Webster's New World Medical Speller/Divider

Compiled by
Joy H. Walworth
and based upon

Webster's
New World Dictionary
of the American Language
Second College Edition

Technical Consultant
Edward M. Chester, M.D.

PRENTICE HALL PRESS
New York, N.Y. 10023

Copyright © 1987 by Simon & Schuster, Inc.

All rights reserved
including the right of reproduction
in whole or in part in any form

Published by New World Dictionaries/
Prentice Hall Press
A Division of Simon & Schuster, Inc.
Gulf + Western Plaza
New York, NY 10023

Dictionary Editorial Offices:
New World Dictionaries
850 Euclid Avenue
Cleveland, Ohio 44114

PRENTICE HALL PRESS, TREE OF
KNOWLEDGE, WEBSTER'S NEW WORLD
and colophons are trademarks of Simon
& Schuster, Inc.

Manufactured in the United States

1 2 3 4 5 6 7 8 9 10

Library of Congress Cataloging in Publication Data

Walworth, Joy H.
  Webster's New World medical speller.

  1. Medicine—Dictionaries. 2. Biology—Dictionaries.
3. Spellers. 4. Homonyms. I. Webster's New World
dictionary of the American language. II. Title.
III. Title: Medical speller/divider. [DNLM: 1. Dictionaries, Medical. W 13 W242w]
R121.W32   1986   610′.3′21   86-23531
ISBN 0-671-55192-2

after words that are spelled differently but pronounced the same.

    rec′ord *n.* re·cord′ *v.*   hy′dro·ceph′a·lous *adj.*
                              hy′dro·ceph′a·lus *n.*

Irregularly formed inflected forms have been entered beneath the entry for the base word; they are entered only if there is actual evidence that they are used as part of the living language. Inflected forms include the plurals of nouns, the parts of verbs, and the comparative and superlative forms of adjectives and adverbs. For verbs, when two forms are given, the first is the past tense and past participle and the second is the present participle. When three forms are given, the first is the past tense, the second is the past participle, and the third is the present participle. As a space-saving feature, inflected forms have been shortened, when possible, to show only those syllables that are different from the base word.

mam′ma	gout′y	ul′cer·ate′	frost′bite′
·mae	·i·er	·at′ed	·bit′
	·i·est	·at′ing	·bit′ten
			·bit′ing

Many of the entries represent the spelling of a word with more than one part of speech. If, for example, an entry contains verb forms, this does not necessarily mean that the word is used only as a verb; it may be used as a noun as well, but if the spelling, stress pattern, and syllabification are exactly the same as for the verb, the word is not entered again separately. If no part-of-speech label is shown, the form given should be accepted as correct for all parts of speech.

Words with identical spellings, words with identical pronunciations, and words with very similar pronun-

INTRODUCTION

This *Speller/Divider of Medical Terms* is an a[l]betical listing of over 30,000 terms in medicine ar[d re]lated fields. It is to be used as a guide to spel[ling,] syllabification, pronunciation, and irregular in[flec]tions. The speller/divider is based on *Webster's [New] World Dictionary, Second College Edition.*

The entry words have been divided into syllab[les.] Each syllable break is indicated by a centered dot, accent mark, or, in some cases, a hyphen. Wheneve[r a] hyphen is used, that hyphen is part of the spelling [of] the word. Accent marks are included as an aid to p[ro]nunciation. The marks are of two kinds—the heavy, darker, mark shows that the syllable preceding it r[e]ceives the main stress or accent; the lighter mark i[n]dicates that the preceding syllable receives less stre[ss] than one getting the main stress, but somewhat mor[e] than an unmarked syllable.

    car′di·ol′o·gy   Gram′-pos′i·tive   lym′pho·cyte′

When the entry is an open compound, no accent marks are shown on any of the elements of that compound, unless one or more of those elements are not entered separately. In such cases the accent marks will appear with the compound.

    tar′dive dyskinesia   diabetus mel·li′tus
        Ep′stein-Barr′ virus

Part-of-speech labels (*n.* - noun; *v.* - verb; *adj.* - adjective) are inserted only in special instances, notably after words that are accented one way as one part of speech and differently as another part of speech, or

ciations all have identifying definitions. When such words, which are likely to cause confusion, are not close together alphabetically, they are cross-referred to each other. The definitions are intended only to aid users in finding the term being sought and are not to be looked upon as true definitions.

    fa′cial   fas′ci·al   os   os
    (*of the face;*   (*of tissue;*   os′sa   o′ra
    SEE *fascial*)   SEE *facial*) (*bone*) (*mouth*)

Variant spellings and alternate forms are often treated as compound entries with an "or" between them and with the preferred form given first when order of preference can be clearly established. The appearance of such forms in alphabetical order usually indicates that usage is nearly equal with preference, if any, going to the first form.

    cu·ret′ *or* ·rette′   nurs′ling *or*   pe·dun′cu·late *or*
                           nurse′ling   ·lat′ed

The word stock in this book is basically a selection from the vocabulary of medicine, but terms have been included from such related, specialized fields as bacteriology, biochemistry, biophysics, immunology, cytology, dentistry, embryology, genetics, anatomy, physiology, pharmacy, surgery, ophthalmology, microbiology, parasitology, psychology, and psychiatry.

This book has been created as a companion volume to *Webster's New World Speller/Divider*, which lists the 33,000 words most used in American English. The *Medical Speller/Divider* is for the use of medical students and secretaries, paramedical personnel, nurses, and anyone working with health-care terminology.

A

Aar′on's sign
a·bac·te′ri·al
a·bac′tus ven′ter
A·ba·die's′ sign
a·baisse′ment
ab·ap′i·cal
ab′ar·thro′sis
ab′ar·tic′u·lar
ab′ar·tic′u·la′tion
a·ba′si·a
a·ba′sic
a·bate′
  ·bat′ed ·bat′ing
a·bate′ment
a·bat′ic
ab·ax′i·al
Ab′der·hal′den's reaction
ab′do·men
ab·dom′i·nal
ab·dom′i·no·car′di·ac′ reflex
ab·dom′i·no·cen·te′sis
  ·ses
ab·dom′i·no·ves′i·cal
ab·du′cens
ab·du′cent
ab·duct′
  (pull away; SEE adduct*)*
ab·duc′tion
ab·duc′tor

ab′em·bry·on′ic
ab′en·ter′ic
Ab′er·ne′thy's fascia
ab·er′rance *or* ·ran·cy
ab·er′rant
ab′er·ra′tion
a·be′ta·lip′o·pro′te·in·e′mi·a
a·bey′ance
ab′i·ent
a·bil′i·ty
  ·ties
ab′i·o·gen′e·sis
  ·ses′
ab′i·o·ge·net′ic
ab′i·og′e·nist
ab′i·og′e·nous
ab′i·o′sis
ab′i·ot′ic
ab′i·o·troph′ic
ab′i·ot′ro·phy
ab·ir′ri·tant
ab′ir·ri·ta′tion
ab′lac·ta′tion
ab·late′
  ·lat′ed ·lat′ing
ab·la′tion
a′ble–bod′ied
a′ble·phar′i·a
a·bleph′a·rous
ab′lu·ent
ab·lu′tion
ab·lu′tion·ar′y
ab·lu′to·ma′ni·a
ab·mor′tal

ab·nor′mal
ab′nor·mal′i·ty
  ·ties
ab·nor′mi·ty
  ·ties
ab·o′rad
ab·o′ral
a·bort′
a·bort′er
a·bor′ti·cide′
a·bor′tient
a·bor′ti·fa′cient
a·bor′tion
a·bor′tion·ist
a·bor′tive
ABO system
a·bor′tus
  ·tus·es
a·bra′chi·a
a·bra′chi·o·ceph′a·lus
a·bra′chi·us
a·brad′ant
ab·rade′
  ·rad′ed ·rad′ing
ab·rad′er
a·bra′sion
ab·ra′sive
ab′re·act′
ab′re·ac′tion
a·bro′si·a
ab·rup′ti·o
ab′scess
ab·scise′
ab·scis′sa
  ·sas *or* ·sae
ab·scis′sion

1

ab·sco′pal
ab′sence
ab′sent *adj.*
ab·sent′ *v.*
ab′sinthe *or* ·sinth
ab′sinth·ism
ab′so·lute′
ab·sorb′
  (take in; SEE
  adsorb)
ab·sorb′a·bil′i·ty
ab·sorb′a·ble
ab·sorb′ance
ab·sor′be·fa′cient
ab·sorb′en·cy
ab·sorb′ent
ab·sorb′er
ab·sorp′ti·om′e·ter
ab·sorp′tion
ab·sorp′tive
ab·stain′
ab·ste′mi·ous
ab·sten′tion
ab·sten′tious
ab·sterge′
  ·sterged′ ·sterg′ing
ab·ster′gent
ab·ster′sion
ab′sti·nence
ab′sti·nent
ab′stract *n.*
ab·stract′ *v.*
ab·strac′tion
ab·ter′mi·nal
a·bu′li·a
a·bu′lic
a·bu′lo·ma′ni·a

a·buse′
  ·bused′ ·bus′ing
a·bus′er
a·bu′sive
a·but′
  ·but′ted ·but′ting
a·but′ment
a·ca′cia
a·cal′ci·co′sis
a′cal·cu′li·a
a·camp′si·a
a·can′tha
a·can′tha·me·bi′a·sis
a·can′thes·the′si·a
a·can′thi·on
a·can′tho·ceph′a·lan
a·can′tho·ceph′a·li′a·sis
  ·ses′
a·can′tho·cyte′
a·can′thoid *or*
  ·thous
ac′an·thol′y·sis
  ·ses′
a·can′tho·lyt′ic
ac′an·tho′ma
  ·mas *or* ·ma·ta
ac′an·tho′sis
  ·ses
ac′an·thot′ic
a·cap′ni·a
a·cap′ni·al
a·cap′nic
a·car′bi·a
a·car′di·a

a·car′di·a·cus
ac′a·ri′a·sis *or* a·car′i·di′a·sis
  ·ses′
a·car′i·cide′
ac′a·rid
a·car′i·no′sis
  ·ses
ac′a·roid′
ac′a·rol′o·gist
ac′a·rol′o·gy
ac′a·rus
  ·a·ri
a·car′y·ote′
a′cat·a·la′si·a *or*
  ·la·se′mi·a
ac′a·thec′tic
ac′a·thex′i·a
ac′a·thex′is
a·cau′dal *or* ·date
ac·cel′er·ant
ac·cel′er·ate′
  ·at′ed ·at′ing
ac·cel′er·a′tion
ac·cel′er·a′tive
ac·cel′er·a′tor
ac·cel′er·om′e·ter
ac·cen′tu·ate′
  ·at′ed ·at′ing
ac·cen′tu·a′tion
ac·cep′tor
ac·cess
ac·ces′si·ble
ac·ces′sion
ac·ces·so′ri·us

ac·ces′so·ry *or*
·sa·ry
·ries
ac′ci·dent
ac′ci·den′tal
ac′ci·dent–prone′
ac·cip′i·ter
ac′cli·mate′
·mat′ed ·mat′ing
ac′cli·ma′tion
ac·cli′ma·ti·za′tion
ac·cli′ma·tize′
·tized′ ·tiz′ing
ac·com′mo·date′
·dat′ed ·dat′ing
ac·com′mo·da′tion
ac·com′mo·dat′ive
ac·com′mo·da′tor
ac·couche′ment
ac′cou·cheur′
ac′cou·cheuse′
ac′cre·men·ti′tion
ac·crete′
·cret′ed ·cret′ing
ac·cre′ti·o cor′dis
ac·cre′tion
ac·cre′tive
ac′cu·ra·cy
ac′cu·rate
Ace bandage
 (*Ace* is a trademark)
a·ce′di·a
a′ce·nes·the′si·a
a·cen′tric
a′ce·pha′li·a

a·ceph′a·lo·bra′chi·a
a·ceph′a·lous *adj.*
a·ceph′a·lus *n.*
ac′er·ate
a·cerb′ *or* a·cer′bic
a·cer′bi·ty
·ties
a·cer′vu·line
a·cer′vu·lus
·li
a·ce′ta
 (*sing* . a·ce′tum)
ac′e·tab′u·lar
ac′e·tab′u·lum
·la *or* ·lums
ac′e·tal′
ac′e·tal′de·hyde′
ac′et·am′ide
ac′e·ta·min′o·phen
ac′et·an′i·lide′ *or*
·lid
ac′e·tate′
ac′e·tat′ed
ac′e·ta·zol′a·mide′
a·ce′tic
ac′e·ti·fi·ca′tion
a·cet′i·fy′
·fied′ ·fy′ing
ac′e·tin
ac′e·to·a·ce′tic acid
ac′e·tom′e·ter *or*
·tim′e·ter
ac′e·tone′
ac′e·ton′ic
a·ce′to·phe·net′i·din
or ·phen′e·tide′

ac′e·tous *or* ·tose′
a·ce′tum
·ta
ac′e·tyl
a·cet′y·late′
·lat′ed ·lat′ing
a·cet′y·la′tion
ac′e·tyl·cho′line
ac′e·tyl·cho′lin·es′ter·ase
ac′e·tyl·co·en′zyme A
a·cet′y·lene′
a·ce′tyl′ic
a·ce′tyl·sal′i·cyl′ic acid
ac′e·tyl·trans′fer·ase
ach′a·la′si·a
ache
 ached ach′ing
a·chei′li·a
a·chei′lous *or* a·chi′lous
a·chei′ro·po′di·a
a·chieve′ment quotient
A·chil′les tendon
a·chil′lo·bur·si′tis
ach′il·lor′rha·phy
a′chlor·hy′dri·a
a′chlor·hy′dric
a·chlo·rop′si·a
a·cho′li·a
a·cho′lic
a′cho·lu′ri·a
a′chon·dro·pla′si·a
a·chon′dro·plas′tic

a·chro′ma
ach′ro·mat′
ach′ro·mat′ic
a·chro′ma·tin
a·chro′ma·tism *or*
 ·chro′ma·tic′i·ty
a·chro′ma·tize′
 ·tized′ ·tiz′ing
a′chro·mat′o·phil
 or a·chro′ma·phil
a·chro′ma·tous
a·chro′mi·a
a·chro′mic *or*
 ·mous
ach′y
 ·i·er ·i·est
a·chy′li·a
a·chy′lous
a·chy′mi·a
a·cic′u·la
 ·lae′
a·cic′u·lar
a·cic′u·late *or*
 ·lat′ed
a·cic′u·lum
 ·lums *or* ·la
ac′id
ac′id·am′i·nu′ri·a
ac′i·de′mi·a
ac′id–fast′
ac′id–form′ing
a·cid′ic
a·cid′i·fi′a·ble
a·cid′i·fi·ca′tion
a·cid′i·fi′er
a·cid′i·fy′
 ·fied′ ·fy′ing

ac′i·dim′e·ter
a·cid′i·met′ric
ac′i·dim′e·try
a·cid′i·ty
 ·ties
a·cid′o·phil *or*
 ·phile′
a′cid·o·phil′ic
ac′i·doph′i·lus milk
ac′i·do′sis
 ·ses
ac′i·dot′ic
a·cid′u·late′
 ·lat′ed ·lat′ing
a·cid′u·la′tion
a·cid′u·lous *or* ·lent
ac′i·dum
ac′i·du′ri·a
ac′i·du′ric
ac′i·form′
ac′i·ne′si·a
ac′i·net′ic
a·cin′i·form′
ac′i·nous *or* ·nose′
ac′i·nus
 ·ni′
a′clad·i·o′sis
 ·ses
ac′la·sis *or* a·cla′
 si·a
a·clas′tic
a·cleis′to·car′di·a
a·clu′sion
ac′me
ac′ne
ac′ne·gen′ic

ac·ne′i·form′ *or* ac′
 ne·form′
ac·ne′mi·a
ac·ni′tis
ac′o·nite′ *or* ac′o·
 ni′tum
a′co·re′a *(absence of
 the pupil)*
a·co′ri·a *(unrelieved
 hunger)*
a·cou′es·the′si·a
a·cou′me·ter
a·cous′tic
a·cous′ti·cal
ac′ous·ti′cian
a·cous′tics
ac·quir′a·ble
ac·quire′
 ·quired′ ·quir′ing
ac′qui·si′tion
ac′ral
a·cra′ni·a
a·cra′ni·al
ac′rid
a·cri·dine′
a·crid′i·ty
ac′ri·fla′vine
ac′ri·mo′ny
 ·nies
a·crit′i·cal
ac′ro·ag·no′sis
ac′ro·an·es·the′si·a
ac′ro·cen′tric
ac′ro·ce·phal′ic *or*
 ·ceph′a·lous
ac′ro·ceph′a·ly
ac′ro·chor′don

ac′ro·cy·a·no′sis
· ses
ac′ro·der′ma·ti′tis
ac′ro·dyn′i·a
ac′ro·es·the′si·a
ac′ro·ker′a·to′sis
· ses
a·cro′le·in
ac′ro·me·gal′ic
ac′ro·meg′a·ly
ac′ro·me·lal′gi·a
a·cro′mi·al
ac′ro·mic′ri·a
a·cro′mi·o·cla·vic′u·lar
a·cro′mi·on
a·crom′pha·lus
ac′ro·phobe′
ac′ro·pho′bi·a
ac′ro·pho′bi·ac′
ac′ro·pho′bic
ac′ro·scle′ro·der′ma
ac′ro·scle·ro′sis
· ses
ac′ro·so′mal
ac′ro·some′
a·crot′ic
ac′ro·tism
ac′ro·troph′o·neu·ro′sis
ac′ry·late′
a·cryl′ic
ac′ry·lo·ni′trile
ACTH
ac′tin
ac·tin′ic
ac·tin′i·form′

ac′tin·ism
ac·tin′i·um
ac′ti·no·der′ma·ti′tis
ac·tin′o·gen
ac·tin′o·graph′
ac·tin′o·lite′
ac′ti·nol′o·gy
ac′ti·no·met′ric
ac′ti·nom′e·try
ac′ti·no·my′ces
ac′ti·no·my·cete′
· ce′tes
ac′ti·no·my·ce′tic
ac′ti·no·my·ce′tous
ac′ti·no·my′cin
ac′ti·no·my·co′sis
· ses
ac′ti·no·my·cot′ic
ac′ti·non′
ac′ti·no·ther′a·py
ac′tion
ac′ti·vate′
· vat′ed · vat′ing
ac′ti·va′tion
ac′ti·va′tor
ac′tive
ac·tiv′i·ty
· ties
ac′tiv·ize′
· ized′ · iz′ing
ac′to·my′o·sin
ac′tu·al
ac′u·fi′lo·pres′sure
a·cu′i·ty
a·cu′mi·nate n.

a·cu′mi·nate′
· nat′ed · nat′ing
ac′u·pres′sure
ac′u·punc′ture n.
ac′u·punc′ture
· tured · tur·ing
a′cus
a·cute′
ac′u·tor′sion
a·cy′a·nop′si·a
a·cy′a·not′ic
a·cy′clic
a′cy·e′sis
ac′yl
ac′y·lase′
ac′yl·a′tion
a·cyl′o·in
a·cys′ti·a
a′dac·tyl′i·a or a·dac′ty·lism
a·dac′ty·lous
a·dac′ty·ly
ad′a·man′tine
ad′a·man·ti·no′ma
· mas or · ma·ta
ad′a·man·to′ma
· mas or · ma·ta
Ad′am's ap′ple
ad′ams·ite′
a·dapt′
a·dapt′a·bil′i·ty
· ties
a·dapt′a·ble
ad′ap·ta′tion
ad′ap·ta′tion·al
a·dapt′er or · dap′tor

a·dap′tion
a·dap′tive
ad′ap·tom′e·ter
ad·ax′i·al
ad′der *(snake)*
add′er *(one who adds)*
ad′dict *n.*
ad·dict′ *v.*
ad·dic′tion
ad·dic′tive
Ad′dis count test
ad′di·son·ism
Ad′di·son's disease
ad′di·tive
ad′dle
 ·dled ·dling
ad′dle·brained′
ad·du′cent
ad′duct *n.*
ad·duct′ *v. (pull toward;* SEE abduct*)*
ad·duc′tion
ad·duc′tive
ad·duc′tor
ad′e·nal′gi·a
ad′e·nase′
a′den·drit′ic *or* a·den′dric
ad′e·nec′to·my
 ·mies
a·de′ni·a
a·den′i·form′
ad′e·nine′
ad′e·ni′tis
ad′e·ni·za′tion

ad′e·no·ac′an·tho′ma
 ·mas *or* ·ma·ta
ad′e·no·car′ci·no′ma
 ·mas *or* ·ma·ta
ad′e·no·cele′
ad′e·no·fi·bro′ma
 ·mas *or* ·ma·ta
ad′e·noid′
ad′e·noi′dal
ad′e·noid·ec′to·my
 ·mies
ad′e·noid·i′tis
ad′e·noids′
ad′e·no·li·po′ma
 ·ma·ta *or* ·mas
ad′e·no′ma
 ·mas *or* ·ma·ta
ad′e·nom′a·tous
ad′e·no·mere′
ad′e·no·my·o′sis
 ·ses
ad′e·no·scle·ro′sis
 ·ses
ad′e·nose′
a·den′o·sine′
ad′e·no′sis
 ·ses
ad′e·no·tome′
ad′e·not′o·my
 ·mies
ad′e·no·vi′rus
ad′e·nyl
a·den′y·late
ad′e·nyl·yl
ad′e·qua·cy

ad′e·quate
a·der′mi·a
a·der′mo·gen′e·sis
 ·ses′
ad·here′
 ·hered′ ·her′ing
ad·her′ence
ad·her′ent
ad·her′er
ad·he′sion
ad·he′si·ot′o·my
 ·mies
ad·he′sive
ad·hib′it
ad′hi·bi′tion
ad′i·a·bat′ic
a·di′a·do′cho·ki·ne′si·a
a·di′a·pho·re′sis
a·di′a·pho·ret′ic
a·di′a·pho′ri·a
ad′i·aph′o·rous
ad′i·ent
A′die's pupil
a·dip′ic
ad′i·po·cele′
ad′i·po·cere′
ad′i·po·cyte′
ad′i·po·gen′ic *or*
 ·pog′e·nous
ad′i·poid
ad′i·pose′
ad′i·po′sis
 ·ses
ad′i·pos′i·ty
a·dip′si·a

ad'i·tus
 ·tus or ·tus·es
ad·join'
ad'junct
ad·junc'tive
ad·just'
ad·just'a·ble
ad·just'er or
 ·jus·tor
ad·jus'tive
ad·just'ment
ad'ju·vant
ad' lib'i·tum
ad·min'is·ter
ad·mis'sion
ad·mit'
 ·mit'ted ·mit'ting
ad·mit'tance
ad·mix'
ad·mix'ture
ad' nau'se·am
ad·ner'val or
 ·neu'ral
ad·nex'a
ad·nex'al
ad'nex·i'tis
ad'o·les'cence
ad'o·les'cent
ad·o'ral
ADP
ad·re'nal
ad·re'nal·ec'to·
 mize'
 ·mized' ·miz'ing
ad·re'nal·ec'to·my
 ·mies

A·dren'al·in
 (trademark)
ad·ren'a·li·ne'mi·a
ad·ren'a·li·nu'ri·a
ad·ren'a·lism
ad·re·nal·i'tis
ad'ren·ar'che
ad·ren·er'gic
ad·ren'o·chrome'
ad·re'no·cor'ti·cal
ad·re'no·cor'ti·co·
 troph'ic or
 ·trop'ic
ad·sorb' *(collect on a
 surface;* SEE absorb*)*
ad·sorb'a·ble
ad·sor'bate
ad·sor'bent
ad·sorp'tion
ad·sorp'tive
a·dult'
a·dul'ter·ant
a·dul'ter·ate'
 ·at·ed ·at·ing
a·dul'ter·a'tion
a·dul'ter·a'tor
ad·um'brate
 ·brat·ed ·brat·ing
ad·um'bra·tion
ad·um'bra·tive
ad·vance'
 ·vanced' ·vanc·ing
ad·vance'ment
ad'ven·ti'tia
ad'ven·ti'tial
ad'ven·ti'tious
ad·verse'

ad·ver'sive
ad·vice' *n.*
ad·vis'a·ble
ad·vise'
 ·vised' ·vis'ing
ad'y·na'mi·a
ad'y·nam'ic
a·ë'des
*For words
 beginning* aeg– *see*
 EG–
ae'lu·rop'sis
ae'o·lo·trop'ic
ae'quo·rin
aer·ate'
 ·at·ed ·at·ing
aer·a'tion
aer'a'tor
aer·if'er·ous
aer'i·fi·ca'tion
aer'i·form'
aer'i·fy'
 ·fied' ·fy'ing
aer'o·bac'ter
aer'obe
aer·o'bic
aer'o·bi·ol'o·gist
aer'o·bi·ol'o·gy
aer·o'bi·um
 ·bi·a
aer'o·cele'
aer'o·col'pos
aer'o·don'ti·a
aer'o·dy·nam'ic
aer'o·dy·nam'ics
aer'o·em'bo·lism
aer'o·gel'

7

a′er·o·gen′
aer′o·gen′e·sis
  ·ses′
aer′o·gen′ic
aer·og′e·nous
aer′o·me·chan′ic
aer′o·me·chan′ics
aer′o·med′i·cal
aer′o·med′i·cine
aer·om′e·ter
aer′o·met′ric
aer·om′e·try
aer′o·neu·ro′sis
  ·ses
aer′o·pha′gi·a
aer′o·phobe′
aer′o·pho′bi·a
aer′o·pho′bi·ac′
aer′o·pho′bic
aer′o·phore′
aer′o·pi·e·so·ther′a·py
aer′o·ple·thys′mo·graph′
aer′o·scope′
aer′o·sol′
aer′o·sol·ize′
  ·ized′ ·iz′ing
aer′o·space′ medicine
aer′o·stat′ic
aer′o·stat′ics
aer′o·ther′a·peu′tics *or* ·ther′a·py
For words beginning aes– *or* aet– *see* ES–, ET–

Aes·cu·la′pi·an
Aes·cu·la′pi·us, staff of
aes·the′si·a
aes·thet′ic *or* ·thet′i·cal
aes·thet′ics
aes′ti·val
ae′ther
ae·the′ri·al
ae′ti·ol′o·gy
a·fe′brile
af·fect′
af′fect′ *n. (emotion;* SEE effect)
af′fect *v.*
af·fect′a·bil′i·ty
  ·ties
af·fect′ed
af·fec′tion
af·fec′tive *(emotional;* SEE effective)
af′fec·tiv′i·ty
af′fer·ent *(bringing inward;* SEE efferent)
af·fil′i·ate *n.*
af·fil′i·ate′
  ·at′ed ·at′ing
af·fil′i·a′tion
af·fil′i·a′tive
af·fin′i·ty
  ·ties
af′fir·ma′tion
af·flict′
af·flic′tion
af·flic′tive
af′flux

af·fu′sion
af′la·tox′i·co′sis
af′la·tox′in
af′ter·birth′
af′ter·brain′
af′ter–care′
af′ter·ef·fect′
af′ter·im′age
af′ter·pains′
af′ter·sen·sa′tion
af′ter·taste′
a′ga·lac′ti·a
a′ga·mete′
a′ga·met′ic
a·gam′ic
a·gam′ma·glob′u·li·ne′mi·a
ag′a·mo·gen′e·sis
  ·ses′
ag′a·mous
a′gar–a′gar
ag·ar′ic
ag′a·tha·na′si·a
age
  aged ag′ing *or* age′ing
a′ged *n.*
a·gen′e·sis
  ·ses′
a′gent
A′gent Or′ange
a·geu′si·a
a·geu′sic
ag′er
ag·glom′er·ate *n.*
ag·glom′er·ate′
  ·at′ed ·at′ing

ag·glom'er·a'tion
ag·glom'er·a'tive
ag·glu'ti·na·ble
ag·glu'ti·nant
ag·glu'ti·nate *adj.*
ag·glu'ti·nate'
 ·nat'ed ·nat'ing
ag·glu'ti·na'tion
ag·glu'ti·na'tive
ag·glu'ti·na'tor
ag·glu'ti·nin
ag·glu·tin'o·gen
ag·glu·tin'o·gen'ic
ag'gra·vate'
 ·vat'ed ·vat'ing
ag·gra·va'tion
ag'gre·gate n.
ag'gre·gate'
 ·gat'ed ·gat'ing
ag·gre·ga'tion
ag·gre·ga'tive
ag·gres'sin
ag·gres'sion
ag·gres'sive
ag'i·tate'
 ·tat'ed ·tat'ing
ag·i·ta'tion
ag'i·ta'tor
a·glos'si·a
ag·lu·ti'tion
ag'nail
ag·no·gen'ic
ag·no'si·a
ag·nos'tic
ag'o·nal
ag'o·nist

ag'o·nize'
 ·nized' ·niz'ing
ag'o·ny
 ·nies
ag'o·ra·pho'bi·a
a·graffe' or ·grafe'
a·gran'u·lo·cyte'
a·gran'u·lo·cyt'ic
a·gran'u·lo·cy·to'sis
 or ·lo'sis
a·graph'i·a
a·graph'ic
a'gri·us
a·gryp'ni·a
a'gue
a·gu'ish
a·his'ta·da'sia
aid or aide n.
aid v.
aid'man
 ·men
AIDS
ail'ment
ai·lu'ro·phobe'
ai·lu'ro·pho'bi·a
ain'hum
air bed
air'borne'
air'bra'sive
air'flow', laminar
air'sick'
air'tight'
air'way'
aitch'bone'
a·kar'y·o·cyte'
ak'a·thi'si·a

a'la
 a'lae
al'a·nine'
a'lar or ·la·ry
a'late or ·lat·ed
al'ba
Al'bar·rán's'
 glands
al·be'do
al·bes'cence
al·bes'cent
al'bi·cans
 ·can'ti·a
al·bin'ic
al'bi·nism
Al·bi'ni's nodules
al·bi'no
 ·nos
al·bi·noid'ism
al·bi·not'ic
Al'bright's
 syndrome
al·bu·gin'e·a
al·bu·gin'e·ous
al·bu'men *(egg white)*
al·bu'men·ize'
 ·ized' ·iz'ing
al·bu'min *(protein)*
al·bu'mi·nate'
al·bu'mi·noid'
al·bu'mi·nous
al·bu'mi·nu·ret'ic
al·bu'mi·nu'ri·a
al·bu'mi·nu'ric
al·bu'mo·scope'
al'bu·mose'

al·bu′te·rol′
Al′cock's canal
al′co·gel′
al′co·hol′
al′co·hol′ic
al′co·hol·ism
al′co·hol·ize′
 ·ized′ ·iz′ing
al′co·hol·om′e·ter
al′de·hyde′
al′de·hy′dic
al′dol
al′do·lase′
al′dose
al·do′ste·rone′
al·do′ste·ron·ism
Al′drich's
 syndrome
al′drin
a·lem′bic
a·lert′
a′leu·ke′mi·a
a′leu·ke′mic
a·leu′ki·a
a′leu·rone′ or ·ron′
a′leu·ron′ic
a·lex′e·ter′ic
a·lex′i·a
a·lex′ic
a·lex′in
a·lex′i·phar′mic or
 ·mac
a·ley′dig·ism
al′gae
 (sing. al′ga)

al′gae·cide′ or
 ·gi·cide′
al′gal
al′ge·fa′cient
al·ge′si·a
al·ge′sic or ·get′ic
al·ge·sim′e·ter
al′ges·the′si·a
al′ges·the′sis
al′gi·cide′
al′gid
al′gin
al′gi·nate′
al·gin′ic acid
al′go·gen′ic
al′go·lag′ni·a
al′go·log′i·cal
al·gol′o·gist
al·gol′o·gy
al′go·met′ric
al′go·pho′bi·a
al′gor
al′i·ble
al′i·cy′clic
al′ien·ate′
 ·at′ed ·at′ing
al′ien·a′tion
al′ien·ism
al′ien·ist
al′i·form′
a·lign′
a·lign′ment
al′i·ment n.
al′i·ment′ v.
al′i·men′tal

al′i·men′ta·ry
 (connected with food;
 SEE elementary)
al′i·men·ta′tion
al′i·men′ta·tive
al′i·men′to·ther′
 a·py
al′i·na′sal
a·line′
 ·lined′ ·lin′ing
a·line′ment
al′i·phat′ic
al′i·quant
al′i·quot
al′i·sphe′noid
a·live′
a·liz′a·rin or ·rine
al′ka·le′mi·a
al′ka·les′cence or
 ·cen·cy
al′ka·les′cent
al′ka·li′
 ·lies′ or ·lis′
al′ka·lim′e·ter
al′ka·lim′e·try
al′ka·line
al′ka·lin′i·ty
al′ka·lin·i·za′tion
al′ka·li·nu′ri·a
al′ka·li·ther′a·py
al′ka·lize′
 ·lized′ ·liz′ing
al′ka·li·za′tion
al′ka·liz′er
al′ka·loid′
al′ka·loid′al

al′ka·lo′sis
·ses
al′kane
al′ka·net′ or al·kan′na
al·kan′nin
al′kap·to·nu′ri·a
al′kene
al′kyl
al′ky·late′
·lat′ed ·lat′ing
al′ky·la′tion
al·kyl′ic
al′kyne or ·kine
al·la·ches·the′si·a
al·lan′to·cho′ri·on
al·lan′to·ic
al·lan′toid
al·lan′to·in
al·lan′to·i·nu′ri·a
al·lan′to·is
al′lan·to′i·des′
al·lay′
·layed′ ·lay′ing
al·lele′
al·le′lic
al·lel′ism
al·le′lo·morph′
al·le′lo·mor′phic
al·le′lo·mor′phism
al′le·lop′a·thy
al′ler·gen
al′ler·gen′ic
al·ler′gic
al′ler·gist
al′ler·gi·za′tion

al′ler·gize′
·gized′ ·giz′ing
al′ler·gy
·gies
al′le·thrin
al·le′vi·ate′
·at′ed ·at′ing
al·le′vi·a′tion
al·le′vi·a′tive
al·le′vi·a′tor
al·le′vi·a·to′ry
al·li·a′ceous
al′li·cin
al·lit′er·a′tion
al·lit′er·a′tive
al′li·um
al′lo·an′ti·bod′y
·ies
al′lo·an′ti·gen
al·log′a·mous
al·log′a·my
al′lo·ge·ne′ic or ·gen′ic
al′lo·graft′
al·lom′er·ism
al·lom′er·ous
al·lom′e·try
al′lo·morph′
al′lo·mor′phic
al′lo·mor′phism
al′lo·path′ or al·lop′a·thist
al′lo·path′ic
al·lop′a·thy
·thies
al′lo·phe′nic
al′lo·plasm

al′lo·plas′mic or ·plas·mat′ic
al′lo·plast′
al′lo·plas′tic
al′lo·plas′ty
al′lo·pol′y·ploid′
al′lo·pol′y·ploi′dy
al′lo·pu′ri·nol′
al′lo·rhyth′mi·a
all-or-none law
al′lo·ster′ic
al′lo·trope′
al′lo·troph′ic
al′lo·trop′ic or ·trop′i·cal
al·lot′ro·py or ·ro·pism
al′lo·type′
al′lo·typ′ic
al′loy n.
al·loy′ v.
all′-trans-ret′i·nal
al′lyl
al·lyl′ic
a·lo′chi·a
al′oe
·oes
al′o·et′ic
a·lo′gi·a
al′o·in
al′o·pe′ci·a
al′o·pe′cic
al′pha
al′pha-ad′re·ner′gic

al′pha-fe′to·pro′tein
al′pha·lyt′ic
al′pha-1-an′ti·tryp′sin
al′pha-to·coph′er·ol
al·pho′sis
·ses
al′ter·a′tive
al′ter e′go
al′ter·nans
al′ter·nate n.
al′ter·nate′
·nat′ed ·nat′ing
al′ter·na′tion
al·ter′na·tive
al′ter·na′tor
al·the′a or ·thae′a
al′um
a·lu′mi·na
a·lu′mi·nate′
a·lu′mi·nif′er·ous
a·lu′mi·nize′
·nized′ ·niz′ing
a·lu′mi·no′sis
a·lu′mi·nous
a·lu′mi·num
al·ve′o·lar
al·ve′o·late or ·lat′ed
al·ve′o·la′tion
al·ve′o·lo·den′tal
al·ve′o·lus
·li′
al′ve·us
·ve·i′
al′vine

al′vus
·vi
Alz′hei′mer's disease
a′ma·crine
am′a·dou′
a·mal′gam
a·mal′gam·a·ble
a·mal′ga·mate′
·mat′ed ·mat′ing
a·mal′ga·ma′tive
a·mal′ga·ma′tion
a·mal′ga·ma′tor
am′a·ni′ta
a·man′ta·dine′ hydrochloride
am′a·ranth′
am′a·tive
am′au·ro′sis
·ses
am′au·rot′ic
am′bi·dex′ter
am′bi·dex·ter′i·ty
am′bi·dex′trous
am′bi·ent
am·big′u·ous
am·bi·gu′i·ty
·ties
am·bi·lat′er·al
am′bi·sex′ual
am′bi·tend′en·cy
am·biv′a·lence
am·biv′a·lent
am′bi·ver′sion
am′bi·vert
am′bly·o′pi·a
am′bly·o′pic

am′bly·o·scope′
am′bo·cep′tor
am′bon or ·bo
am′bos
Am′bu bag (trade name)
am′bu·lance
am′bu·lant
am′bu·late′
·lat′ed ·lat′ing
am′bu·la′tion
am′bu·la·to′ry
a·me′ba
·bas or ·bae
am′e·bi′a·sis
·ses′
a·me′bic or ·ban
a·me′bi·cide′
a·me′bi·form′
a·me′bo·cyte′
a·me′boid
a·mel′ei·a (apathy)
a·me′li·a (absence of a limb)
a·mel′i·fi·ca′tion
a·mel′io·rant
a·mel′io·rate′
·rat′ed ·rat′ing
a·mel′io·ra′tion
a·mel′io·ra′tive
am′e·lus
a·me′na·bil′i·ty
a·me′na·ble
a·me′ni·a
a·men′or·rhe′a or ·rhoe′a

a·men'or·rhe'al *or*
 ·ic
a'ment
a·men'tia
am·er·ic'i·um
Am'es·lan'
am'e·tro'pi·a
am'e·tro'pic
am'i·dase'
am'ide
a·mid'ic
am'i·din *(soluble starch)*
am'i·dine' *(nitrogen base)*
a·mi'do
a·mi'do·gen
am'i·done'
a·mim'i·a
am'i·nate'
 ·nat·ed ·nat'ing
a·mine'
a·mi'nic
a·mi'no
a·mi'no·ben·zo'ic acid
a·mi'no·py'rine
a·mi'no·thi'a·zole'
a'mi·to'sis
 ·ses
a'mi·tot'ic
am'me·ter
am'mine
am·mo·ne'mi·a
am·mo'ni·a
am·mo'ni·ac'
am·mo'ni·a·cal
am·mo'ni·ate *n.*
am·mo'ni·ate
 ·at'ed ·at'ing
am·mo'ni·a'tion
am·mo'nic
am·mo'ni·fi·ca'tion
am·mo'ni·fy'
 ·fied' ·fy'ing
am·mo'ni·um
am'mo·no'
am·ne'sia
am·ne'si·ac' *or* ·ne'sic
am·nes'tic
am'ni·o·cen·te'sis
 ·ses
am'ni·on
 ·ons *or* ·ni·a
am'ni·o·ni'tis
am'ni·o·scope'
am'ni·os'co·py
am'ni·ot'ic *or* ·ni·on'ic
am'o·bar'bi·tal'
A'–mode'
am'o·di'a·quin *or* ·quine'
a·moe'ba
 ·bas *or* ·bae
am'oe·bi'a·sis
 ·ses'
a·moe'bic *or* ·ban
a·moe'bo·cyte'
a·moe'boid
a·mok'
a'morph
a·mor'phi·a
a·mor'phism
a·mor'phous *(shapeless)*
a·mor'phus
 ·phi *or* ·phus·es *(shapeless form)*
a·mo'ti·o
a·mox'i·cil'lin
AMP
amp
am'per·age
am'pere
am·phet'a·mine'
am·phi·ar·thro'sis
 ·ses
am'phi·as'ter
am·phib'i·ous
am'phi·bol'ic
am'phi·chro'ic
am'phi·coe'lous *or* ·ce'lous
am'phi·mix'is
am'phi·the'a·ter *or* ·tre
am·phit'ri·chous *or* ·chate
am'pho·phil'
am'pho·phil'ic
am·pho'ric
am·pho·ter'ic
am·pho·ter'i·cin
am·pi·cil'lin
am'pli·fi·ca'tion
am'pli·fi'er
am'pli·fy'
 ·fied' ·fy'ing
am'pli·tude'

am′pul *or* ·pule *or* ·poule
am·pul′la
  ·lae
am′pul·la′ceous
am·pul′lar
am′pu·tate′
  ·tat′ed ·tat′ing
am′pu·ta′tion
am′pu·ta′tor
am′pu·tee′
a·muck′
a·mu′si·a
Am′us·sat's′ valves
a·myc′tic
a′my·e′li·a
a·myg′da·la
  ·lae
a·myg′da·lin
a·myg′da·loid′ *or* a·myg′da·loid′al
am′yl
am′y·la′ceous
am′yl·ase′
am′yl·ene′
a·myl′ic
a·myl′in
a·myl′o·gen
am′y·lo·gen′ic
am′y·loid′
am′y·loi·do′sis
  ·ses
am′y·lol′y·sis
  ·ses′
am′y·lo·lyt′ic
am′y·lo·pec′tin
am′y·lop′sin

am′y·lose′
am′y·lum
a·my′o·to′ni·a
a′my·o·tro′phi·a
a′my·o·troph′ic
a·myx′i·a
an′a
an′a·bae′na
a·nab′a·sis
an′a·bat′ic
an′a·bi·o′sis
  ·ses
an′a·bi·ot′ic
an′a·bol′ic
a·nab′o·lin
a·nab′o·lism
a·nab′o·lite′
a·nac′la·sis
  ·ses′
an′a·clas′tic
an′a·cli′sis
  ·ses
an′a·clit′ic
an′a·crot′ic
a·nac′ro·tism
a·nae′mi·a
an·aer′obe
an′aer·o′bic
an·a′er·o·bi·o′sis
  ·ses
an·a′er·o·bi·ot′ic
an′aer·o′bism
an′aer·o′bi·um
  ·bi·a
an′aes·the′sia
an′aes·thet′ic
an·aes′the·tist

an·aes′the·tize′
  ·tized′ ·tiz′ing
an′a·go′ge *or* ·go′gy
an′a·gog′ic *or* ·gog′i·cal
an′a·ku′sis
a′nal
an′a·lep′tic
an′a·lep′sis
an′al·ge′si·a
an′al·ge′sic
an·al′gi·a
an·al′gic
a·nal′o·gize′
  ·gized′ ·giz′ing
a·nal′o·gous
an′a·logue′ *or* ·log′
a·nal′o·gy
  ·gies
a·nal′y·sand′
a·nal′y·sis
  ·ses′
an′a·lyst
an′a·lyt′ic *or* ·lyt′i·cal
an′a·lyz′a·ble
an′a·lyze′
  ·lyzed′ ·lyz′ing
an′a·lyz′er
an·am·ne′sis
  ·ses
an′am·nes′tic
an′a·mor′pho·scope′
an′a·mor′pho·sis
  ·ses′

an′a·phase′
an·a′phi·a
an·aph′ro·dis′i·a
an·aph′ro·dis′i·ac′
an′a·phy·lac′tic
an′a·phy·lac′to·gen
an′a·phy·lac′to·gen′e·sis
an′a·phy·lac′to·gen′ic
an′a·phy·lax′is
·es
an′a·pla′si·a
an′a·plas′tic
an′a·plas′ty
an·ap′tic
an·ar′thri·a
an′a·sar′ca
an′a·sar′cous
an′a·stal′sis
an′a·stal′tic
an·as′tig·mat′
an·as′tig·mat′ic
a·nas′to·le′
a·nas′to·mose′
·mosed′ ·mos′ing
a·nas′to·mo′sis
·ses
a·nas′to·mot′ic
an′a·tom′i·cal or
·tom′ic
a·nat′o·mist
a·nat′o·mi·za′tion
a·nat′o·mize′
·mized′ ·miz′ing
a·nat′o·my
·mies

a·nat′to
an·cho′ne
an′chor·age
an·chu′sin
For words beginning anchylo– and ancylo– see ANKYLO–
an′cil·lar′y
an·cip′i·tal or ·cip′i·tous
an′co·nad′
an′co·nal or ·ne·al
an·co′ne·us
·ne·i′
an′cy·los·to·mi′a·sis
·ses′
An′der·nach's ossicle
An′der·sen's syndrome
An′der·son splint
an′dro·gen
an′dro·gen′ic
an·drog′e·nous
an′dro·gyne
an·drog′y·nous
an·drog′y·ny
an′droid
An·drom′e·da strain
an′dro·stane′
an′dro·stene′
an·dros′ter·one′
an′ec·dot·al

a·ne′mi·a
a·ne′mic
an′e·mo·pho′bi·a
an′en·ce·phal′ic
an′en·ceph′a·lous
adj.
an′en·ceph′a·lus *n.*
an′en·ceph′a·ly
·lies
an′er·ga′si·a
an′er gas′tic
an·er′gic
an′er·gy
an′er o′bic
an′er·oid
an′e·ryth′ro·pla′si·a
an′e·ryth′ro·plas′tic
an′es·the′si·a
an′es·the′si·ol′o·gist
an′es·the′si·ol′o·gy
an′es·thet′ic
an·es′the·tist
an·es′the·ti·za′tion
an·es′the·tize′
·tized′ ·tiz′ing
a·net′ic
an·e′ti·o·log′ic or
·log′i·cal
an′e·to·der′ma
an′eu·rysm or ·rism
an′eu·rys′mal or ·ris′mal
an·frac′tu·os′i·ty
an·frac′tu·ous
an′gel dust
an·gel′i·ca

An'ge·luc'ci's syndrome
an'gi·ec'to·my
an'gi·i'tis
an·gi'na
an'gi'nal *or* an'gi·nose'
angina pec'to·ris
an'gi·noid'
an'gi·o·blast'
an'gi·o·blas'tic
an'gi·o·gram'
an'gi·o·graph'ic
an'gi·og'ra·phy
an'gi·oid'
an'gi·ol'o·gy
an'gi·o'ma
·ma·ta *or* ·mas
an'gi·o·ma·to'sis
·ses
an'gi·om'a·tous
an'gi·o·plas'ty
an'gi·o·poi·e'sis
·ses
an'gi·o·poi·et'ic
an'gi·o·sar·co'ma
·mas *or* ·ma·ta
an'gi·o·scope'
an'gi·o·ten'sin
an·gi'tis
an'gle
an'gor
an·gos·tu'ra
ang'strom
an'gu·lar
an'gu·late *n.*
an'gu·late'
·lat'ed ·lat'ing
an'gu·la'tion
ang'u·lus
·li'
an'he·do'ni·a
an'he·don'ic
an'hi·dro'sis
·ses
an'hi·drot'ic
an·his'tic *or* ·his'trous
an·hy'drase
an·hy'dride
an·hy'drous
an'ile
an'i·line
an'i·lin·ism *or* ·i·lism
a·nil'i·ty
an'i·ma
an'i·mal'cu·lar
an'i·mal'cule
an'i·mal'cu·lum
·la
an'i·mate'
·mat'ed ·mat'ing
an'i·mate *adj.*
an'i·ma'tion
an'i·ma·tism
an'i·mism
an'i·mis'tic
an'i·mos'i·ty
·ties
an'i·mus
(*animosity;* SEE anomous)
an'i·on
an'i·on'ic
an'i·sa·ki'a·sis
·ses'
an'ise
an'i·seed'
an·i·sei·ko'ni·a
an·i·sei·kon'ic
a·nis'ic acid
an'i·so·co'ri·a
an·i·so·gam'ete
an·i'sog·a·mous
an·i'sog·a·my
an·i·so·met'ric
an·i'so·me·tro'pi·a
an·i·so·me·trop'ic
an·i·so·trop'ic
an·i·sot'ro·py *or* ·ro·pism
an'kle·bone'
an'ky·lose'
·losed' ·los'ing
an'ky·lo'sis
·ses
an·ky·los'to·ma
an·ky·los'to·mi'a·sis
·ses'
an'ky·lot'ic
an'ky·roid'
an'la·ge
·gen *or* ·ges
an·nat'to
an·neal'
an·neal'er
an·nec'tent
an'ne·lid
an·nel'i·dan

an′nu·lar
an′nu·lar′i·ty
an′nu·late or ·lat′ed
an′nu·la′tion
an′nu·lus
 ·li′ or ·lus·es
an·o′dal or ·od′ic
an′ode
an′o·dize′
 ·dized′ ·diz′ing
an′o·don′ti·a
an′o·dyne′
an′o·dyn′i·a
an′o·dyn′ic
a·noi′a
a·nom′a·lism
a·nom′a·lis′tic
a·nom′a·lous
a·nom′a·ly
 ·lies
a·no′mi·a
a·nom′ic
an′o·mie or ·my
an′o·mous
 (without shoulders;
 SEE animus)
a·noph′e·les′
a·noph′e·line′
an′o·pho′ri·a
an′o·rec′tal
an′o·rec′tum
an′o·rex′i·a ner·
 vo′sa
an′o·rex′ic or
 ·rec′tic or ·ret′ic
an·os′mi·a
an·os′mic

an′ox·e′mi·a
an′ox·e′mic or
 ·e·mat′ic
an·ox′i·a
an·ox′ic
an′sa
 ·sae
ansa cer′vi·ca′lis
an′sate
an′si·form′
an·sot′o·my
An′ta·buse′
 (trademark)
ant·ac′id
an·tag′o·nism
an·tag′o·nist
an·tag′o·nis′tic
an·tag′o·nize′
 ·nized′ ·niz′ing
ant·al·ge′sic or
 ·al′gic
ant·al′ka·li′
 ·lies′ or ·lis′
ant·al′ka·line
ant·ar·thrit′ic
an·taz′o·line′
 hydrochloride
an′te·bra′chi·al
an′te·bra′chi·um
an′te·ced′ent
an′te ci′bum
an′te·flect′
an′te·flex′ion
an′te·mor′tem
an′te·na′tal
an′te·par′tal
an′te par′tum

an′te·py·ret′ic
an·te′ri·or
an′ter·o·grade′
an′ter·o·in·fe′ri·or
an′te·ver′sion
an′te·vert′
ant′hel·min′tic
an′tho·cy′a·nin or
 ·cy′an
an′thra·ce′mi·a
an′thra·cene′
an·thra′ci·a
an′thra·coid′
an′thra·co·ne·cro′
 sis
 ·ses
an′thra·co′sis
 ·ses
an′thra·lin
an′thra·nil′ic acid
an′thra·qui·none′
an′thrax
 ·thra·ces′
an′thro·po·cen′tric
an′thro·poid′
an′thro·poi′dal
an′thro·po·log′i·cal
 or ·log′ic
an′thro·pol′o·gist
an′thro·pol′o·gy
an′thro·pom′e·ter
an′thro·po·met′ric
 or ·met′ri·cal
an′thro·pom′e·try
an′thro·poph′a·
 gous
an′thro·poph′a·gy

an'thro·po·phil'ic
an'ti·a·bor'ti·fa'cient
an'ti·ad·ren·er'gic
an'ti·an'ti·bod'y
 ·bod·ies
an'ti·an'ti·tox'in
an'ti·anx·i'e·ty
an'ti·ar·rhyth'mic
an'ti·bac·te'ri·al
an'ti·bi·o'sis
 ·ses
an'ti·bi·ot'ic
an'ti·bod'y
 ·bod·ies
an'ti·bra'chi·um
an'ti·car'i·o·gen'ic
 or ·car'i·us
an'ti·cat'a·lyst
an'ti·cath'ode
an'ti·chlor'
an'ti·cho·les'ter·e'mic
an'ti·cho'lin·er'gic
an'ti·cho'lin·es·ter·ase'
an'ti·cli'nal
an'ti·co·ag'u·lant
an'ti·co·ag'u·lin
an·ti'cus
an'ti–D'
an'ti·de·pres'sant
an'ti·di·u·ret'ic
an'ti·dot'al
an'ti·dote'
an'ti·drom'ic
an'ti·e·met'ic

an'ti·fe'brile
an'ti·gen
an'ti·gen'ic
an'ti·ge·nic'i·ty
an'ti·he'lix
 ·hel'i·ces' or
 ·he'lix·es
an'ti·his'ta·mine'
an'ti·his'ta·min'ic
an'ti·hy'per·lip'o·pro'te·in·e'mic
an'ti·hy'per·on'
an'ti·hy'per·ten'sive
an'ti·hy'po·ten'sive
an'ti–ic·ter'ic
an'ti·lew'is·ite
an'ti·lith'ic
an'ti·ma·lar'i·al
an'ti·mere'
an'ti·me·tab'o·lite'
an'ti·mo'ni·al
an'ti·mon'ic
an'ti·mo'nous or
 an'ti·mo'ni·ous
an'ti·mo'ny
an'ti·mo·nyl'
an'ti·neu·tri'no
 ·nos
an'ti·neu'tron
an·tin'i·on
an'ti·ox'i·dant
an'ti·path'ic
an·tip'a·thy
 ·thies'
an'ti·pe'ri·od'ic
an'ti·per'i·stal'sis
 ·ses

an'ti·per'spir·ant
an'ti·phlo·gis'tic
an·tip'o·dal
an'ti·pode'
 an·tip'o·des'
an·tip'o·de'an
an'ti·pro'ton
an'ti·py·ret'ic
an'ti·py'rine
an'ti·ra·chit'ic
an'ti·scor·bu'tic
an'ti·sep'sis
 ·ses
an'ti·sep'tic
an'ti·sep'ti·cize'
 ·cized' ·ciz'ing
an'ti·se'rum
 ·rums or ·ra
an'ti·so'cial
an'ti·spas·mod'ic
an'ti·the'nar
an'ti·tox'ic
an'ti·tox'in
an·ti'tra·gus
 ·gi
an'ti·tus'sive
an'ti·ven'in
an'ti·viv'i·sec'tion
an'ti·viv'i·sec'tion·ist
an'tral
an'trum
 ·tra or ·trums
an·u're·sis
 ·ses
an·u·ret'ic
an·u'ri·a

an·u′ric
a′nus
   a′nus·es *or* a′ni
an′vil
anx·i′e·ty
   ·ties
anx′i·o·lyt′ic
a·or′ta
   ·tas *or* ·tae
a·or′tic *or* ·tal
a·or′to·cor′o·nar′y
a·or′to·gram′
a′or·tog′ra·phy
ap′a·thet′ic
ap′a·thy
   ·thies
ap′a·tite′
APC
a·pel′lous
ap′en·ter′ic
a·pe′ri·ent
a′pe·ri·od′ic
a′per·tu′ra
   ·rae
ap′er·ture
a′pex
   a′pex·es *or* ap′i·ces
a·pex′i·graph′ *or* ·o·graph′
Ap′gar score
a·pha′ki·a *or* ·ci·a
a·pha′kic *or* ·cic
a·pha′si·a
a·pha′sic *or* ·si·ac′
a·pho′ni·a
a′phon′ic

aph′ro·dis′i·ac′
aph′tha
   ·thae
aph·tho′sis
   ·ses
aph′thous
ap′i·cal
ap′i·ces′
   (sing. a′pex)
ap′la·nat′ic
a·pla′si·a
a·plas′tic
ap·ne′a *or* ·noe′a
ap·ne′ic *or* ·noe′ic
ap′o·chro′mat
ap′o·chro·mat′ic
a·poc′o·pe′
ap′o·cop′tic
ap′o·crine′
a·poc′y·na′ceous
a·poc′y·nin
ap′o·dal *or* ·dous
ap′o·en′zyme
ap′o·ge′an *or* ·ge′al
ap′o·gee′
a·po′lar
ap′o·lep′sis
ap′o·lip′o·pro′tein
ap′o·mor′phine
ap′o·neu·ro′sis
   ·ses
ap′o·neu·rot′ic
a·poph′y·se′al *or* ap′o·phys′i·al
a·poph′y·sis
   ·ses′

ap′o·plec′tic *or* ·plec′ti·cal
ap′o·plec′toid
ap′o·plex′y
ap′o·pro′tein
ap′o·re·pres′sor
a·pos′ta·sis
   ·ses′
a·poth′e·car′y
   ·ies
ap′o·them′
ap′o·zem′ *or* ap·oz′e·ma *or* ap′o·zeme′
ap′pa·ra′tus
   ·tus *or* ·tus·es
ap·par′ent
ap·pen′dage
ap′pen·dec′to·my
   ·mies
ap·pen′di·ci′tis
ap·pen·dic′u·lar
ap·pen′dix
   dix·es *or* ·di·ces′
ap′per·ceive′
   ·ceived′ ·ceiv′ing
ap′per·cep′tion
ap′per·cep′tive
ap′pe·stat′
ap′pe·tence
ap′pe·ten·cy
   ·cies
ap′pe·tite′
ap′pe·ti′tive
ap′pla·na′tion
ap′pla·nom′e·ter
ap·pli′ance

ap'pli·ca·bil'i·ty
ap'pli·ca·ble
ap'pli·ca'tion
ap'pli·ca'tive
ap'pli·ca'tor
ap·ply'
  ·plied' ·ply'ing
ap·pos'a·ble
ap·pose'
  ·posed' ·pos'ing
  (to place near; SEE
  oppose*)*
ap'po·site
ap·po·si'tion
ap·pre·hen'sion
ap·pre·hen'sive
ap·proach'
ap·prox'i·mal
ap·prox'i·mate *adj.*
ap·prox'i·mate'
  ·mat'ed ·mat'ing
ap·prox'i·ma'tion
a·prax'i·a
a·prax'ic or
  ·prac'tic
a'pron
ap'ti·tude'
a'pus
a'py·ret'ic
aq'ua
  ·uas or ·uae
aqua for'tis
aq'ua·punc'ture
aqua pu'ra
aqua re'gi·a
aq'ue·duct'
aq'ue·duc'tal

a'que·ous
aqueous humor
ar'a–A
a·rab'i·nose'
a·rach'nid
a·rach'ni·dan
a·rach'noid
Ar·an'–Du·chenne'
  disease
A·ran'ti·us' body
ar'a·ro'ba
ar'bor
  ·bo·res'
ar·bo're·al
ar·bo're·ous
ar'bo·res'cence
ar'bo·res'cent
ar'bor·i·za'tion
ar'bor vi'tae
ar'bo·vi'rus
ARC
arc
  arced *or* arcked
  arc'ing *or* arck'ing
ar·cade'
ar·ca'num
  ·na *or* ·nums
arch
ar'chen·ter'ic
ar'chen·ter·on'
  ·ter·a
ar'che·spore' *or* ar'-
  che·spo'ri·um
ar'che·spo'ri·al
ar'che·typ'al *or*
  ·typ'i·cal
ar'che·type'

ar'chi·blast'
arch'o·plasm *or*
  ·i·plasm
ar'cu·ate
ar'cu·a'tion
ar'cus
ar'dent
ar'dor
ar'e·a
  ·as *or* ·ae'
a're·na·vi'rus
  ·rus·es
a·re'o·la
  ·lae' *or* ·las
a·re'o·lar
a·re'o·late
a·re'o·la'tion
a·re'o·le'
  ·les'
ar'ge·ma
ar·gen'taf'fin *or*
  ·fine'
ar'gen·taf'fi·no'ma
ar·gen'tic
ar'gen·tine
ar·gen'tum
  ·ti
ar'gi·nase'
ar'gi·nine'
ar'gi·ni'no·suc·cin'-
  ic·ac'i·du'ri·a
ar'gon
ar·gyr'i·a
ar·gyr'ic
ar'gy·rism
a·rhyth'mi·a
a·rhyth'mic

Ar'i·as–Stel'la reaction
a·ri'bo·fla'vin·o'sis
  ·ses
a·ris'to·gen'ics
ar'ma·men·tar'i·um
  ·i·a
ar'ma·ture
arm'pit'
ar'ni·ca
Ar'nold's canal
a·ro'ma
ar'o·mat'ic
a·ro'ma·ti·za'tion
a·ro'ma·tize'
  ·tized' ·tiz'ing
a·rous'al
a·rouse'
  ·roused' ·rous'ing
ar'o·yl
ar'rache·ment'
ar·range'
  ·ranged' ·rang'ing
ar·range'ment
ar·rec'tor
  ar'rec·to'res
ar·rest'
ar·rhe'no·blas·to'ma
  ·mas or ·ma·ta
ar·rhyth'mi·a
ar·rhyth'mic or ·mi·cal
ar'sa·nil'ic acid
ar'sen·ate'
ar'se·ni'a·sis
ar'sen·ic' n.

ar·sen'ic adj.
ar·sen'i·cal
ar'se·nide'
ar·se'ni·ous
ar·se'niu·ret'ed or ·ret'ted
ar·sen'o·blast'
ar'se·nous
ar·sine'
ars·phen'a·mine'
ar'te·fact'
ar·te'ri·a
  ·ae'
ar·te'ri·al
ar·te'ri·al·i·za'tion
ar·te'ri·al·ize'
  ·ized' ·iz'ing
ar·te'ri·o·fi·bro'sis
  ·ses
ar·te'ri·o·gram'
ar·te'ri·og'ra·phy
ar·te'ri·o'la
  ·lae
ar·te'ri·o'lar
ar·te'ri·ole'
ar·te'ri·o·lo·scle·ro'sis
  ·ses
ar·te'ri·o·lo·scle·rot'ic
ar·te'ri·o·pres'sor
ar·te'ri·o·scle·ro'sis
  ·ses
ar·te'ri·o·scle·rot'ic
ar·te'ri·o·tome'
ar·te'ri·ot'o·my

ar·te'ri·ot'o·ny
ar·te'ri·o·ve'nous
ar·te'ri·o·ver'sion
ar·te'ri·o·ver'ter
ar·te·ri'tis
ar'ter·y
  ·ies
ar·thrag'ra
ar·thral'gia
ar'threm·pi·e'sis
  ·ses
ar·thrit'ic
ar·thri'tis
ar·throc'a·ce
ar'thro·cen·te'sis
  ·ses
ar'thro·em·pi·e'sis
  ·ses
ar'thro·pod'
ar·throp'o·dal or ·dan or ·dous
ar'thro·scope'
ar·thros'co·py
ar'thro·spore'
ar'throus
ar·throx'e·sis
Ar'thus' reaction
ar·tic'u·lar
ar·tic'u·la're
ar·tic'u·late adj.
ar·tic'u·late'
  ·lat'ed ·lat'ing
ar·tic'u·la'ti·o
  ·la'ti·o'nes
ar·tic'u·la'tion
ar·tic'u·la'tor
ar·tic'u·la·to'ry

21

ar′ti·fact′
ar′ti·fi′cial
ar′ti·fi′ci·al′i·ty
  ·ties
ar′y·ep′i·glot′tic
ar·yt′e·no·ep′i·glot′tic
ar′y·te′noid
as′a·fet′i·da or ·foet′i·da
as·bes′tine
as·bes′tos or ·tus
as·bes·to′sis
  ·ses
A′–scan
as′ca·ri′a·sis
  ·ses′
as′ca·rid
  as′ca·rids or as·car′i·des′
as′car·is
  as·car′i·des′
as·cend′ing
as·cer·tain′ment
asc″hel·minth
Asch′off's bodies
as′ci
  (sing. as′cus)
as·ci′a
as·ci′tes
as·ci′tic
as′co·carp′
as′co·car′pous
as′co·go′ni·um
  ·ni·a
As·co′li's test
as′co·my·cete′

as′co·my·ce′tous
a·scor′bate
a·scor′bic acid
as′co·spore′
as′cus
  ·ci
a·se′mi·a
a·sep′sis
  ·ses
a·sep′tic
a·sex′u·al
a·sex′u·al′i·ty
a·sex′u·al·i·za′tion
a′si·a′li·a
A′sian flu
A′si·at′ic cholera
a·sid′er·o′sis
a·sit′i·a
a·so′cial
as·par′a·gin·ase′
as·par′a·gine′
as·par′tame′
as·par′tic acid
a′spe·cif′ic
as′pect
as′per·gil·lo′sis
  ·ses
as′per·gil′lus
  ·li
as·phyx′i·a
as·phyx′i·al
as·phyx′i·ant
as·phyx′i·ate′
  ·at′ed ·at′ing
as·phyx′i·a′tion
as·phyx′i·a′tor
as·pid′i·um

as′pi·rate n.
as′pi·rate′
  ·rat′ed ·rat′ing
as′pi·ra′tion
as′pi·ra·tor
as·pir′a·to′ry
as′pi·rin
as′sa·fet′i·da or ·foet′i·da
as·sault′
as′say n.
as·say′ v.
as′si·dent
as·sim′i·la·ble
as·sim′i·late′
  ·lat′ed ·lat′ing
as·sim′i·la′tion
as·sim′i·la′tive or ·la·to′ry
as·sist′ant
as·so′ci·a·ble
as·so′ci·ate n.
as·so′ci·ate′
  ·at′ed ·at′ing
as·so′ci·a′tion
as·so′ci·a′tion·ism
as·so′ci·a′tive
as′so·nance
as·sort′ment
a·sta′si·a
a·stat′ic
as·ta′tine′
as′ter
as·te′ri·on
  ·ri·a
as·ter·ix′is
a·ster′nal

as·ter·oid'
as·the'ni·a
as·then'ic
as'the·nom'e·ter
as'the·no'pi·a
as'the·nop'ic
asth'ma
asth·mat'ic *or*
  ·mat'i·cal
as·tig·mat'ic
a·stig'ma·tism
as'tig·ma·tom'e·ter
  or ·tig·mom'e·ter
a·stom'a·tous
as'tra·gal
as·trag'a·lar
as·trag'a·lus
  ·li'
as'tral'
as·trict'
as·tric'tion
as·tringe'
  ·tringed' ·tring·ing
as·trin'gen·cy
as·trin'gent
as'tro·cyte'
as'tro·cyt'ic
as·trog'li·a
as'tro·sphere'
a·sy'lum
a'sym·met'ri·cal *or*
  ·met'ric
a·sym'me·try
  ·tries
a·symp'to·mat'ic
as'ymp·tot'ic
a·syn'cli·tism

a·syn'er·gy
a·sys'to·le
a'sys·tol'ic
a·tac'tic
a·tac'ti·form'
at'a·rac'tic *or*
  ·rax'ic
at'a·rax'i·a *or*
  ·rax'y
at'a·vism
at'a·vist
at'a·vis'tic *or* a·
  tav'ic
a·tax'i·a *or* ·tax'y
a·tax'ic
at'e·brin
at'e·lec'ta·sis
  ·ses'
a·te'li·a
a·te'li·ot'ic
at'e·lo·car'di·a
a·ther'man·cy
a·ther'ma·nous
ath'e·ro·em'bo·lus
  ·bo·li
ath'er·o'ma
  ·mas *or* ·ma·ta
ath'er·o·ma·to'sis
  ·ses
ath'er·o'ma·tous
ath'er·o·scle·ro'sis
  ·ses
ath'er·o·scle·rot'ic
ath'e·toid
ath'e·to'sis
  ·ses
ath'lete's' foot

ath·let'i·cism
at·lan'tad
at·lan'tal
at·lan'to·ax'i·al
at'las
at'lo·ax'loid
at·mom'e·ter
at'mos·phere'
at'mos·pher'ic *or*
  ·pher'i·cal
a·to'ci·a
at'om
a·tom'ic
at'om·i·za'tion
at'om·ize'
  ·ized' ·iz'ing
at'om·iz'er
at'o·my
  ·mies
a·ton'ic
a'to·nic'i·ty
at'o·ny
at'o·pen
at'o·py
ATP
at'ra·bil'ious *or*
  ·bil'iar
a·tre'si·a
a·tret'ic *or* ·tre'sic
a'tri·al
at·tri'o·nec'tor
a'tri·o·nod'al
a'tri·o·ven·tric'u·lar
a'tri·um
  ·tri·a *or* ·tri·ums
a·tro'phi·a
a·troph'ic

23

at′ro·pho·der′ma
at′ro·phy
  ·phied ·phy·ing
at′ro·pine′ *or* ·pin
at·ro′pin·i·za′tion
at·tach′ment
at·tack′
at·taint′
  ·taint′ed ·taint′ing
at·tend′ant
at·tend′ing
at·ten′tion
at·ten′u·ant
at·ten′u·ate′
  ·at′ed ·at′ing
at·ten′u·ate *adj.*
at·ten′u·a′tion
at·ten′u·a′tor
at′tic
at′ti·cot′o·my
at′ti·tude′
at′ti·tu′di·nal
at·tol′lens
at·tract′ant
at·trac′tion
at·tri′tion
a·typ′i·a
a·typ′i·cal *or* ·typ′ic
Aub′–Du·bois′ table
au′dile
au′di·o-fre′quen·cy
au′di·o·gen′ic
au′di·o·gram′
au′di·o·log′i·cal
au′di·ol′o·gist

au′di·ol′o·gy
au′di·om′e·ter
au′di·o·met′ric
au′di·om′e·trist
au′di·om′e·try
au′di·o-vis′u·al
au′di·phone′
au·di′tion
au′di·tive
au′di·to′ry
Au′er's bodies
aug′ment *n.*
aug·ment′ *v.*
aug′men·ta′tion
aug·ment′or
au′la
au′ra
  ·ras *or* ·rae
au′ral
  (*of the ear;* SEE oral)
au′ric
au′ri·cle
au·ric′u·la
  ·las *or* ·lae′
au·ric′u·lar
au·ric′u·late
au·ric′u·lo·ven·tric′u·lar
au′ri·form′
au′ris
  ·res
au′ro·ther′a·py
au′rous
au′rum
aus·cult′

aus′cul·tate′
  ·tat′ed ·tat′ing
aus′cul·ta′tion
aus′cul·ta′tor
aus·cul′ta·to′ry
Aus·tral′ia antigen
au′ta·coid′
au·te′cious
au′then·tic′i·ty
au′tism
au·tis′tic
au′to-an′a·lyz′er
au′to-an′ti·bod′y
  ·bod′ies
au′to-an′ti·gen
au′to·ca·tal′y·sis
  ·ses′
au·toch′thon
  ·thons *or* ·tho·nes′
au·toch′tho·nous *or* ·tho·nal
au′to·clave′
  ·claved′ ·clav′ing
au′to·coid′
au·toe′cious
au·toe′cism
au′to·e·rot′ic
au′to·er′o·tism *or* ·e·rot′i·cism
au·tog′a·mous
au·tog′a·my
au′to·gen′e·sis
  ·ses′
au′to·ge·net′ic
au·tog′e·nous *or* au′to·gen′ic
au′to·hyp·no′sis

au·to·hyp·not′ic
au·to·im·mune′
au·to·im·mu′ni·ty
au·to·in·fec′tion
au·to·in·oc·u·la′tion
au·to·in·tox·i·ca′tion
au·to·ki·net′ic
au·tol′o·gous
au·tol′y·sate
au·tol′y·sin
au·tol′y·sis
·ses
au·to·lyt′ic
au·to·lyze′
·lyzed′ ·lyz′ing
au·to·mat′ic
au·tom′a·tism
au·tom′a·tist
au·tom′a·ti·za′tion
au·tom′a·tize
·tized′ ·tiz′ing
au·tom′a·ton′
·tons′ or ·ta
au·to·nom′ic
au·ton′o·mous
au·ton′o·my
au·to–ox·i·da′tion
au·to·plas′tic
au·to·plas′ty
au′top·sy
·sies
au′top′sy
·sied ·sy·ing
au·to·ra′di·o·graph′
or ·gram′

au·to·ra′di·o·
graph′ic
au·to·ra′di·og′ra·
phy
·phies
au·to·so′mal
au′to·some′
au·to·sug·ges′tion
au·to·tox·e′mi·a or
·ae′mi·a
au·to·tox′ic
au·to·tox′in
au·to·troph′
au·to·troph′ic
au·tox′i·da′tion
au·tox′i·da′tive
aux·e′sis
·ses
aux·et′ic
aux·il′ia·ry
aux′o·troph′
aux′o·troph′ic
a·vail′a·bil′i·ty
av′er·age
a·ver′sion
a·ver′sive
a′vi·an
a′vi·a′tion medicine
av′i·din
a·vid′i·ty
a′vir′u·lent
a′vi·ta·min·o′sis
·ses
A′vo·ga′dro's law
a·void′a·ble
a·void′ance
av′oir·du·pois′

a·vul′sion
a·wake′
a·woke′ or
a·waked′
a·waked′
a·wak′ing
a·wak′en
a·xe′nic
ax′es
(sing. ax′is)
ax′i·al
ax·il′la
·lae or ·las
ax′il·la′ry
ax′il·lo·fem′o·ral
ax′i·o·cer′vi·cal
ax′i·on
ax′is
·es
ax·og′e·nous
(originating in an axon; SEE exogenous)
ax′on
a′ya·huas′ca
A·yer′za's
syndrome
a′yur·ve′da
a′yur·ve′dic
az′a·thi′o·prine′
a·zed′a·rach′
a·ze′o·trope′
a′ze·o·trop′ic
az′ide or az′id
az′o
az′o·ben′zene
az′ole

az′ote
az′o·te′mi·a
az′o·te′mic
az′o·tize′
  ·tized′ ·tiz′ing
a·zo′to·bac′ter
Az′tec two–step
az′u·ro·phil or
  ·phile′
az′y·gos
  (unpaired part)
az′y·gous
  (unpaired)

B

Bab′bitt metal
bab′ble
  ·bled ·bling
ba·be′si·a
ba·be′si·o′sis or ba·be·si′a·sis
  ·ses
Ba·bin′ski's reflex
ba′by
  ·bies
bac′ca
  ·cae
bac′cate
bac′ci·form′
Bach′mann's bundle
bac·il·lar′y or ba·cil′lar
bac′il·le′mi·a
ba·cil′li
  (sing. ba·cil′lus)
ba·cil′li·form′
ba·cil′lin
bac′il·lo′sis
  ·ses
ba·cil′lus
  ·li
bac′i·tra′cin
back′ache′
back′bone′
back′cross′
back′flow′
back′pack′
back′rest′
back′scat′ter·ing or ·scat′ter
back′side′
back′up′ or back′–up′
back′ward or ·wards
bac·te·re′mi·a
bac·te′ri·a
  (sing. bac·te′ri·um)
bac·te′ri·al
bac·te·ri·ci′dal
bac·te·ri·cide′
bac′ter·id
bac·te′ri·o·cin
bac·te′ri·o·log′ic or ·log′i·cal
bac·te′ri·ol′o·gist
bac·te′ri·ol′o·gy
bac·te′ri·ol′y·sis
bac·te′ri·o·lyt′ic
bac·te′ri·o·phage′
bac·te′ri·o·pha′gi·a
bac·te′ri·o·phag′ic
bac·te′ri·oph′a·gy
bac·te′ri·op′so·nin
bac·te′ri·o·rho·dop′sin
bac·te′ri·os′co·py
bac·te′ri·o·sta′sis
  ·ses
bac·te′ri·o·stat′
bac·te′ri·o·stat′ic
bac·te′ri·um
  ·ri·a
bac·te′ri·u′ri·a
bac·te′ri·u′ric
bac·te·ri·za′tion
bac·te·rize′
  ·rized′ ·riz′ing
bac′te·roid′ or bac′te·roid′al
bac′te·roi′des
ba·cu′li·form′
bad
  worse worst
bag·as·so′sis
  ·ses
bal′ance
  ·anced ·anc·ing
ba·lan′ic
bal′a·ni′tis
bal′a·no·plas′ty
bal′a·no·pos·thi′tis
bal′a·no·pre·pu′tial
bal′a·nor·rha′gi·a
bal′an·ti·di′a·sis
  ·ses′
bald′head′ed

ball′-and-sock′et joint
bal·lis′tic
bal·lis′tics
bal·lis′to·car′di·o·gram′
bal·lis′to·car′di·o·graph′
bal·loon′
bal·lotte′ment
balm
bal′ne·al
bal′ne·ol′o·gy
bal′sam
bal·sam′ic
band′age
  ·aged ·ag·ing
Band′-Aid′ *(trademark)*
ban′dy·leg′ged
bane
bang
bank, human-milk
bar
bar′ag·no′sis
bar·bas′co
  ·cos *or* ·coes
bar′ber's itch
bar′bi·tal′
bar·bi′tu·rate
bar·bi·tu′ric acid
bar′bo·tage′
bare
  bared bar′ing
  (uncover; SEE bear*)*
bare′foot′
bare′foot′ed

bar′es·the′si·a
bar′es·the′si·om′e·ter
bar′i·at′ric
bar′i·at′rics
bar′ic
bar′i·um
bar′o·cep′tor
bar′og·no′sis
bar′o·re·cep′tor
bar′o·scope′
bar′o·scop′ic
bar′o·ti′tis
Barr body
bar′rel-chest′ed
bar′ren
bar′ri·er
bar′tho·lin·i′tis
Bar′tho·lin's glands
bar′to·nel·li′a·sis *or* ·nel·lo′sis
  ·ses
bar′y·la′li·a
ba·ry′ta
ba·ryt′ic
bas′ad
bas′al
base
  bas′es
base line
base′plate′
bas′es *(sing.* base*)*
ba′ses *(sing.* ba′sis*)*
ba′si·al
bas′ic

ba·sic′i·ty
ba′si·cra′ni·al
ba·sid′i·al
ba·sid′i·o·my′cete
ba·sid′i·o·my·ce′tous
ba·sid′i·o·spore′
ba·sid′i·o·spo′rous
ba·sid′i·um
  ·i·a
bas′i·fy′
  ·fied′ ·fy′ing
bas′i·lar *or* ·lar′y
ba′si·lat′er·al
ba·sil′ic *or* ·sil′i·cal
bas′i·lo′ma
  ·mas *or* ·ma·ta
ba′sin
ba′si·on
ba·sip′e·tal
ba′sis
  ·ses
ba′si·sphe′noid
bas′ket
ba′so·phile′ *or* ·phil′
ba′so·phil′i·a
ba′so·phil′ic
bas′si·net′
bas′tard
bath
  baths
bathe
  bathed bath′ing
bath′robe′
bath′tub′
bath′y·an·es·the′sia

27

bath'y·es·the'sia
bath'y·hyp'es·the'sia
ba·tra'cho·tox'in
bat'tered–child' syndrome
bat'ter·y
·ies
bat'tle fatigue
bat'tle scarred'
Bau'de·locque's' method
Bau·mé's' scale
bay'o·net leg
BCG (vaccine)
bdel'lo·vib'ri·o
bead'ing
beak'er
beam
bear
  bore, borne or born,
  bear'ing
  (carry; see bare)
bear'a·ble
beat
  beat beat'en
  beat'ing
beau'ty spot
be·bee'rine
bed'bug'
bed'clothes'
bed'ding
bed'fast'
bed'lam
bed'lam·ite'
bed lin'en

bed'pan'
bed'rail'
bed rest
bed'rid'den or ·rid'
bed'side'
bed'sore'
bed'–wet'ting
bees'wax'
be·have'
  ·haved' ·hav'ing
be·hav'ior
be·hav'ior·al
be·hav'ior·al·ism
be·hav'ior·ism
be·hav'ior·ist
be·hav'ior·is'tic
bel
belch
bel'la·don'na
bell'–crowned'
Bell's palsy
bel'ly
  ·lies
bel'ly·ache'
bel'ly·band'
bel'ly·but'ton
bel'o·noid
bel'o·no·ski·as' co·py
ben·act'y·zine'
Ben'a·dryl
  (trademark)
bends, the
Ben'e·dict's test
Ben'e·dikt's syndrome
be·nign'

be·nig'nant
ben·zal'de·hyde'
ben'zal·ko'ni·um chloride
Ben'ze·drine'
  (trademark)
ben'zene
ben'zi·dine'
ben'zine
ben·zo·ate'
ben'zo·caine'
ben·zo'ic
ben'zo·in
ben'zol
ben'zo·phe·none'
ben'zo·py'rene or benz'py'rene
ben'zo·yl
ben'zyl
ben'zyl·pen'i·cil'lin
ber'ber·ine'
ber'ga·mot
ber'i·ber'i
berke'li·um
Ber·nard's' duct
Ber·noul'li's principle
ber'ry
  ·ries
ber·serk'
Ber'til·lon' system
be·ryl'li·o'sis
be·ryl'li·um
bes'ti·al'i·ty
  ·ties
bes'tial·ize'
  ·ized' ·iz'ing

be′ta
be·ta–ad′re·ner′gic
beta blocker
be′ta·cism
be′ta·ine′
be′ta·naph′thol
be′ta·tron′
be′ta·zole
  hydrochloride
be′tel
be·than′e·chol
  chloride
bet′ter
Betz's cells
bev *or* Bev
bev′el
  ·eled *or* ·elled
  ·el·ing *or* ·el·ling
be′zoar
bhang
bi′ar·tic′u·lar
bi′aur′al
bi′au·ric′u·late *or*
  ·u·lar
bi·ax′i·al
bi·bas′ic
bi′-bi·va′lent
bib′li·o·mane′
bib′li·o·ma′ni·a
bib′li·o·ma′ni·ac
bib′u·lous
bi·cam′er·al
bi·cap′su·lar
bi·car′bon·ate
bi·cel′lu·lar

bi·ceph′a·lous *or*
  bi′ce·phal′ic
bi′ceps
  ·ceps *or* ·ceps·es
Bi·chat's′ canal
bi·chlo′ride
bi·chro′mate
bi·cip′i·tal
bi′con·cave′
bi′con·vex′
bi′corn *or* bi·cor′-
  nu·ate
bi·cor′po·rate
bi·cus′pid *or*
  ·cus′pi·date′
bi·den′tate
bi·det′
bid′u·ous
bi′fid
bi·fid′i·ty
bi′flex
bi·fo′cal *adj.*
bi′fo′cal *n.*
bi′fo′cals
bi′forked′
bi′form′
bi′fur·cate′
  ·cat′ed ·cat′ing
bi′fur·ca′tion
big
  big′ger big′gest
bi·gem′i·nal
bi·gem′i·ny
bi·lat′er·al
bi·lat′er·al·ism
bile′stone′
bil·har′zi·a

bil′har·zi·a′sis
bil·har′zic
bil′i·ar′y
bil′i·gen′e·sis
bil′i·gen′ic
bil′ious
bil′i·ra′chi·a
bil′i·ru′bin
bil′i·ru′bi·ne′mi·a
bil′i·ru′bi·nu′ri·a
bil′i·u′ri·a
bil′i·ver′din
bil·lo′bate *or* bi·
  lobed′ *or* bi·lo′-
  bat·ed
bi·lob′u·lar
bi·loc′u·lar *or*
  ·u·late
bi·ma′nous
bi·man′u·al
bi·mes′tri·al
bi·mod′al
bi·mo·dal′i·ty
bi·mo·lec′u·lar
bi′na·ry
  ·ries
bin·au′ral
bind
  bound bind′ing
bind′er
Bi·net′–Si′mon test
bin·oc′u·lar
bin·oc′u·lar′i·ty
bi·no′mi·al
bin·ov′u·lar

bi·nu′cle·ate *or*
  ·cle·at′ed *or* ·cle·ar
bi′o·ac·cu′mu·la′tion
bi′o·ac′ti·va′tion
bi′o·ac′tive
bi′o·an′a·lyst
bi′o·as′say
bi′o·as′tro·nau′tics
bi′o·a·vail′a·bil′i·ty
bi′o·cat′a·lyst
bi′o·cat′a·lyt′ic
bi′o·ce·no′sis *or*
  ·coe·no′sis *or*
  ·ce′nose
bi′o·chem′i·cal
bi′o·chem′ist
bi′o·chem′is·try
bi′o·ci′dal
bi′o·cide′
bi′o·clean
bi′o·cli·mat′ic
bi′o·cli·ma·tol′o·gy
bi′o·com·pat′i·bil′i·ty
bi′o·de·grad′a·bil′i·ty
bi′o·de·grad′a·ble
bi′o·deg′ra·da′tion
bi′o·e·col′o·gy
bi′o·e·lec′tric *or*
  ·lec′tri·cal
bi′o·e·lec·tron′ics
bi′o·en·er·get′ics
bi′o·en′gi·neer′
bi′o·e·quiv′a·lence

bi′o·eth′i·cal
bi′o·eth′ics
bi′o·feed′back′
bi′o·fla′vo·noid′
bi′o·gen′e·sis
  ·ses′
bi′o·ge·net′ic *or*
  ·net′i·cal
bi·o·gen′ic
bi′o·haz′ard
bi′o·haz′ard·ous
bi′o·im′plant
bi′o·in′stru·men·ta′tion
bi′o·ki·net′ics
bi′o·log′i·cal *or*
  ·log′ic
bi′o·log′i·cals
bi·ol′o·gist
bi·ol′o·gy
bi′o·lu′mi·nes′cence
bi′o·lu′mi·nes′cent
bi·ol′y·sis
  ·ses′
bi′o·lyt′ic
bi′o·ma·te′ri·al
bi′o·me·chan′i·cal
bi′o·me·chan′ics
bi′o·med′i·cal
bi′o·med′i·cine
bi′o·me·te·or·ol′o·gy
bi′o·met′ric *or*
  ·met′ri·cal
bi′o·met′rics
bi·om′e·try

bi′o·mi′cro·scope′
bi′o·mol′e·cule′
bi·on′ic
bi·on′ics
bi′o·nom′ics
bi′o·phys′i·cal
bi′o·phys′i·cist
bi′o·phys′ics
bi′o·plasm
bi′op′sy
  ·sies
bi′o·re·vers′i·ble
bi′o·rhythm′
bi′o·rhyth·mic′i·ty
bi′o·rhyth′mics
bi·os′co·py
bi′o·sta·tis′ti·cal
bi′o·sta·tis′tics
bi′o·syn′the·sis
  ·ses′
bi′o·syn·thet′ic
bi′o·sys′te·mat′ic
bi′o·sys′te·mat′ics
bi·o′ta
bi′o·tech·nol′o·gy
bi′o·te·lem′e·try
bi′o·ther′a·py
bi·ot′ic *or* ·ot′i·cal
bi′o·tin
bi′o·tox′in
bi′o·trans′for·ma′tion
Bi·ots′ breathing
bi′o·type′
bi′o·typ′ic
bi′o·var

30

bi′o′vu·lar
bip′a·ra
  ·ras *or* ·rae
bi′pa·rcn′tal
bi′pa·ri′e·tal
bip′a·rous
bi·par′tite
bi′ped
bi·ped′al
bi·per′i·den
bi·phen′yl
bi·po′lar
bi′po·lar′i·ty
bi′po·ten′ti·al′i·ty
bi·ra′mous
bi′re·frac′tive
bi′re·frin′gence
bi′re·frin′gent
birth control
birth′mark′
birth′rate′ *or* birth rate
bis·ac′o·dyl
bis′a·cro′mi·al
bis′cuit–bake′
bi·sect′
bi·sec′tion
bi·sec′tion·al
bi·sex′u·al
bi·sex′u·al′i·ty *or* bi·sex′u·al·ism
bis·fer′i·ous
bis·il′i·ac
bis′muth
bis′muth·al
bis·mu′thic

bis′muth′ism
bis′mu·tho′sis
bis′tou·ry
  ·ries
bi·sul′fate
bi·sul′fide
bi·sul′fite
bi·tar′trate
bite
  bit, bit′ten *or* bit, bit′ing
bite′–block′
bite′lock′
bi·tem′po·ral
bite′plate′
bite′wing′
Bi′tot's spots
bit′ten
  (alt. pp. of bite)
bit′ter
bit′ters
bi·tu′men
bi·tu′mi·noid
bi·tu′mi·no′sis
bi·tu′mi·nous
bi′u·ret
bi·va′lence *or* ·len·cy
bi·va′lent
bi′ven·ter
bi′ven·tric′u·lar
Bjer′rum's sign
black′–and–blue′
black′damp′
Black Death
black eye

black′head′
black lung (disease)
black′out′
black′strap′ mo·las′ses
black′wa′ter fever
black wid′ow
blad′der
blad′der·worm′
blain
blanc fixe
blanch
bland
blan′ket
blast
blas·te′ma
  ·mas *or* ·ma·ta
blas·te′mic *or* blas′te·mat′ic
blas′to·coele′ *or* ·cele′
blas′to·coe′lic
blas′to·cyst′
blas′to·cyte′
blas′to·cy·to′ma
  ·mas *or* ·ma·ta
blas′to·derm′
blas′to·der′mic
blas′to·disc′ *or* ·disk′
blas′to·gen′e·sis
blas·to′ma
  ·mas *or* ·ma·ta
blas′to·mere′
blas′to·mer′ic
blas′to·my·cete′
blas′to·my·co′sis

blas′to·pore′
blas′to·sphere′
blas′to·spore′
blas′tu·la
 ·las *or* ·lae
blas′tu·lar
blas′tu·la′tion
bleach
blear
blear′y
 ·i·er ·i·est
blear′y–eyed′ *or* blear′eyed′
bleb
bleb′by
bleed
 bled bleed′ing
bleed′er
blem′ish
blen′nad·e·ni′tis
blen′no·gen′ic
blen′noid
blen′nor·rha′gi·a
blen′nor·rhe′a
blen′nor·rhe′al
blen·nos′ta·sis
blen′no·stat′ic
blen′no·tho′rax
blen·nu′ri·a
bleph′ar·ad·e′ni·tis
bleph′a·ral
bleph′a·rec′to·my
 ·mies
bleph′a·rism
bleph′a·ri′tis
bleph′a·ro·ad′e·ni′tis

bleph′a·ro·ath′er·o′ma
 ·mas *or* ·ma·ta
bleph′a·ron′cus
 ·ci
bleph′ar·o·plas′ty
bleph′a·ro·ple′gi·a
bleph′a·rot′o·my
blind gut
blind spot
blink
blis′ter
blis′ter·y
bloat
block·ade′
block′age
block′er
blood bank
blood count
blood group
blood′less
blood′let′ting
blood′mo·bile′
blood plasma
blood platelet
blood poisoning
blood pressure
blood′–red′
blood′root′
blood′shot′
blood′stained′
blood′stream′
blood′suck′er
blood sugar
blood type
blood′y
 ·i·er ·i·est

blood′y
 ·ied ·y·ing
blotch
blotch′y
 ·i·er ·i·est
blow
 blew blown blow′ing
blow′fly′
 ·flies′
blow′pipe′
blue baby
blue′bot′tle
blue dev′ils
blue′jack′
blue′stone′
blue vitriol
Blum′berg's sign
blur
 blurred blur′ring
blur′ry
blush
Blyth's test
B′–mode′
Bo′as' point
Boch′da·lek's′ ganglion
bod′y
 ·ies
body bag
boil
bo′lo·graph′
bo′lo·graph′ic
bo·lom′e·ter
bo′lo·met′ric
bo′lus
 ·lus·es

32

bom·bard' v.
bom'be·sin
bond
bone'let
bon'y
  ·i·er ·i·est
boost'er shot
bo·rac'ic
bo'rate
  ·rat·ed ·rat·ing
bo'rax
bor'bo·ryg'mus
  ·mi
bor'der
bor'der·line'
bor'de·tel'la
Bor·det'–Gen·gou'
  bacillus
bo'ric
bo'rism
born
  (brought into being)
borne *or* born
  (alt. pp. of bear*)*
bor'ne·ol'
Born'holm disease
bo'ron
bor·rel'i·a
boss
bos'se·lat·ed
bos'se·la'tion
bot
Bo·tal'lo's duct
botch
bot'fly'
  ·flies'

bot'ry·oid'al *or*
  bot'ry·oid'
bot'ry·o·my·co'sis
bott
bot'tle
  ·tled ·tling
bot'u·li·form
bot'u·lin
bot'u·li'nus
bot'u·lism
bou·gie'
bouil'lon
bound
bound'a·ry
  ·ries
bou·quet'
bout
bou·ton·niere' *or*
  ·nière'
bou·tons' ter·mi·
  naux'
  (sing. bou·ton'
  ter·mi·nal'*)*
bow
bow'el
bow'leg·ged
box'ing
box'–note'
Boyle's law
brace
  braced brac'ing
brace'let
bra'chi·al
bra·chi·al'gi·a
bra'chi·o·ce·phal'ic
bra'chi·o·cru'ral
bra'chi·o·cu'bi·tal

bra'chi·um
  ·chi·a
brach'y·ce·phal'ic
  or ·ceph'a·lous
brach'y·ceph'a·ly
brach'y·cra'ni·al *or*
  ·cra'nic
brach'y·cra'ny
brach'y·dac·tyl'ic
  or ·dac'ty·lous
brach'y·dac'ty·ly
brach'y·fa'cial
brach'y·gna'thi·a
brack'et
brad'y·car'di·a
brad'y·car'di·ac'
brad'y·di·as'to·le
brad'y·ki·ne'si·a
brad'y·ki·net'ic
brad'y·kin'in
brad'y·la'li·a
brad'y·lex'i·a
brad'y·lo'gi·a
brad'y·pha'si·a
brad'y·phra'si·a
brad'y·phre'ni·a
brad'y·pne'a
brad'y·sper'ma·tism
brad'y·stal'sis
Braille *or* braille
  Brailled *or* brailled
  Brail'ling *or* brail'
    ling
brain'case'
brain death
brain fever
brain'pan'

brain'sick'
brain'stem'
brain'wash'
brain wave
branch
bran'chi·al
bran'chi·ate
brash
braz'ing
breadth
break
  broke bro'ken
  break'ing
break'bone' fever
break'down'
break'out'
break'through'
breast'bone'
breast'–feed'
  –fed' –feed'ing
breath *n.*
breath'a·ble
Breath'a·lyz'er
  (trademark)
breathe
  breathed breath'ing
breath'er
breath'less
breath'y
breech
breg'ma
  ·ma·ta
breg·mat'ic
Breus mole
bridge
  bridged bridg'ing
bridge'a·ble

bridge'work'
bri'dle
brie
Bright's disease
bril'liant green
Brill's disease
brise·ment'
brit'tle
Brix scale
broach
broad'–spec'trum
Bro'die's ligament
broke
  (pt. of break*)*
bro'mate
  ·mat·ed ·mat·ing
bro'me·lain *or* bro·mel'in
bro'mic
bro'mide
bro'mi·nate'
  ·nat·ed ·nat·ing
bro'mi·na'tion
bro'mine
bro'mism
bro'mize
  ·mized ·miz·ing
Bromp'ton's cock'tail'
bron'chi
  (sing. bron'chus)
bron'chi·al
bron'chi·ec'ta·sis
bron·chil'o·quy
bron'chi·o·gen'ic
bron'chi·ole'
bron'chi·o·lec'ta·sis

bron'chi·o·lus
  ·li'
bron·chit'ic
bron·chi'tis
bron'chi·um
  ·chi·a
bron'cho·cele'
bron'cho·con·stric'tor
bron'cho·di·la'tor
bron·chog'ra·phy
bron·chop'a·thy
bron'choph'o·ny
bron'cho·pleu'ral
bron'cho·pneu·mo'ni·a
bron'cho·pul'mo·nar'y
bron'chor·rha'gi·a
bron'cho·scope'
bron'cho·scop'ic
bron·chos'co·py
bron'cho·ste·no'sis
  ·ses
bron·chos'to·my
bron·chot'o·my
bron'cho·ve·sic'u·lar
bron'chus
  ·chi
broth
brow'lift'
Brown'i·an movement
brown lung (disease)
bru·cel'la

bru·cel′lar
bru·cel·lo′sis
bru′cine
bruise
   bruised bruis′ing
bruisse·ment′
bruit
brux′ism
bu′ba
bub′ble
  ·bled ·bling
bu′bo
  ·boes
bu·bo·nal′gi·a
bu·bon′ic plague
bu·bon′o·cele′
bu·car′di·a
buc′ca
  ·cae
buc′cal
buc′ci·na′tor
buc′co·clu′sal
buc′co·clu′sion
buc′co–oc·clu′sal
buck′le
  ·led ·ling
buck′tooth′
  ·teeth′
buck′toothed′
buc·ne′mi·a
bud
  bud′ded bud′ding
Buer′ger's disease
buff′er
bu′fo·ten′ine
bug′ger·y
bulb

bul′bar
bul′bi
  (sing. bul′bus)
bul′bi·form′
bul·bi′tis
bul′bo·u·re′thral
bul′bous
  (of a bulb)
bul′bus
  ·bi
  (rounded mass)
bu·lim′i·a
bu·lim′ic
bulk
bul′la
  ·lae
bul′late
bul·lo′sis
bul′lous
bump
bun′dle
  ·dled ·dling
bundle branch
  block
bun′ion
bun′ion·ette′
Bun′sen burner
buph·thal′mos
bur *or* burr
  (in dentistry)
bu·rette′ *or* ·ret′
burke
  burked burk′ing
Bur′kitt's
  lymphoma

burn
  burned *or* burnt
  burn′ing
bur′nish
bur′nish·er
burn′out′ *n.*
burnt
  (alt. pt. & pp. of
  burn)
burp
burr *or* bur
  (in speech)
bur′sa
  ·sae *or* ·sas
bur′sal
bur·sec′to·my
bur′si·form′
bur·si′tis
bur′so·lith
bur·sop′a·thy
burst
  burst burst′ing
bu′ta·caine′ sulfate
bu′tane
bu′ta·nol′
bu′tene
butt
but′ter·fly′
  ·flies′
but′tock
but′ton·hole′
bu′tyl
bu′tyl·ene′
bu′ty·ra′ceous
bu′ty·rate′
bu·tyr′ic
bu′ty·rin

35

bu′ty·roid
bu′ty·rous
by′pass′
by′prod′uct *or* by′–prod′uct
bys′si·no′sis
  ·ses
bys′si·not′ic

C

Cab′ot's ring bodies
ca·chec′tic *or* ·chex′ic
ca·chet′
ca·chex′i·a *or* ·chex′y
cach′in·nate′
  ·nat·ed ·nat·ing
cach′in·na′tion
ca·chou′
cac′o·dyl
cac′o·dyl′ic
ca·co·ë′thes *or* ·e′thes
cac′o·geu′si·a
cac·os′mi·a
ca·cu′men
  ·cu′mi·na
ca·dav′er
ca·dav′er·ic
ca·dav′er·ine
ca·dav′er·ous
cad′mi·o′sis

cad′mi·um
ca·du′ca
ca·du′ce·us
  ·ce·i′
ca·du′ci·ty
For words beginning cae– *see also* CE–
cae′cum
  ·ca
Cae·sar′e·an *or* ·i·an
cae′si·um
ca·fé au lait′ spot
caf′feine *or* ·fein
caf′fein·ism
cais′son disease
caj′e·put *or* ·u·put
Cal′a·bar′ bean
ca·lage′
cal′a·mine′
cal′a·mus
  ·mi′
cal·ca′ne·al *or* ne·an
cal·ca′ne·o·dyn′i·a
cal·ca′ne·o·val′go·
  ca′vus
cal·ca′ne·um
  ·ne·a
cal·ca′ne·us
  ·ne·i′
cal′car
  cal·car′i·a
cal′ca·rate′
cal′car′e·ous
cal′ca·rine

cal·ca′ri·u′ri·a
cal·ce′mi·a
cal′ces
  (*sing.* calx)
cal′cic
cal′ci·co′sis
cal·cif′er·ol′
cal·cif′er·ous
cal·cif′ic
cal′ci·fi·ca′tion
cal′ci·fy′
  ·fied′ ·fy′ing
cal′ci·na′tion
cal′cine
  ·cined ·cin·ing
cal′cin·o′sis
  ·ses
cal′ci·o·ki·ne′sis
cal′ci·o·ki·net′ic
cal′ci·pec′tic *or* ·pex′ic
cal′ci·pex′is *or* cal′ci·pex′y
cal′ci·phy·lac′tic
cal′ci·phy·lax′is
cal′ci·priv′i·a
cal′ci·priv′ic
cal′cite
cal′ci·to′nin
cal′ci·um
cal′ci·u′ri·a
cal′co·spher′ite
calc′spar′
cal′cu·lif′ra·gous
cal′cu·lo′sis
  ·ses
cal′cu·lous *adj.*

cal′cu·lus
·li′ or ·lus·es
(stony mass; SEE calculous)
cal′e·fa′cient
cal′e·fac′tion
cal′en·ture
calf
calves
calf′–bone′
cal′i·ber or ·bre
cal′i·brate′
·brat′ed ·brat′ing
cal′i·bra′tion
cal′i·bra′tor
cal′i·ce′al or cal′y·ce′al
cal′i·cec′ta·sis or ca′li·ec′ta·sis
·ses′
ca′li·ces′
(sing. ca′lix)
cal′i·cle
ca·lic′u·lus
·li′
cal′i·for′ni·um
cal′i·per
cal′i·say′a bark
cal′is·then′ic or ·then′i·cal
cal′is·then′ics
ca′lix
ca′li·ces′
cal′li·per
cal·lo′sal
cal·los′i·ty
·ties

cal·lo′sum
·sa
cal′lous adj.
cal′lus n.
·lus·es
calm′a·tive
cal·mod′u·lin
cal′o·mel′
ca·lor′ic
cal′o·rie
cal·o·rif′ic
ca·lor′i·gen′ic or ·ge·net′ic
cal′o·rim′e·ter
cal′o·ri·met′ric or ·met′ri·cal
cal′o·rim′e·try
Ca·lo′ri's bursa
cal′o·ry
·ries
cal′se·ques′trim
cal·u′ster·one′
cal·var′i·al or ·var′i·an
cal·var′i·um
·var′i·a
Cal·vé′–Per′thes disease
cal·vi′ti·es′
calx
calx′es or cal′ces
ca′ly·ces′
(sing. ca′lyx)
ca′ly·cine or ca·lyc′i·nal
ca·lyc′u·lus
·li′

ca·lym′ma·to·bac·te′ri·um
·ri·a
ca′lyx
ca′lyx·es or ca′ly·ces′
cam′bi·um
·bi·ums or ·bi·a
cam′er·a
·er·as or ·er·ae′
camera lu′ci·da
camera lu′ci·das
camera ob·scu′ra
camera ob·scu′ras
cam′i·sole′
cam′o·mile′
cam′phene
cam′phol
cam′phor
cam′phor·ate′
·at′ed ·at′ing
cam·phor′ic
cam·pim′e·ter
cam·pim′e·try
cam·pot′o·my
camp′to·dac′ty·ly
camp′to·me′li·a
camp′to·me′lic
camp′to·the′cin
cam′sy·late
ca·nal′
can′a·lic′u·late or ·lic′u·lat′ed or ·lic′u·lar
can′a·lic′u·lus
·li′

ca·na′lis
·les
ca·nal′i·za′tion
ca·nal′ize
·ized ·iz·ing
can′cel·lous or
·cel·late or
·cel·lat′ed
can′cer
can·cer·a′tion
can′cer·i·gen′ic or
·cer·o·gen′ic
can′cer·ous
can′cer·pho′bi·a or
can′cer·o·pho′bi·a
can′cri·form′
can′croid′
can′crum
can′cra
can·de′la
can′di·da al′bi·cans
can′di·dal
can′di·di′a·sis
·ses′
can′di·did
can′di·din
can′dy strip′er
ca·nes′cent
ca′nine
ca·ni′ti·es
can′ker
can′ker·ous
can′na·bin
can′na·bis
can′nu·la
·lae′ or ·las

can′nu·lar or
·nu·late
can′thal
can·thar′i·des′
(sing. can′tha·ris)
can·thec′to·my
can·thi′tis
can·thor′rha·phy
can·thot′o·my
can′thus
·thi
caou·tchouc′
ca·pac′i·tance
ca·pac′i·tate′
·tat′ed ·tat′ing
ca·pac′i·tive
ca·pac′i·tor
ca·pac′i·ty
·ties
cap·il·la′ceous
cap·il·lar′ec·ta′si·a
cap·il·la′ri·a
cap·il·la·ri′a·sis
·ses′
cap·il·la·ri′tis
cap·il·lar′i·ty
cap′il·lar′y
·ies
cap′i·tal
cap′i·tate′
cap′i·ta′tion
cap′i·ta′tum
cap·i·tel′lum
ca·pit′u·lar
ca·pit′u·lum
·la
ca·pote·ment′

cap′re·o·my′cin
sulfate
cap′ric acid
cap′ro·ate′
ca·pro′ic acid
ca·pryl′ic acid
cap·sa′i·cin
cap′si·cum
cap′sid
cap′so·mer or
·mere′
cap′su·la
·lae′
cap′su·lar
cap′su·late′ or
·lat′ed
cap′su·la′tion
cap′sule
·suled ·sul·ing
cap′su·li′tis
cap′sul·ize′
·ized′ ·iz′ing
cap′su·lor′rha·phy
cap′ture
car′a·geen′
car′a·mel
ca·ram′i·phen
car′a·way′
car′ba·mate′
car·bam′ic acid
car′bam·ide′
car′ba·moyl·trans
fer·ase′
car·ban′i·on
car·bar′sone′
car′ba·zole′
car′ben·i·cil′lin

car′bide
car′bi·nol′
car′bo·cy′clic
car′bo·hy′drase
car′bo·hy′drate
car′bo·late′
　·lat′ed ·lat′ing
car·bol′ic acid
car′bo·lism
car′bo·lize′
　·lized′ ·liz′ing
car′bon
car′bo·na′ceous
car′bo·nate′
　·at′ed ·at′ing
car′bon·ate n.
car′bon·a′tion or
　·a·ta′tion
carbon black
carbon disulfide
car·bon′ic
car·bon′ic–ac′id
　gas
car′bon·if′er·ous
car·bon′i·um
car′bon·i·za′tion
car′bon·ize′
　·ized′ ·iz′ing
car′bo·nom′e·ter
car′bo·nom′e·try
car′bon·yl′
car′bon·yl′ic
car′box′y·he′mo·
　glo′bin
car·box′yl
car·box′yl·ase′

car·box′yl·ate′
　·at′ed ·at′ing
car·box′yl·a′tion
car′box′yl·ic
car′bun·cle
car·bun′cu·lar
car′ci·nec′to·my
car′ci·no·em′bry·
　on′ic
car·cin′o·gen
car′ci·no·gen′e·sis
　or ·no·ge·nic′i·ty
car′ci·no·gen′ic
car′ci·no′ma
　·mas or ·ma·ta
car′ci·no·ma·to′sis
　·ses
car′ci·nom′a·tous
car′da·mom or
　·mon
car′di·al′gi·a
car′di·ec′ta·sis
　·ses′
car′di·o·ac·cel′er·a′
　tor
car′di·o·a·or′tic
car′di·o·cele′
car′di·o·di′la·tor
car′di·o·dyn′i·a
car′di·o·gram′
car′di·o·graph′
car′di·o·graph′ic
car′di·og′ra·phy
car′di·oid′
car′di·o·in·hib′i·
　to′ry
car′di·ol′o·gist

car′di·ol′o·gy
car′di·o·my·op′a·
　thy
car′di·o·pul′mo·
　nar′y
car′di·o′ta·chom′e·
　ter
car′di·o·vas′cu·lar
car′di·o·ver′sion
car′di·o·vert′er
car·di′tis
care
　cared car′ing
ca·ri′bi
car′ies
ca·ri′na
　·nas or ·nae
ca·ri′nal
car′i·nate′ or
　·nat′ed
car′i·ous
car′i·so·pro′dol
car·min′a·tive
car′mine
car′ni·fi·ca′tion
car′ni·fy′
　·fied′ ·fy′ing
car′ni·tine′
car·nos′i·ty
ca′ro
　car′nes
car·ot′e·nase′
car′o·tene′ or ·tin
car′o·te·ne′mi·a or
　·te·no′sis
ca·rot′e·noid′ or
　·rot′i·noid′

ca·rot′id
car′o·ti·nase′
ca·rot′o·dyn′i·a or
 ·rot′i·dyn′i·a
car′pal
car·pa′le
 ·li·a
car·phen′a·zine ma·
 le′ate
car·phol′o·gy
car′pi
 (sing. car′pus)
car′po·met′a·car′
 pal
car′po·pe′dal
car′pus
 ·pi
 (wrist; SEE corpus)
car′ra·geen′ or
 ·gheen′
car′ri·er
Car·ri·ón′s′ disease
car′sick′
car′ti·lage
car′ti·lag′i·nous
car′ti·la′go
 ·lag′i·nes
car′un·cle
ca·run′cu·la
 ·lae′
ca·run′cu·lar or
 ·lous or ·late
car′va·crol′
carv′er

*For words
beginning* cary–
and caryo– *see*
KARY– *and*
KARYO–
cas·cade′
 ·cad′ed ·cad′ing
cas·car′a sa·gra′da
cas·ca·ril′la
ca′se·ase′
ca′se·ate′
 ·at′ed ·at′ing
ca′se·a′tion
case′book′
ca′se·fy′
 ·fied′ ·fy′ing
case his′to·ry (or
 stud′y)
ca′sein
ca′se·in′o·gen
ca′se·ose′
ca′se·ous
case′worm′
Ca·so′ni′s reaction
casque
casqued
cas·sette′
cas′sia
cast brace
Cas·tel·lan′i′s paint
Cas·tile′ (or cas·
 tile′) soap
cas′tor oil
cas′trate
 ·trat·ed ·trat·ing
cas·tra′tion

cas′u·al·ty
 ·ties
cas′u·is′tics
ca·tab′a·sis
 ·ses′
cat′a·bat′ic
cat′a·bol′ic
ca·tab′o·lism
ca·tab′o·lite′
ca·tab′o·lize′
 ·lized′ ·liz′ing
cat′a·caus′tic
cat′a·crot′ic
ca·tac′ro·tism
cat′a·di·crot′ic
cat′a·di′cro·tism
cat′a·gen
cat′a·gen′e·sis
cat′a·lase′
cat′a·lec′tic
cat′a·lep′sy
 ·sies
cat′a·lep′tic
cat′a·lep′ti·form′ or
 ·lep′toid
ca·tal′y·sis
 ·ses′
cat′a·lyst
cat′a·lyt′ic
cat′a·lyze′
 ·lyzed′ ·lyz′ing
cat′a·lyz′er
cat′a·me′ni·a
cat′a·me′ni·al
cat·am·ne′sis
cat·am·nes′tic
cat′a·pha′si·a

ca·taph'o·ra
  (lethargy)
cat'a·pho·re'sis
  ·ses
cat'a·pho'ri·a
  (visual abnormality)
cat'a·phor'ic
cat'a·pla'si·a
  ·si·ae
cat'a·plasm
cat'a·plas'tic
cat'a·ract'
cat'a·rac'to·gen'ic
cat·ta·ri'a
ca·tarrh'
ca·tarrh'al or
  ·tarrh'ous
cat'ar·rhine'
cat'a·stal'sis
cat'a·stal'tic
ca·tas'ta·sis
  ·ses'
cat'a·to'ni·a
cat'a·ton'ic
cat'a·tri·crot'ic
cat'a·tri'cro·tism
cat'a·tro'pi·a
catch
  caught catch'ing
catch'ment area
cat'e·chin
cat'e·chol'
cat'e·chol'a·mine'
cat'e·chol·am'in·er'
  gic
cat'e·chu'

cat'e·gor'i·cal or
  ·gor'ic
cat'e·go·ry
  ·ries
cat'e·nate'
  ·nat'ed ·nat'ing
cat'e·na'tion
ca·ten'u·late
cat'gut'
ca·thar'sis
  ·ses
ca·thar'tic or
  ·thar'ti·cal
ca·thect'
ca·thec'tic
ca·thep'sin
cath'e·ter
cath'e·ter·i·za'tion
cath'e·ter·ize'
  ·ized' ·iz'ing
cath'e·ter·o·stat'
ca·thex'is
cath'ode
cath'ode–ray' tube
ca·thod'ic
ca·thol'i·con'
cat'i·on
cat'i·on'ic
cat'lin or ·ling
ca·top'tric
ca·top'trics
CAT scan
CAT scan'ner
CAT scan'ning
cau·ca'sian or cau·
  cas'ic

cau'da
  ·dae
cau'dad
cau'dal
  (near the tail)
cau'date or ·dat·ed
cau'dle
  (drink)
caul
cau'li·flow'er ear
cau'mes·the'si·a
caus'al
cau·sal'gi·a
cau·sa'tion
caus'a·tive
cause
  caused caus'ing
caus'tic
caus·tic'i·ty
cau'ter·ant
cau'ter·i·za'tion
cau'ter·ize'
  ·ized' ·iz'ing
cau'ter·y
  ·ies
cav'a·scope'
cav'ern
ca·ver'na
  ·nae
cav'er·ni'tis
cav'er·no'ma
  ·mas or ·ma·ta
cav'er·no·si'tis
cav'ern·ous
cav'i·ta'tion
ca·vi'tis

cav′i·ty
  ·ties
ca′vum
  ·va
ca′vus
CEA
ce·as′mic
ce′cal
ce·cec′to·my
ce·ci′tis
ce′co·pli·ca′tion
ce′co·sig′moi·dos′to·my
ce′cum
  ·ca
cef′a·drox′il
ce·faz′o·lin
ce·la′tion
ce′li·lac′
ce′li·ec′to·my
ce′li·o·cen·te′sis
  ·ses
ce′li·o′ma
  ·mas or ·ma·ta
ce′li·o·my′o·si′tis
ce′li·op′a·thy
ce′li·o·scope′
ce·li′tis
cell
cel′la
  ·lae
  (enclosure; SEE
  sella)
celled
cell membrane
cel·loi′din
cel′lo·phane′

cel′lu·la
  ·lae
cel′lu·lar
cel′lu·lar′i·ty
cel′lu·lase′
cel′lule
cel′lu·li·ci′dal
cel′lu·lif′u·gal
cel′lu·lip′e·tal
cel′lu·lite′
cel′lu·li′tis
cel′lu·lo·fi′brous
cel′lu·lose′
cel′lu·los′ic
cel′lu·lous
*For words
beginning* celo—
see CELIO—,
COELO—
ce′lom *or* ce·lo′ma
ce′lo·scope′
ce′lo·zo′ic
ce·ment′
ce′men·ta′tion
ce·men′ti·cle
ce·men′to·blast′
ce·men′to·cla′si·a
ce·men′to′ma
  ·mas *or* ·ma·ta
ce·men′tum
  ·ta
ce′nes·the′sia
ce′nes·the′sic
ce′nes·the′sis
*For words
beginning* ceno—
see COENO—

ce′no·gen′e·sis
ce′no·ge·net′ic
ce·no′sis
ce′no·site
ce·not′ic
cen′sor
cen′sor·ship′
cen′ter
cen′ti·grade′
cen′ti·gram′
cen′tile
cen′ti·li′ter
cen′ti·me′ter
cen′ti·pede′
cen′ti·poise′
cen′trad
cen′tral
cen·tra′tion
cen′tren·ce·phal′ic
cen′tric *or* ·tri·cal
cen·tric′i·put
cen·trif′u·gal
cen·trif′u·gal·i·za′tion
cen·trif′u·gal·ize′
  ·ized′ ·iz′ing
cen·trif′u·ga′tion
cen′tri·fuge′
  ·fuged′ ·fug′ing
cen′tri·ole′
cen·trip′e·tal
cen′tro·mere′
cen′tro·mer′ic
cen′tro·some′
cen′tro·som′ic
cen′tro·sphere′

cen′trum
  ·trums or ·tra
ceph′a·lad′
ceph′a·lal′gi·a
ceph′a·lex′in
ceph′al·hy′dro·cele′
ce·phal′ic
ceph′a·lin
ce·phal′o·cele′
ceph′a·lo·gy′ric
ceph′a·lom′e·ter
ceph′a·lo·met′rics
ceph′a·lom′e·try
ceph′a·lor′i·dine′
ceph′a·lo·spo′rin
ceph′a·pi′rin
ce·ra′ceous
ce·ram′ics
ce·ram′i·dase′
cer′a·mide′
ce′rate
ce′rat·ed
For words beginning cerato– see KERATO–
cer·ca′ri·a
  ·ae′
cer·clage′
cer′cus
  ·ci
cer′e·a flex′i·bil′i·tas
cer′e·bel′lar
cer′e·bel′li·fu′gal
cer′e·bel·lip′e·tal
cer′e·bel′lum
  ·lums or ·la

cer′e·bral
cer′e·brate′
  ·brat·ed ·brat′ing
cer′e·bra′tion
cer′e·bro′ma·la′ci·a
cer′e·bro·side′
cer′e·bro·spi′nal
cer′e·bro·spi′nant
cer′e·bro·vas′cu·lar
cer′e·brum
  ·brums or ·bra
cere′cloth′
Ce·ren′kov radiation
cer′e·sin or ce′rin
ce′ric
ce′ri·um
ce′ro·plas′tic
ce′ro·plas′tics
ce′ro·plas′ty
ce·rot′ic
ce′rous
cer′ti·fi′a·ble
cer′ti·fi·ca′tion
cer′ti·fy′
  ·fied′ ·fy′ing
ce·ru′men
ce·ru′mi·nal
ce·ru′mi·nous
ce′ruse
cer′vi·cal
cer′vi·ces′
  (*sing.* cer′vix)
cer′vi·ci′tis
cer′vix
  ·vi·ces′ or ·vix·es

Ce·sar′e·an or ·i·an
ce′si·um
ces·sa′tion
ces′tode
ces·to·di′a·sis
  ·ses′
ces′toid
cet′al·ko′ni·um chloride
ce′tyl·py′ri·din′i·um chloride
ce′vi·tam′ic acid
chae′ta
  ·tae
chafe
  chafed chaf′ing
Cha′gas′ disease
cha·go′ma
chain
cha·la′si·a or ·sis
cha·la′zi·on
  ·zi·a or ·zi·ons
chal·co′sis
chal′ice cell
chal·i·co′sis
chalk′stone′
chal′lenge
  ·lenged ·leng·ing
chal′one
cha·lyb′e·ate
cham′ber
cham′fer
cham′o·mile′
chan′cre
chan′croid
chan′crous

43

change of life
chan′nel
· ·neled *or* ·nelled
· ·nel·ing *or* ·nel·ling
chap
· chapped *or* chapt
· chap′ping
chap′pa
char′ac·ter
char′ac·ter·is′tic
char′ac·ter·i·za′tion
char′ac·ter·ol′o·gy
char′coal′
Char·cots′ joint
charge
char′la·tan
char′la·tan·ism
char′la·tan·ry
· ·ries
char′ley horse
chart
char′ta
· ·tae
chauf·fage′
chaul·moo′gra
check′–bite′
check′up′
cheek′bone′
chees′y
· ·i·er ·i·est
chei·li′tis
chei′lo·plas′ty
chei·los′chi·sis
· ·ses′
chei·lo′sis
· ·ses

chei′lo·sto·mat′o·plas′ty
chei′rog·nos′tic
chei′ro·plas′ty
che′late
· ·lat·ed ·lat·ing
che·la′tion
che′la·tor
che·lic′er·a
· ·ae′
che′loid
chem′a·bra′sion
chem′ex·fo′li·a′tion
chem′i·cal
chem′i·co·phys′i·cal
chem′i·lu′mi·nes′cence
chem′i·lu′mi·nes′cent
che·mise′
chem′i·sorb′
chem′i·sorp′tion
chem′ist
chem′is·try
· ·tries
che′mo·au′to·troph′
che′mo·au′to·troph′ic
che′mo·au·tot′ro·phy
· ·phies
che′mo·cau′ter·y
che′mo·cep′tor
che′mo·dec·to′ma
· ·mas *or* ·ma·ta
che′mo·nu·cle·ol′y·sis

che′mo·or′ga·no·troph′ic
che′mo·pro′phy·lac′tic
che′mo·pro′phy·lax′is
· ·es
che′mo·re·cep′tive
che′mo·re·cep′tor
che·mo′sis
· ·ses
chem·os·mo′sis
· ·ses
chem′os·mot′ic
che′mo·stat′
che′mo·ster′i·lant
che′mo·sur′ger·y
che′mo·syn′the·sis
· ·ses′
che′mo·syn·thet′ic
che′mo·tac′tic
che′mo·tax′is
· ·tax′es
che′mo·ther′a·peu′tic
che′mo·ther′a·pist
che′mo·ther′a·py *or*
· ·ther′a·peu′tics
che′mo·trop′ic
chem·ot′ro·pism
chest′ed
Cheyne′–Stokes′ asthma
chi·as′ma
· ·ma·ta
chi·as′ma·typ′y
chick′en breast

chicken pox
chief of staff
chig′er
chig′oe
  ·oes
chil′blain′
chil′blained′
child
  chil′dren
child′bear′ing
child′bed′
child′birth′
child′hood′
child′ing
Chil′e saltpeter
chill
chi′lo·mas·ti·gi′a·sis
  or ·mas′tix·i′a·sis
  or ·mas′to′sis
chi′lo·plas′ty
chi′lo·pod′
chi·me′ra or
  ·mae′ra
chi·mer′ism
chin′cap′
chinch bug
chine
Chin·ese′ res′tau·rant syndrome
chip
  chipped chip′ping
chip′–blow′er
chi′ral
chi·ral′i·ty
chi′ro·plas′ty
chi·rop′o·dist
chi·rop′o·dy

chi′ro·prac′tic
chi′ro·prac′tor
chi·rur′geon
chi·rur′ger·y
chi·rur′gi·cal
chi′–square′
chi′tin
chi′tin·ase
chi′tin·ous
chi′to·san
chlam′y·de′mi·a
chla·myd′i·a
chla·myd′o·spore′
chla·myd′y·o′sis
chlo·as′ma
  ·ma·ta
chlor·ac′ne
chlo′ral
chlo′ra·mine′
chlor′am·phen′i·col′
chlo′rate
chlor·cy′cli·zine′
chlor′dane or ·dan
chlor′di·az′e·pox′ide
chlo·rel′la
chlo·rel′lin
chlo·re′mi·a
chlor·hex′i·dine
chlor·hy′dri·a
chlo′ric
chlo′ride
chlo′ri·du′ri·a
chlo′ri·nate′
  ·nat′ed ·nat′ing

chlo′ri·na′tion
chlo′rine
chlo′rite
chlo·rit′ic
chlo′ro·bu·ta′nol
chlo′ro·form′
chlo′ro·form′ism
chlo·ro′ma
  ·mas or ·ma·ta
Chlo′ro·my·ce′tin
  (trademark)
chlo′ro·phyll′ or
  ·phyl′
chlo′ro·phyl′lose or
  ·phyl′lous
chlo′ro·pic′rin
chlo′ro·pri′vic
chlo′ro·quine
chlo·ro′sis
  ·ses
chlo′ro·thi′a·zide′
chlo·rot′ic
chlo′rous
chlor·pic′rin
chlor·prom′a·zine′
chlor′tet·ra·cy′cline
chlor·thal′i·done′
cho′a·na
  ·nae′
choke
  choked chok′ing
chol′a·gog′ic
chol′a·gogue′
cho·lan′gi·o·gas·tros′to·my
cho·lan′gi·og′ra·phy

cho·lan′gi·o·ma
·mas *or* ·ma·ta
chol′an·git′ic
chol′an·gi′tis
cho′late
cho·le·cal·cif′er·ol
chol′e·cyst
chol′e·cys·tec′to·my
·mies
chol′e·cys′tic
chol′e·cys′tis
chol′e·cys·ti′tis
chol′e·cys′to·gas·tros′to·my
chol′e·cys′to·je·ju·nos′to·my
chol′e·cys′to·kin′in
cho·led′o·cho·du′o·de·nos′to·my
cho·led′o·chor·rha·phy
cho·led′o·chous
adj.
cho·led′o·chus *n.*
cho·le′ic
cho′le·lith
cho′le·li·thi′a·sis
·ses′
cho′le·li·thot′o·my
cho·le′mi·a
cho·le′mic
chol′er·a
chol′e·ra′ic
(*of cholera*; SEE choleric)
cholera mor′bus

chol′er′e·sis
·ses′
cho′le·ret′ic
chol′er·ic
(*irritable*; SEE choleraic)
cho′le·sta′sis
·ses
cho′le·stat′ic
cho·les′ter·e′mi·a
cho·les·ter′ic
cho·les′ter·ol′
cho′lic acid
cho′line
cho′lin·er′gic
cho′lin·es′ter·ase′
chon′dral
chon·drec′to·my
chon′dri·o·some′
chon·dri′tis
chon′dro·cal′ci·no′sis
·ses
chon′droid
chon·dro·li·po′ma
·mas *or* ·ma·ta
chon·dro′ma
·mas *or* ·ma·ta
chon′dro·po·ro′sis
·ses
Cho·part's′ joint
chor′da
·dae
chor′dal
chor′dee
chor·di′tis
chor·dot′o·my

cho·re′a
cho·re′ic
cho·re′i·form′
cho′ri·o·al′lan·to′ic
cho′ri·o·al′lan·to′is
cho′ri·o·cap′il·la′ris
cho′ri·on′
cho′ri·on′ic
cho′ri·o·ret′i·ni′tis
cho′roid *or* cho′ri·oid′
cho·roi′dal
cho·roi′do·ret′i·ni′tis
Christ′mas factor
chro·maf′fin
chro′maf·fi·no′ma
·mas *or* ·ma·ta
chro′mate
chro·mat′ic
chro·mat′i·cism
chro′ma·tid
chro′ma·tin
chro′ma·tin–neg′a·tive
chro′ma·tism
chro·mat′o·gram′
chro·mat′o·graph′
chro·mat′o·graph′ic
chro′ma·tog′ra·phy
chro·ma·tol′y·sis
·ses′
chro·mat′o·lyt′ic
chrome
chro′mic
chro′mi·dro′sis
or chrom′hi·dro′sis

chro'mi·um
chro'mo·gen
chro'mo·gen'e·sis
 ·ses'
chro'mo·gen'ic
chro'mo·mere'
chro'mo·mer'ic
chro'mo·my·co'sis
 ·ses
chro'mo·ne'ma
 ·ma·ta
chro'mo·ne'mal
chro'mo·phil'
chro'mo·phil'ic
chro'mo·phore'
chro'mo·phor'ic
chro'mo·plas'tid or
 ·plast'
chro'mo·pro'te·in
chro'mo·so'mal
chro'mo·some'
chro'mous
chro'nax·ie or
 ·nax·y
 ·ies
chron'ic
chro·nic'i·ty
chron'o·graph'
chron'o·log'i·cal or
 ·log'ic
chron'o·scope'
chron'o·trop'ic
chro·not'ro·pism
chrys'a·ro'bin
Chvos'tek's sign
chy·la'ceous or
 chy'lous

chy'lan·gi·o'ma
 ·mas or ·ma·ta
chyle
chy'li·fac'tion or
 ·li·fi·ca'tion
chy'li·fac'tive
chy·lif'er·ous
chy'li·form'
chy'lo·mi'cron
chyme
chym'ist
chy'mo·pa·pa'in
chy'mo·tryp'sin
chy'mo·tryp'tic
chy'mous
cic'a·trice
cic'a·tri'cial
cic'a·tri'cle
cic'a·trix
 ci·cat'ri·ces'
cic'a·tri·za'tion
cic'a·trize'
 ·trized' ·triz'ing
ci'gua·te'ra
cil'i·a
 (sing. cil'i·um)
cil'i·ar'y
cil'i·ate or ·at·ed
cil'i·o·spi'nal
ci·lo'sis
cim'bi·a
ci·met'i·dine'
ci'mex
 cim'i·ces'
cin·cho'na
cin·chon'ic
cin·chon'i·dine'

cin'cho·nine'
cin'cho·nism
cin'cho·nize'
 ·nized' ·niz'ing
cinc'ture
cin'e·ole'
cin'e·ra'di·og'ra·
 phy
ci·ne're·a
ci·ne're·al
cin'gu·late or
 ·lat'ed
cin'gu·lot'o·my or
 ·gu·lum·ot'o·my
cin'gu·lum
 ·la
cin·nam'ic
cin'na·mon
cir·ca'di·an
cir'ci·nate'
cir'cle
cir'cuit
cir'cu·lar
cir'cu·late'
 ·lat'ed ·lat'ing
cir'cu·la'tion
cir'cu·la'tive
cir'cu·la'tor
cir'cu·la·to'ry
cir'cum·cise'
 ·cised' ·cis'ing
cir'cum·ci'sion
cir'cum·cor'ne·al
cir'cum·duc'tion
cir'cum'fer·ence
cir'cum·flex'
cir'cum·flex'ion

cir'cum·lo·cu'tion
cir'cum·loc'u·to'ry
cir'cum·scribe'
cir'cum·scrip'tus
cir'cum·stan'ti·al'i·ty
cir'cum·stan'ti·ate'
  ·at·ed ·at·ing
cir'cum·val'late *adj.*
cir'cum·vo'lute
cir'cum·vo·lu'tion
cir·rho'sis
  ·ses
cir·rhot'ic
cir'rose *or* ·rous
cir'rus
  ·ri
cir'soid
cir·som'pha·los
cir'so·tome'
cis
cis'sa
cis'tern
cis·ter'na
  ·nae
cis·ter'nal
cis'tron
cit'ral
cit'rate
cit'ric
ci·trul'line
ci·trul'li·nu'ri·a
cit'ta
cit·to'sis
cla·dis'tic
cla·dis'tics *n.*

clad'o·spo·ri·o'sis
clair·voy'ance
clair·voy'ant
clam'my
  ·mi·er ·mi·est
clamp
clap
cla·pote·ment' *or*
  cla'po·tage'
Clap'ton's line
cla·rif'i·cant
clar'i·fi·ca'tion
clar'i·fi'er
clar'i·fy'
  ·fied' ·fy'ing
Clarke's cells
Clark's rule
clas·mat'o·cyte'
clas·mat'o·den·dro'sis
clas·ma·to'sis
clasp
clas'sic
clas'si·cal
clas'si·fi'a·ble
clas'si·fi·ca'tion
clas'si·fy'
  ·fied' ·fy'ing
clas'tic
clas'to·gen
clas'to·gen'ic
clas'to·thrix'
clath'rate
clau'di·cant
clau'di·ca'tion
clau'di·ca·to'ry
claus'tro·pho'bi·a

claus'tro·pho'bic
claus'trum
  ·tra
cla'va
  ·vae
cla'vate
cla·va'tion
clav'i·cle
cla·vic'u·lar
cla'vus
  ·vi
claw'hand'
clean room
cleanse
  cleansed cleans'ing
clear'ance
clear'–cell'
  carcinoma
clear'er
clear'ing
cleav'age
cleav'er
cleft
cleft palate
clei'do·cra'ni·al
clei·dot'o·my
clench
clep'to·ma'ni·a
click
cli·mac'ter·ic *or* cli'mac·ter'i·cal
cli·mac'tic *or*
  ·mac'ti·cal
cli'ma·to·ther'a·py
cli'max
clin'ic
clin'i·cal

cli·ni′cian
clin′i·co·path′o·log′ic
cli′no·dac′ty·lous
cli′no·dac′ty·ly
cli·nom′e·ter
cli′no·met′ric *or* ·met′ri·cal
cli·nom′e·try
cli′no·scope′
clip
  clipped clip′ping
clis′e·om·e·ter
clit′i·on
clit′o·ral *or* cli·tor′ic
cli′to·ri·dec′to·mize′
  ·mized′ ·miz′ing
clit′o·ri·dec′to·my *or* ·o·rec′to·my
clit′o·ris
clit′o·ri′tis
cli′vus
  ·vi
clo·a′ca
  ·cae *or* ·cas
clo·a′cal
clo·fi′brate
clo′mi·phene′ citrate
clon′al
clone *or* clon
  cloned clon′ing
clon′ic
clo·nic′i·ty
clon′i·co·ton′ic

clon′ism
clon′o·graph′
clo′no·spasm
clo′nus
close
  closed clos′ing
clos·trid′i·al
clos·trid′i·um
  ·trid′i·a
clo′sure
clot
  clot′ted clot′ting
cloud′y
  ·i·er ·i·est
clo′ven
  (alt. pp. of cleave*)*
clove oil
clox′a·cil′lin sodium
club
  clubbed club′bing
club′foot′
  ·feet′
club′foot′ed
club′hand′
clump
clu·ne′al
clu′nis
  clu′nes
clus′ter
clut′ter·ing
cly′sis
clys′ter
cne′mi·al
cne′mis
co·ac′er·vate′
  ·vat′ed ·vat′ing

co·ac′er·vate *n.*
co·ac′er·va′tion
co′ad·ap·ta′tion
co·ad′u·nate
co·ad′u·na′tion
co′ag·glu′ti·na′tion
co·ag′u·la·bil′i·ty
co·ag′u·la·ble
co·ag′u·lant
co·ag′u·lase′
co·ag′u·late′
  ·lat′ed ·lat′ing
co·ag′u·la′tion
co·ag′u·la′tive
co·ag′u·la′tor
co·ag′u·lum
  ·la
co′a·lesce′
  ·lesced′ ·lesc′ing
co′a·les′cence
co′a·les′cent
coal tar
co′apt *v.*
co′ap·ta′tion
co·arc′tate
co·arc·ta′tion
coarse
  (not fine; SEE course*)*
coat′ed
co·bal′a·min
co′balt
co·bal′tic
co·bal′tous
co′bra
co′ca
co·caine′ *or* ·cain′

co·cain'ism
co·cain'ize
  ·ized ·iz·ing
co·car·cin'o·gen'
coc'ci
  (sing. coc'cus)
coc·cid'i·al
coc·cid'i·oi'dal
coc·cid'i·oi'do·my·
  co'sis
  ·ses
coc·cid'i·o'sis
  ·ses
coc·cid'i·um
  ·i·a
coc'co·bac·il'lar·y
coc'co·ba·cil'lus
  ·li
coc'coid
coc'cus
  coc'ci
coc·cyg'e·al
coc'cy·gec'to·my
coc'cy·go·dyn'i·a
coc'cyx
  coc·cy'ges
coch'i·neal'
coch'le·a
  ·ae *or* ·as
coch'le·ar
coch·le·a're
coch'le·ate *or*
  ·at'ed
coch'le·i'tis
coch'le·o·ves·tib'u·
  lar
cock·ade'

cock'eye'
cock'roach'
cock'tail'
co'coa but'ter
co'con·trac'tion
coc'to·la'bile
coc'to·sta'bile *or*
  ·ble
code
  cod'ed cod'ing
co'deine *or* ·cin *or*
  co·de'la
co'dex
  co'di·ces'
cod'–liv'er oil
co'don
*For words
beginning* coe– *see
also* CE–
co'ef·fi'cient
coe·len'ter·ate'
coe·len'ter·on'
  ·ter·a
coe'li·ac'
coe'lom
  coe'loms *or* coe·lo'
  ma·ta
coe·lom'ic
coe·nes·the'sia *or*
  ·the'sis
coe'no·cyte'
coe·nu'rus
  ·ri
co'en·zy·mat'ic
co·en'zyme
co'fac'tor
co'ge·ner

cog·ni'tion
cog'ni·tive
co·here'
  ·hered' ·her'ing
co·her'ence *or*
  ·her'en·cy
co·her'ent
co·he'sion
co·he'sive
Cohn'heim's areas
co'ho·bate'
  ·bat'ed ·bat'ing
co'hort
co'hosh
coil
coin test
co'i·tal
co·i'tion
co·i'tus
coitus in'ter·rup'
  tus
col
co'la
  (tree)
co'la
  (sing. co'lon)
co·la'tion
col'chi·cine'
col'chi·cum
cold'–pack' *v.*
cold pack
cold sore
co·lec'to·my
  ·mies
co'li·bac·il·le'mi·a
co'li·bac·il·lo'sis
  ·ses

co′li·bac·il·lu′ri·a
co′li·ba·cil′lus
 ·li
col′ic
col′i·ca
col′ick·y
co′li·cys·ti′tis
col′i form′
co′li·pli·ca′tion
co′li·punc′ture
co·lis′tin
co·li′tis
co′li·tox·e′mi·a
co′li·tox′in
col′la·gen′
col·lag′e·nase′
col·lag′e·na′tion
col′la·gen′ic
col lag′e·no·gen′ic
col·lag′e·nous
col·lapse′
 ·lapsed′ ·laps′ing
col·lap′so·ther′a·py
col′lar·bone′
col·lat′er·al
col′lect
Col′les' ligament
col·lic′u·lus
 ·li′
col′li·gate′
 ·gat′ed ·gat′ing
col′li·ga′tion
col′li·ga′tive
col′li·mate′
 ·mat′ed ·mat′ing
col′li·ma′tion
col′li·ma′tor

col′li·qua′tion
col·liq′ua·tive
col·lo′di·on
col′loid
col·loi′dal
col′lop
col′lum
 ·la *(neck;* SEE column)
col·lyr′i·um
 ·i·a *or* ·i·ums
col·o·bo′ma
 ·mas *or* ·ma·ta
co′lo·cen·te′sis
 ·ses
co′lo·cho′le·cys·tos′to·my
co′lo·cly′sis
co′lo·clys′ter
co′lon
 ·lons *or* ·la
co′lon·al′gi·a
co·lon′ic
col′o·ni·za′tion
co·lon′o·scope′
co·lo·nos′co·py
col′o·ny
 ·nies
co′lop·to′sis
col′or·a′tion
col′or·blind′
col′o·rec′tal
col′o·rec′tum
col′or·im′e·ter
col′or·i·met′ric
col′or·im′e·try
col′o·scope′

col·los′to·my
 ·mies
co·los′trum
co·lot′o·my
 ·mies
col·pal′gi·a
col·pec′to·my
col′peu·ryn′ter
col·peu′ry·sis
 ·ses′
col·pi′tis
col′po·cele′
col′po·cys′to·cele′
col′po·hy′per·pla′si·a
col′por·rhex′is
 ·rhex′es
col′po·scope′
col·po·scop′ic
col·pos′co·py
col·pot′o·my
col′po·xe·ro′sis
co·lum′bi·um
col′u·mel′la
 ·mel′lae
col′umn
 (vertical structure;
 SEE collum)
co·lum′na
 ·nae
col·um′nar
col′um·ni·za′tion
co′ma
co′ma·tose′
com·bi·na′tion
com·bine′
 ·bined′ ·bin′ing

com·bus'ti·ble
com·bus'tion
com'e·do'
  com'e·do'nes or
  com'e·dos'
co'mes
  com'i·tes'
com'fort·a·ble
com'ma bacillus
com·men'sal
com·men'sal·ism
com·mi'nute'
  ·nut'ed ·nut'ing
com'mi·nu'tion
com'mis·su'ra
  ·rae
com·mis'su·ral
com'mis·sure'
com·mit'
  ·mit'ted ·mit'ting
com·mit'ment
com·mit'tal
com·mix'
com·mix'ture
com·mode'
com'mon cold
com mu'ni·ca·bil'
  i·ty
com·mu'ni·ca·ble
com·mu'ni·cate'
  ·cat'ed ·cat'ing
com·mu'ni·ca'tion
com·mu'ni·ty
  ·ties
com·pact' *adj.*
com·pac'tion
com·par'a·tive

com·par'a·tor'
com·pare'
  ·pared' ·par'ing
com·part'men'tal·i·
  za'tion
com·pat'i·bil'i·ty
com·pat'i·ble
Com'pa·zine
  (trademark)
com'pen·sate'
  ·sat'ed ·sat'ing
com'pen·sa'tion
com·pen'sa·to'ry
com'pe·tence or
  ·ten·cy
com'pe·tent
com'pe·ti'tion
com·pim'e·ter
com·plaint'
com'ple·ment *n.*
com'ple·men'ta·ry
  or ·men'tal
com'ple·men·ta'tion
com·plete'
  ·plet'ed ·plet'ing
com'plex *n.*
com·plex' *adj.*
com·plex'ion
com·plex'ioned
com·pli'ance or
  ·an·cy
com'pli·cate'
  ·cat'ed ·cat'ing
com'pli·ca'tion
com·po'nent
com·pose'
  ·posed' ·pos'ing

com·pos'ite
com'po·si'tion
com'pos men'tis
com·pound' *n.*
com'pound' *v.*
com'pre·hen'sion
com'press *n.*
com·press' *v.*
com·pres'sion
com·pres'sor
com'pro·mised'
com·pul'sion
com·pul'sive
com·pul'so·ry
co·na'ri·um
co·na'tion
co·na'tion·al
con'a·tive
con'ca·nav'i·lin
con·cat'e·nate'
  ·nat'ed ·nat'ing
con·cat'e·na'tion
con·cave' *n.*
con·cav'i·ty
  ·ties
con·ca'vo–con·
  cave'
con·ca'vo–con·vex'
con·ceive'
  ·ceived' ·ceiv'ing
con'cen·trate'
  ·trat'ed ·trat'ing
con'cen·tra'tion
con·cen'tric or
  ·cen'tri·cal
con'cen·tric'i·ty
con'cept

con·cep'tion
con·cep'tion·al
con·cep'tive
con·cep'tu·al
con·cep'tus
  ·tus·es *or* ·ti
con'cha
  ·chae
con·chi'tis
con·choi'dal
con·cho·tome'
con·cli·na'tion
con·coc'tion
con·com'i·tant
con·cord'ance
con·cord'ant
con'cre·ment
con·cres'cence
con·crete'
con·cre'tion
con·cuss'
con·cus'sion
con·cus'sive
con·den·sate' *n.*
con·den·sa'tion
con·dense'
  ·densed' ·dens'ing
con·dens'er
con·di'tion
con·di'tion·al
con·di'tioned
con'dom
con·duct'ance
con·duc'tion
con·duc·tiv'i·ty
con·duc'tor
con·duit

con'dy·lar
con'dyle
con'dy·loid
con'dy·lo'ma
  ·ma·ta
con'dy·lot'o·my
con'dy·lus
  ·li
cone
cone'nose'
con·fab'u·late'
  ·lat'ed ·lat'ing
con·fab'u·la'tion
con·fab'u·la·to'ry
con·fec'tion
con·fi·den'ti·al'i·ty
con·fig'u·ra'tion
con·fig'u·ra'tion·al
con·fig'u·ra'tion·ism
con·fig'u·ra'tive
con·fine'ment
con'flict *n.*
con'flu·ence
con'flu·ent
con'flux
con·fo'cal
con·for·ma'tion
con·form'er
con·fron·ta'tion
con·fu'sion
con·fu'sion·al
con·geal'
con·ge·la'tion
con'gen·er
con·ge·ner'ic *or*
  con·gen'er·ous

con·gen'i·tal
con·gest'
con·ges'tion
con·ges'tive
con·glo'bate
  ·bat·ed ·bat·ing
con'glo·ba'tion
con·globe'
con·glom'er·ate *n.*
con·glom'er·ate'
  ·at'ed ·at'ing
con·glom'er·a'tion
con·glu'ti·nant
con·glu'ti·nate'
  ·nat'ed ·nat'ing
con·glu'ti·na'tion
con·glu'ti·na'tive
Con'go red
con'i·cal
co·nid'i·al *or* ·i·an
co·nid'i·o·phore'
co·nid'i·um
  ·i·a
co·ni·ine' *or* co'
  nine'
co'ni·o·lymph'sta·
  sis
co'ni·o·phage'
co'ni·o'sis
  ·ses
co'ni·um
co'ni·za'tion
con·join'
con'joint'
con'ju·gant
con'ju·ga'ta
  ·tae

con′ju·gate n., adj.
con′ju·gate′
·gat·ed ·gat′ing
con′ju·ga′tion
con′ju·ga′tor
con·junc·ti′va
·vas or ·vae
con·junc·ti′val
con·junc′tive
con·junc′ti·vi′tis
con·junc′ti·vo′ma
con·junc′ti·vo·
plas′ty
con′nate
con·nect′
con·nec′tion
con·nec′tive
con·nec′tor or
·nect′er
co′noid or co·noi′
dal
con′san·guin′e·ous
or con·san′guine
con′san·guin′i·ty
con′science
con′scious
con·sen′su·al
con·sent′
con·ser′va·tive
con′serve n.
con·sis′ten·cy
·cies
con·sis′tent
con·sol′i·dant
con·sol′i·date′
·dat·ed ·dat′ing
con·sol′i·da′tion

con′spe·cif′ic
con′stan·cy
con′stant
con′stel·la′tion
con·stel′la·to′ry
con′sti·pate′
·pat·ed ·pat′ing
con′sti·pa′tion
con′sti·tu′tion
con′sti·tu′tion·al
con·strict′
con·stric′tion
con·stric′tive
con·stric′tor
con·stringe′
·stringed′
·string′ing
con·struc′tive
con·sult′
con·sult′ant
con′sul·ta′tion
con·sume′
·sumed′ ·sum′ing
con·sump′tion
con·sump′tive
con′tact
con·tac′tant
con·ta′gion
con·ta′gious
con·ta′gi·um
·gi·a
con·tain′
con·tam′i·nant
con·tam′i·nate′
·nat·ed ·nat′ing
con·tam′i·na′tion
con·tam′i·na′tive

con·tam′i·na′tor
con′tent n.
con′ti·gu′i·ty
·ties
con·tig′u·ous
con′ti·nence or
·ti·nen·cy
con′ti·nent
con′ti·nu′i·ty
·ties
con·tin′u·ous
con·tin′u·ous–flow′
e·lec′tro·phor′e·sis
con·tor′tion
con′tour
con′tra–ap′er·ture
con′tra·cep′tion
con′tra·cep′tive
con·tract′ v.
con·tract′i·bil′i·ty
con·tract′i·ble
con·trac′tile
con′trac·til′i·ty
con·trac′tion
con·trac′tive
con·trac′tor
con·trac′ture
con′tra·fis·su′ra or
con′tra·fis′sure
con′tra·in′di·cant
con·tra·in′di·cate′
·cat·ed ·cat′ing
con′tra·in′di·ca′tion
con′tra·in·dic′a·tive
con′tra·lat′er·al
con·trast′ v.
con′trast n.

con'tre·coup'
con·trib'u·to'ry
con·trol'
 ·trolled' ·trol'ling
con·trude' v.
con·tuse'
 ·tused' ·tus'ing
con·tu'sion
con'u·lar
co'nus
 ·ni
con'va·lesce'
 ·lesced' ·lesc'ing
con'va·les'cence
con'va·les'cent
con·vec'tion
con·vec'tive
con·vec'tor
con·ven'ience
con·ven'tion·al
con·verge'
 ·verged' ·verg'ing
con·ver'gence
con·ver'gen·cy
 ·cies
con·ver'gent
con·ver'sion
con·vex' adj.
con'vex n.
con·vex'i·ty
 ·ties
con·vex'o–con·cave'
con·vex'o–con·vex'
con·vex'o–plane'
con'vo·lute'
 ·lut'ed ·lut'ing

con'vo·lu'tion
con·vul'sant
con·vulse'
 ·vulsed' ·vuls'ing
con·vul'sion
con·vul'sion·ar·y
con·vul'sive
Coo'ley's anemia
Coombs' test
co·or'di·nate or
 co–or'di·nate n.
co·or'di·nate' or
 co–or'di nate'
 ·nat'ed ·nat'ing
co·or'di·na'tion or
 co–or'di·na'tion
co·pai'ba
COPD
cope
 coped cop'ing
co'pe·pod'
co'pi·ous
cop·i·o'pi·a
co·pol'y·mer
co·pol'y·mer'ic
co·po·lym·er'i·za'tion
co·po·lym'er·ize'
 ·ized' ·iz'ing
cop'per
cop'per·as
cop'per·head
cop·rem'e·sis
cop'ro·an'ti·bod'y
 ·bod'ies
cop'ro·la'li·a
cop'ro·lith

cop·rol'o·gy
cop·roph'a·gous
cop·roph'a·gy or
 cop'ro·pha'gi·a
cop'ro·phil'i·a
cop'ro·phil'i·ac
cop'ro·por'phy·rin
cop'ro·zo'a
cop'u·la
 ·las
cop'u·late'
 ·lat'ed ·lat'ing
cop'u·la'tion
cop'u·la·to'ry
cor
cor·a·cid'i·um
 ·i·a
cor'a·co·a·cro'mi·al
cor'a·coid'
cord
cord'al
cor'date
cord'ed
cor'dial
cord'i·form'
cor'dis
cor·di'tis
cor·don sa·ni·taire'
cor·dot'o·my
core
 cored cor'ing
 (central part; SEE
 corps)
cor'e·cli'sis
 ·ses
cor·ec'tome
cor·ec·to'pi·a

co′-re·la′tion
co·-rel′a·tive
cor′e·mor·pho′sis
co′re·pres′sor
co′ri·an′der
Co′ri cycle
co′ri·um
 ·ri·a
corn
cor′ne·a
cor′ne·al
cor′ne·i′tis
cor′ne·o·i·ri′tis
cor′ne·ous
cor·nic′u·late
cor′ni·fi·ca′tion
cor′ni·fied′
corn′starch′
corn sugar
corn syrup
cor′nu
 ·nu·a
cor′nu·al
co·ro′na
 ·nas or ·nae
cor′o·nad
co·ro′nal *adj.*
cor′o·nar′y
 ·nar′ies
cor′o·na·vi′rus
cor′o·ner
co·ros′co·py
co·rot′o·my
cor′po·ra
 (sing. cor′pus*)*
cor′po·ral
cor·po′re·al

corps
 (group of people;
  SEE core*)*
corpse
 (dead body)
corps′man
 ·men
cor′pu·lence *or*
 ·len·cy
cor′pu·lent
cor′pus
 ·po·ra
 (body; SEE carpus*)*
cor′pus cal·lo′sum
 cor′po·ra cal·lo′sa
cor′pus·cle *or* cor·
 pus′cule
cor·pus′cu·lar
cor·pus′cu·lum
 ·la
cor′pus de·lic′ti
 cor′po·ra de·lic′ti
cor′pus lu′te·um
 cor′po·ra lu′te·a
cor′pus stri·a′tum
 cor′po·ra stri·a′ta
cor·rect′
cor·rect′a·ble
cor·rec′tion
cor·rec′tive
cor′re·late′
 ·lat′ed ·lat′ing
cor′re·la′tion
cor·rel′a·tive
cor′re·spond′ence
Cor′ri·gan's pulse
cor′ri·gent
cor′rin

cor·rob′o·rant
cor·rode′
 ·rod′ed ·rod′ing
cor·ro′sion
cor·ro′sive
cor·ro′sive sub′li·
 mate′
cor′ru·ga′tor
cor′tex
 cor′ti·ces′
Cor′ti, canal of
cor′ti·cal
cor′ti·cate *or*
 ·cat′ed
cor′ti·cif′u·gal
cor′ti·cip′e·tal
cor′ti·co·ad·re′nal
cor′ti·co·bul′bar
cor′ti·coid′
cor′ti·cose′
cor′ti·co·ste′roid
cor′ti·co·ste′rone or
cor′ti·co·troph′ic *or*
 ·trop′ic
cor′ti·co·tro′phin *or*
 ·tro′pin
cor′tin
cor′ti·sol′
cor′ti·sone′
co·run′dum
cor′us·ca′tion
co·rym′bi·form′ *or*
 cor′ym·bose′
co·ry′ne·bac·te′
 ri·um
 ·ri·a
co·ry′ne·form′

co·ry′za
co·sen′si·tize′
  ·tized′ ·tiz′ing
cos·met′ic
cos′ta
  ·tae
cos′tal
cos·tal′gi·a
cos′tate
cos·tec′to·my
cos′tive
cos′to·cla·vic′u·lar
cos′to·ster′nal
co·throm′bo·plas′tin
co·trans′port
co′–tri·mox′a·zole′
Cotte's operation
cot′ton bat′ting
cot′y·le·don
cot′y·le·don·ous or
  ·don·al
cot′y·loid′
couch′ing
cough drop
cou·lomb′
cou′ma·rin
coun′sel
  ·seled or ·selled
  ·sel·ing or ·sel·ling
coun′se·lor or
  ·sel·lor
count
count′er
  (one that counts)
coun′ter
  (opposite)

coun′ter·act′
coun′ter·ac′tion
coun′ter·ac′tive
coun′ter·ex·ten′sion
coun′ter·in′di·cate′
  ·cat′ed ·cat′ing
coun′ter·ir′ri·tant
coun′ter·ir′ri·ta′tion
coun′ter·o′pen·ing
coun′ter·pul·sa′tion
coun′ter·shock′
coun′ter·trans·fer′ence
coup
cou′ple
coup′ling
course
  (duration; SEE
  coarse*)*
court plaster
Cour·voi·sier's′ law
co·va′lence
co·va′lent
co·var′i·ance
cov′er
cov′er·glass′ or
  ·slip′
cow′age or ·hage
cow′per·i′tis
Cow′per's glands
cow′pox′
cox′a
  ·ae
cox′al
cox·al′gi·a
cox·al′gic

cox·i′tis
  ·it′i·des′
cox′o·fem′o·ral
Cox·sack′ie virus
CPR
crab louse
crack
cra′dle
cramp
cra′ni·ad
cra′ni·al
cra′ni·ec′to·my
cra′ni·o·cele′
cra′ni·o·clei′do·dys′os·to′sis
  ·ses
cra′ni·ol′o·gy
  ·gies
cra′ni·om′e·ter
cra′ni·om′e·try
cra′ni·o·punc′ture
cra′ni·o·sa′cral
cra′ni·o·tome′
cra′ni·ot′o·my
  ·mies
cra′ni·um
  ·ni·ums or ·ni·a
crap′u·lence
crap′u·lent
crap′u·lous
cra′ter
cra·ter′i·form′
cra′ter·i·za′tion
cra·vat′ bandage
crav′ing
craze
  crazed craz′ing

cra′zy bone
cream
crease
cre·at′i·nase′
cre′a·tine′
cre·at′i·nine′
cre·at′i·nu′ri·a
cre·mas′ter
cre′mas·ter′ic
cre′mate
  ·mat·ed ·mat·ing
cre·ma′tion
cre′na
  ·nae
cre′nate or ·nat·ed
cre·na′tion
cren′a·ture
cre′o·sol′
cre′o·sote′
crep′i·tant
crep′i·tate′
  ·tat·ed ·tat·ing
crep′i·ta′tion
crep′i·tus
cre·pus′cu·lar
cres′cent
cres·cen′tic
cre′sol
crest
cre·syl′ic
cre′tin
cre′tin·ism
cre′ti·noid
cre′tin·ous
Creutz′feldt–Ja′kob
  disease
crev′ice

cre·vic′u·lar
crib death
crib′rate
crib·ra′tion
crib′ri·form′
cri′brum
  ·bra
cri′coid
cri′co·thy·rot′o·my
cri du chat
  syndrome
Crig′ler–Naj·jar′
  syndrome
Crile's clamp
crim′i·nal
crim′i·nol′o·gy
cri′no·gen′ic
crip′ple
  ·pled ·pling
cri′sis
  ·ses
cris·pa′tion
cris′ta
  ·tae
cri·te′ri·on
  ·ri·a or ·ri·ons
cri·thid′i·a
crit′i·cal
cro′cus
  cro′cus·es or cro′ci
Crohn's disease
cro′mo·lyn sodium
crook′back′
crook′backed′
Crookes' tube
cross′bite′
cross′–dress′

cross′–eye′
cross′–eyed′
cross′–fer′tile
cross′–fer·til·i·za′
  tion
cross′–fer′ti·lize′
  ·lized′ ·liz′ing
cross′ing–o′ver
cross′match′ing
cross′o′ver
cross′–re·ac·tiv′i·ty
cross section
cross′–sec′tion v.
cross′–sec′tion·al
cross′–tol′er·ance
cross′way′
crot′a·lid
cro·taph′i·on
crotch
cro′ton
cro·ton′ic acid
croup
croup′ous
croup′y
crown
crown′work′
cru′ces
  (sing. crux)
cru′cial
cru′ci·ate
cru′ci·ble
cru′ci·form′
cru′or
cru′ral
crus
  cru′ra
crust

crus'ta
  ·tae
crutch
crux
  crux'es *or* cru'ces
cry'al·ge'si·a
cry'an·es·the'si·a
cry·es·the'si·a
cry'o·an'al·ge'si·a
cry'o·bi·ol'o·gist
cry'o·bi·ol'o·gy
cry'o·cau'ter·y
cry'o·crit
cry'o·gen
cry'o·gen'ic
cry'o·gen'ics
cry'o·glob'u·li·ne'mi·a
cry·om'e·ter
cry·on'ic
cry·on'ics
cry'o·probe'
cry·os'co·py
cry'o·stat'
cry'o·sur'ger·y
cry'o·sur'gi·cal
cry'o·ther'a·py
crypt
crypt'al
cryp·ten'a·mine'
cryp'tic
cryp'to·coc·co'sis
  ·ses
cryp'to·crys'tal·line
cryp'to·gen'ic

cryp'to·mer'o·ra·chis'chi·sis
  ·ses'
cryp'tom·ne'si·a
cryp'tom·ne'sic
cryp'to·pine'
cryp·tor'chism *or* ·tor'chi·dism
cryp'to·xan'thin
cryp'to·zy'gous
crys'tal
crys'tal·lin
  (eye globulin)
crys'tal·line
  (like crystal)
crys'tal·liz'a·ble
crys'tal·li·za'tion
crys'tal·lize'
  ·lized' ·liz'ing
crys'tal·lo·graph'ic *or* ·graph'i·cal
crys'tal·log'ra·phy
crys'tal·loid'
crys'tal·loi'dal
crys'tal·lu'ri·a
C'–ter'mi·nal
CT scan
cu'beb
cu'bi·form'
cu'bit
cu'bi·tal
cu'bi·tus
  ·ti'
cu'boid
cu·boi'dal
cu·cur'bit

cue
  cued cu'ing *or* cue'ing
cuff
cui·rass'
cul'–de–sac'
cul'do·cen·te'sis
  ·ses
cul'do·scope'
cul·dos'co·py
cu·lic'id
cu'li·cide'
cu·lic'i·fuge'
cu'li·cine
cul'men
  ·mi·na
cul'ti·vate'
  ·vat'ed ·vat'ing
cul'ti·va'tion
cul'ture
  ·tured ·tur·ing
cu·mu·la'tive
cu'mu·lus
  ·li'
cu'ne·ate *or* ·at'ed
cu·ne'i·form'
cun'ni·lin'gus
cup
  cupped cup'ping
cu'po·la
cu'pric
cu'prous
cu'prum
cu'pu·la
  ·lae'
cu'pu·lar
cu'pu·lo·gram'

cu′pu·lom′e·try
cur′a·bil′i·ty
cur′a·ble
cu·ra′re or ·ri
cu·ra′rine
cu·ra′ri·za′tion
cu·ra′rize
·rized ·riz·ing
cur′a·tive
curd
cure
cured cur′ing
cu·ret′ or ·rette′
·ret′ted ·ret′ting
cu·ret′tage′ or cu·rette′ment or cu·ret′ment
cu′rie
Cu·rie's′ law
cu′ri·um
cur′rent
cur·va·tu′ra
·rae
cur′va·ture
curve
curved curv′ing
cur′vi·lin′e·ar or ·lin′e·al
Cush′ing's disease
cush′ion
cusp
cus′pate or cus′pat·ed or cusped
cus′pid
cus′pi·date′ or ·dat′ed

cut
cut cut′ting
cu·ta′ne·ous
cut′down′ n.
cut′i·cle
cu·tic′u·la
·lae′
cu·tic′u·lar
cu′ti·re·ac′tion
cu′tis
cu′tes′ or ·tis·es
cut′off′
cu·vette′
cy·an′am·ide′ or ·am·id
cy′a·nate′
cy′a·ne′mi·a
cy·an′ic
cy′a·nide′
cy′an·met·he′mo·glo′bin
cy′an·o·co·bal′a·min
cy′an·o·gen
cy′a·no·hy′drin
cy′a·noph′i·lous
cy′a·nop′si·a or ·a·no′pi·a
cy′a·no′sis
·ses
cy′a·not′ic
cy′a·nu′rate
cy′a·nu′ric acid
cy′ber·net′ic
cy′ber·net′ics
cy′cla·mate′

cyc′lar·thro′sis
·ses
cy′cla·zo′cine
cy′cle
·cled ·cling
cy·clec′to·my
cy′clic or cy′cli·cal
cy′cli·cot′o·my
cy′cli·zine′
cy′cloid
cy·cloi′dal
cy′clo·pho′ri·a
cy′clo·phos′pha·mide′
cy′clo·pi·a
cy′clo·ple′gi·a
cy′clo·ple′gic
cy′clo·pro′pane
cy′clops
cy·clo′sis
·ses
cy′clo·spo′rin or ·rine
cy′clo·thyme′ or cy′clo·thym′i·ac
cy′clo·thy′mi·a
cy′clo·thy′mic
cy·clot′o·my
cy′clo·tron′
cyl′in·der
cy·lin′dri·cal or ·lin′dric
cyl′in·droid′
cyl′in·dru′ri·a
cyl·lo′sis
cy′mo·graph′
cyn′ic spasm

cyr·tom′e·ter
cyr·to′sis
  ·ses
cyst
cys′tad·e·no′ma
  ·mas *or* ·ma·ta
cys′ta·thi′o·nine
cys′ta·thi′o·nin·u′ri·a
cys·tec·ta′si·a *or* cys·tec′ta·sy
cys·tec′to·my
  ·mies
cys′te·ine
cys′tic
cys′ti·cer′coid
cys′ti·cer·co′sis
  ·ses
cys′ti·cer′cus
  ·cer′ci
cys′ti·form′
cys′tine
cys′ti·stax′is
cys·ti′tis
cys′to·ad′e·no′ma
  ·mas *or* ·ma·ta
cys′to·car′ci·no′ma
  ·mas *or* ·ma·ta
cys′to·cele′
cys′to·dyn′i·a
cys′to·fi·bro′ma
  ·mas *or* ·ma·ta
cyst′oid
cys′to·lith′
cys′to·li·thec′to·my
cys′to·li·thi′a·sis
  ·ses′

cys′to·lith′ic
cys′to·li·thot′o·my
cys·to′ma
  ·mas *or* ·ma·ta
cys′to·mor′phous
cys′to·sar·co′ma
  ·mas *or* ·ma·ta
cys′to·scope′
  ·scoped′ ·scop′ing
cys′to·scop′ic
cys·tos′co·py
  ·pies
cys·tot′o·my
  ·mies
cyt′a·phe·re′sis
  ·ses
cy·tar′a·bine
cy′ti·dine
cy′to·ar′chi·tec·ton′ic *or* ·tec′·tu·ral
cy′to·ar′chi·tec′ture
cy′to·chem′is·try
cy′to·chrome′
cy′to·ci′dal
cy′to·cide′
cy·toc′la·sis
cy′to·clas′tic
cy′to·dis′tal
cy′to·ge·net′ic *or* ·net′i·cal
cy′to·ge·net′i·cist
cy′to·ge·net′ics
cy·tog′e·nous
cy′to·gly′co·pe′ni·a *or* ·glu′co·pe′ni·a
cy′to·his′to·log′ic

cy′to·his·tol′o·gy
cy′to·ki·ne′sis
cy′to·ki·net′ic
cy′to·ki′nin
cy′to·log′ic *or* ·log′i·cal
cy·tol′o·gist
cy·tol′o·gy
  (cell biology; SEE sitology*)*
cy·tol′y·sin
cy·tol′y·sis
  ·ses′
cy′to·lyt′ic
cy′to·meg′a·lo·vi′rus
cy·tom′e·ter
cy·tom′e·try
cy′to·mor·phol′o·gy
cy′to·path′o·gen′e·sis
cy′to·path′o·ge·net′ic
cy′to·path′o·gen′ic
cy′to·path′o·ge·nic′i·ty
cy′to·path′o·log′ic *or* ·log′i·cal
cy′to·pa·thol′o·gist
cy′to·pa·thol′o·gy
cy′to·plasm
cy′to·plas′mic
cy′to·plast′
cy′to·sine′
cy′to·some′
cy′to·stat′ic
cy′to·tax′is

61

cy·to·tox′ic
cy·to·tox′in
cy·to·troph′o·blast′
cy·to·troph′o·blas′tic
cyt′u·la
cy·tu′ri·a

D

dac′ry·o·ad′e·nal′gi·a *or* dac′ry·ad′e·nal′gi·a
dac′ry·o·ad′e·ni′tis *or* dac′ry·ad·e·ni′tis
dac′ry·o·cyst
dac′ry·o·cys·tec′to·my
dac′ry·o·cys·ti′tis
dac′ry·o·cys′to·blen′nor·rhe′a
dac′ry·o·lith′
dac′ry·o′ma
 ·mas *or* ·ma·ta
dac′ry·on
 ·ry·a
dac′ry·ops
dac′tyl
dac·tyl′o·gram′
dac·ty·log′ra·phy
dac·ty·lol′o·gy
dac·ty·los′co·py
dac′ty·lus
 ·li′

dag′ga
dai′ly
Da′kin's solution
dal′ton
Dal′ton·ism
dam
damp
dan′a·zol′
dance
D and C
dan′der
dan′druff
dan′thron
dap′sone
dark′–a·dapt′ed
dark′–field microscope
dar·to′ic *or* dar′toid
dar′tos
Dar′win·ism
Dar′win·ist
Dar′win·is′tic
da·sym′e·ter
da′ta
 (*sing.* da′tum)
da′ta base *or* da′ta·base′ *or* bank·
date
 dat′ed dat′ing
da′tum
 ·ta
da·tu′ra
daugh′ter
day care
day′dream′

daz′zle
 ·zled ·zling
de·ac′ti·vate′
 ·vat′ed ·vat′ing
de·ac′ti·va′tion
dead′en
dead′ly
 ·li·er ·li·est
deaf′–and–dumb′
deaf′en
de·af′fer·en·tate′
 ·tat′ed ·tat′ing
de·af′fer·en·ta′tion
deaf′–mute′
de·al′co·hol′i·za′tion
de·am′i·dase′
de·am′i·di·za′tion
de·am′i·nase′
de·am′i·nate′
 ·nat′ed ·nat′ing
de·am′i·na′tion
de·am′i·ni·za′tion
de·am′i·nize′
 ·nized ·niz′ing
death′bed′
death rate
death rat′tle
death′watch′
de·bil′i·tant
de·bil′i·tate′
 ·tat′ed ·tat′ing
de·bil′i·ta′tion
de·bil′i·ty
 ·ties
de·bouch′
dé·bouch·ment′

dé·bride′
·brid′ed ·brid′ing
dé·bride′ment
de·bris′ *or* dé·bris′
de·caf′fein·at′ed
dec′a·gram′
de·cal′ci·fi·ca′tion
de·cal′ci·fi′er
de·cal′ci·fy′
·fied′ ·fy′ing
dec′a·li′ter
dec′a·me·ter
de·cant′
de′can·ta′tion
de·cap′i·tate′
·tat′ed ·tat′ing
de·cap′i·ta′tion
de·cap′i·ta·tor
de·car′bon·i·za′tion
de·car′bon·ize′
·ized′ ·iz′ing
de·car′box·y·late′
·lat′ed ·lat′ing
de·car′box·y·la′tion
dec′a·vi′ta·min
de·cay′
de·cease′
·ceased′ ·ceas′ing
de·ceit′
de·cel′er·ate′
·at′ed ·at′ing
de·cel′er·a′tion
de·cel′er·a′tor
de·cen′ter
de·cen′tra′tion
de·cer′e·brate′
·brat′ed ·brat′ing

de·cer′e·brate *adj.*
de·cer′e·bra′tion
de·cer′e·brize′
·brized′ ·briz′ing
dec′i·bel′
de·cid′u·a
·ae
de·cid′u·al
de·cid′u·ous
dec′i·gram′
dec′i·li′ter
dec′i·me′ter
dec′li·na′tion
de·cline′
·clined′ ·clin′ing
de·coct′
de·coc′tion
de·col′late′
·lat′ed ·lat′ing
de·col·la′tion
de·col′or
de·col′or·ant
de·col′or·a′tion
de·col′or·i·za′tion
de·col′or·ize′
·ized′ ·iz′ing
de·com′pen·sate′
·sat′ed ·sat′ing
de·com′pen·sa′tion
de·com·pos′a·ble
de·com·pose′
·posed′ ·pos′ing
de·com·po·si′tion
de·com·press′
de·com·pres′sion
de·con·di′tion
de·con·gest′ant

de′con·tam′i·nate′
·nat′ed ·nat′ing
de′con·tam′i·na′tion
de′con·trol′
·trolled′ ·trol′ling
de·cor′ti·cate′
·cat′ed ·cat′ing
de·cor′ti·ca′tion
de·crease′
·creased′ ·creas′ing
dec′re·ment
de·crep′it
de·crep′i·tate′
·tat′ed ·tat′ing
de·crep′i·ta′tion
de·crep′i·tude′
de·cre·scen′do
·dos
de·cres′cent
de·cu′bi·tal
de·cu′bi·tus
·ti′
de·cus′sate
·sat·ed ·sat·ing
de·cus·sa′tion
de·dif′fer·en′ti·ate′
·at′ed ·at′ing
de·dif′fer·en′ti·a′tion
deer′fly′
·flies′
def′e·cate′
·cat′ed ·cat′ing
def′e·ca′tion
de′fect *n.*
de·fec′tive

de·fense′ mechanism
de·fen′sive
def′er·ens
def′er·ent
def′er·en′tial
de′fer·ves′cence
de·fib′ril·late′
　·lat′ed ·lat′ing
de·fib′ril·la′tion
de·fib′ril·la′tor
de·fi′cien·cy
　·cies
de·fi′cient
def′i·cit
def′i·ni′tion
de·fin′i·tive
de·fla′tion
de·flect′
de·flec′tion
def′lo·ra′tion
de·flow′er
de·for·ma′tion
de·formed′
de·form′i·ty
　·ties
de·gen′er·a·cy
de·gen′er·ate adj., n.
de·gen′er·ate′
　·at′ed ·at′ing
de·gen′er·a′tion
de·gen′er·a·tive
de′glu·ti′tion
de·grad′a·ble
deg′ra·da′tion

de·grade′
　·grad′ed ·grad′ing
de·gree′
de·gust′
de′gus·ta′tion
de·hisce′
　·hisced′ ·hisc′ing
de·his′cence
de·his′cent
de·hu′man·i·za′tion
de·hu′man·ize′
　·ized′ ·iz′ing
de′hu·mid′i·fi·ca′tion
de′hu·mid′i·fi′er
de′hu·mid′i·fy′
　·fied′ ·fy′ing
de·hy′drate
　·drat·ed ·drat·ing
de′hy·dra′tion
de·hy′dro·gen·ase′
de·hy′dro·gen·ate′
　·at′ed ·at′ing
de·hy′dro·gen·a′tion
de·hy′dro·gen·ize′
　·ized′ ·iz′ing
de·hyp′no·tize′
　·tized′ ·tiz′ing
de·i′on·i·za′tion
de·i′on·ize′
　·ized′ ·iz′ing
dé·jà vu′
de·jec′ta
de·jec′tion

For words beginning deka– *see* DECA–
de·lac·ta′tion
de·lam′i·nate′
　·nat′ed ·nat′ing
de·lam′i·na′tion
de′–lead′
del′e·te′ri·ous
de·le′tion
de·lim′it
de·lim′i·tate′
　·tat′ed ·tat′ing
de·lim′i·ta′tion
de·lin′quen·cy
　·cies
de·lin′quent
del′i·quesce′
　·quesced′ ·quesc′ing
del′i·ques′cence
del′i·ques′cent
del′i·ra′tion
de·lir′i·ous
de·lir′i·um
　·i·ums *or* ·i·a
delirium tre′mens
del′i·tes′cence
del′i·tes′cent
de·liv′er
de·liv′er·y
　·ies
del′o·mor′phous
de·louse′
　·loused′ ·lous′ing
del′phi·nine′
del′ta

del'toid
de·lu'sion
de·lu'sion·al
de·lu'sive
de·mar'cate
 ·cat·ed ·cat·ing
de'mar·ca'tion or ·ka'tion
deme
de·ment'ed
de·men'tia
dementia prae'cox
 de·men'ti·ae prae·co'ces
dem'i·lune'
de·min'er·al·i·za'tion
demise' n.
de'mo·graph'ic
de·mog'ra·phy
de·mo'ni·ac
de·mon'stra·ble
dem'on·strate'
 ·strat·ed ·strat'ing
dem'on·stra'tor
de·mul'cent
de·my'e·lin·ate'
 ·at·ed ·at·ing
de·my'e·li·na'tion or ·lin·i·za'tion
de·na'tur·ant
de·na'tur·a'tion
de·na'ture
 ·tured ·tur·ing
den'dri·cep'tor
den'dri·form'
den'drite

den·drit'ic or ·drit'i·cal
den'droid
den'dron
 ·drons or ·dra
de·ner'vate
 ·vat·ed ·vat·ing
de'ner·va'tion
den'gue
de·ni'al
den'i·da'tion
den'i·grate'
 ·grat·ed ·grat·ing
den'i·gra'tion
de·ni'trate
 ·trat·ed ·trat·ing
de·ni'tra'tion
de·ni'tri·fi·ca'tion
de·ni'tri·fy'
 ·fied' ·fy'ing
dens
 den'tes
den·sim'e·ter
den'si·tom'e·ter
den'si·tom'e·try
den'si·ty
 ·ties
den'tal
den·tal'gi·a
den'tate
den·ta'tion
den'ti·buc'cal
den'ti·cle
den·tic'u·late or ·lat'ed
den·tic'u·la'tion
den'ti·form'

den'ti·frice'
den·tig'er·ous
den'tin or ·tine
den'ti·no·gen'e·sis
 ·ses'
den'ti·no·gen'ic
den'ti·noid'
den'ti·no'ma
 ·mas or ·ma·ta
den'tist
den'tist·ry
den·ti'tion
den'to·fa'cial
den'toid
den'to·sur'gi·cal
den'tu·lous
den'ture
de·nu'cle·at'ed
de·nu'date
 ·dat·ed ·dat·ing
de'nu·da'tion
de·nude'
 ·nud'ed ·nud'ing
de·o'dor·ant
de·o'dor·i·za'tion
de·o'dor·ize'
 ·ized' ·iz'ing
de·o'dor·iz'er
de·os'si·fi·ca'tion
de·ox'i·da'tion
de·ox'i·dize'
 ·dized' ·diz'ing
de·ox'y·cor'ti·cos'ter·one'
de·ox'y·gen·ate'
 ·at·ed ·at'ing

de·ox′y·ri′bo·nu′cle·ase′
de·ox′y·ri′bo·nu·cle′ic acid
de·ox′y·ri′bo·nu′cle·o·side′
de·ox′y·ri′bo·nu′cle·o·tide′
de·ox′y·ri′bose
de·pend′ence
de·pend′en·cy
  ·cies
de·pend′ent
de·per′son·al·i·za′tion
de·per′son·al·ize′
  ·ized′ ·iz′ing
de′pig·men·ta′tion
dep′i·late′
  ·lat′ed ·lat′ing
dep′i·la′tion
dep′i·la′tor
de·pil′a·to′ry
  ·ries
de·plete′
  ·plet′ed ·plet′ing
de·ple′tion
de′plu·ma′tion
de·plume′
  ·plumed′ ·plum′ing
de·po′lar·i·za′tion
de·po′lar·ize′
  ·ized′ ·iz′ing
de·po′lar·iz′er
de·pos′it
de′pot
dep′ra·va′tion
de·press′
de·pres′sant
de·pressed′
de·pres′si·ble
de·pres′sion
de·pres′sive
de·pres′so·mo′tor
de·pres′sor
dep′ri·va′tion or de·priv′al
de·prive′
  ·prived′ ·priv′ing
de·pro′gram
  ·grammed or ·gramed
  ·gram·ming or ·gram·ing
dep′side
depth psychology
dep′u·rate′
  ·rat′ed ·rat′ing
de·range′
  ·ranged′ ·rang′ing
de·range′ment
de·re′al·i·za′tion
de·re′ism
de·re·is′tic
der′i·va′tion
de·riv′a·tive
de·rive′
  ·rived′ riv′ing
der′ma
der′ma·bra′sion
der′mal
der′ma·ti′tis
  der′ma·tit′i·des′

der′ma·to·au′to·plas′ty
der′ma·to·fi·bro′ma
  ·mas or ·ma·ta
der′mat·o·gen
der′ma·to·glyph′ics
der′ma·to·log′i·cal or ·log′ic
der′ma·tol′o·gist
der′ma·tol′o·gy
der′ma·tol′y·sis
der′ma·tome′
der′ma·to·phyte′
der′ma·to·plas′ty
der′ma·to′sis
  ·ses
der′mic
der′mis
der′moid
der·mop′a·thy
der′mo·phyte′
der′mo·vas′cu·lar
de·sal′i·nate′
  ·nat′ed ·nat′ing
de·sal′i·na′tion or ·i·ni·za′tion
de·sal′i·nize′
  ·nized′ ·niz′ing
de·sat′u·ra′tion
De·sault′s′ apparatus
des′ce·me·ti′tis
des′ce·met′o·cele′
Des·ce·met′s′ membrane
de·scend′ing
de·scen′sus

de·scent'
de·scrip'tive
de·sen'si·ti·za'tion
de·sen'si·tize'
 ·tized' ·tiz'ing
de·sen'si·tiz'er
de·sex'
desex'u·al·i·za'tion
de·sex'u·al·ize'
 ·ized' ·iz'ing
des'ic·cant
des'ic·cate'
 ·cat'ed ·cat'ing
des'ic·ca'tion
des'ic·ca'tive
des'ic·ca'tor
des·mi'tis
des·mog'ra·phy
des'moid
des·mo'ma
 ·mas or ·ma·ta
des'mo·some'
des·mot'o·my
de·so'cial·i·za'tion
de·so'cial·ize'
 ·ized' ·iz'ing
de·sorb'
de·sorp'tion
des·ox'y·cor'ti·cos'ter·one'
des·pre'ci·ate'
 ·at'ed ·at'ing
des·pre'ci·a'tion
de·spu'mate
 ·mat·ed ·mat·ing
des'pu·ma'tion

des'qua·mate'
 ·mat'ed ·mat'ing
des·qua·ma'tion
des·quam'a·tive or
 ·a·to'ry
de·stain'
de·struc'tive
de·stru'do
de·sub'li·ma'tion
de·sulf'hy'drase or
 de·sul'fu·rase'
de·syn'chro·nize'
 ·nized' ·niz'ing
de·tach'ment
de·tect'a·ble or
 ·i·ble
de·tec'tor
de·terge'
 ·terged' ·terg'ing
de·ter'gen·cy or
 ·gence
de·ter'gent
de·te'ri·o·rate'
 ·rat'ed ·rat'ing
de·te'ri·o·ra'tion
de·ter'mi·nant
de·ter'mi·na'tion
de·ter'mine
 ·mined ·min·ing
de·ter'min·er
de·ter'min·ism
de·ter'sive
det'o·na'tion
de·tor'sion
de·tox' v.
de'tox n., adj.

de·tox'i·cate'
 ·cat'ed ·cat'ing
de·tox'i·ca'tion
de·tox'i·fi·ca'tion
de·tox'i·fy'
 ·fied' ·fy'ing
de·tri'tal
de·tri'tion
de·tri'tus
de·trun'cate
 ·cat·ed ·cat·ing
de'trun·ca'tion
de·tru'sor
de·tu·mes'cence
de·tu·mes'cent
deu'ter·an·ope'
deu'ter·an·o'pi·a
deu'ter·a·nop'ic
deu'ter·ate'
 ·at'ed ·at'ing
deu·te'ri·um
deu'ter·on'
deu'to·plasm or
 deu'ter·o·plasm'
deu'to·plas'mic or
 deu'ter·o·plas'mic
de·vas'cu·lar·i·za'tion
de·vel'op·ment
de·vel'op·men'tal
de'vi·an·cy or ·ance
de'vi·ant
de'vi·ate'
 ·at'ed ·at'ing
de'vi·ate n.
de'vi·a'tion
de·vice'

dev'il's grip
de·vi'om·e·ter
de·vi'tal·i·za'tion
de·vi'tal·ize'
·ized' ·iz'ing
dev'o·lu'tion
dev'o·lu'tion·ist
de·worm'
dex'a·meth'a·sone'
dex'ie or dex'y
·ies *slang term*
dex'ter
dex·ter'i·ty
dex'ter·ous
dex'trad
dex'tral
dex·tral'i·ty
·ties
dex'tran
dex'trase
dex'trin or ·trine
dex'trin·u'ri·a
dex'tro
dex'tro·am·phet'a·mine'
dex'troc·u·lar'i·ty
dex'tro·glu'cose
dex'tro·ro·ta'tion
dex'tro·ro'ta·to'ry
or ·ro'ta·ry
dex'trorse
dex'trose
dex'trous
di'a·be'tes in·sip'i·dus
diabetes mel·li'tus
di'a·bet'ic

di'a·bet'o·gen'ic
di'a·be·tog'e·nous
di'a·bro'sis
di'a·brot'ic
di'a·caus'tic
di·ac'id
di·a·clast'
di·ac'ri·nous
di·ac'ri·sis
·ses'
di'a·crit'ic or
·crit'i·cal
di'ac·tin'ic
di'ac·tin·ism
di'ag·nos'a·ble
di'ag·nose'
·nosed' ·nos'ing
di'ag·no'sis
·ses
di'ag·nos'tic
di'ag·nos·ti'cian
di'a·gram'
·gramed' or
·grammed'
·gram'ing or
·gram'ming
di'a·graph'
di'a·ki·ne'sis
·ses
di'al
di·al'y·sance
di·al'y·sate
di·al'y·sis
·ses'
di'a·lyt'ic
di'a·lyz'a·ble

di'a·lyze'
·lyzed' ·lyz'ing
di'a·lyz'er
di'a·mag'net
di'a·mag·net'ic
di'a·mag'net·ism
di·am'e·ter
di·am'e·tral
di'a·met'ric
di'a·met'ri·cal
di·am'ine
di'a·pa'son
di'a·pe·de'sis
·ses
di'a·pe·det'ic
di'a·per
di·aph'a·no·scope'
di·aph'a·nos'co·py
di'a·pho·re'sis
·ses
di'a·pho·ret'ic
di'a·phragm'
di'a·phrag'ma
·ma·ta
di'a·phrag·mat'ic
di'a·phys'e·al or
·i·al
di·aph'y·sis
·ses'
di'ap·o·phys'i·al
di'a·poph'y·sis
·ses'
di'ar·rhe'a or
·rhoe'a
di'ar·rhe'al or
·rhe'ic
di·ar'thric

di·ar·thro′sis
·ses
di·as′chi·sis
·ses′
di′a·stase′
di′a·stat·ic
di′a·ste′ma
·ste′ma·ta
di′a·ste·mat′ic
di·as′ter
di·as′to·le′
di·a·stol′ic
di·as′tral
di·a·tes′sa·ron′
di′a·ther′mal or
·ther′mic
di′a·ther′man·cy
di′a·ther′ma·nous
di′a·ther′mic
di′a·ther′my
di·ath′e·sis
·ses′
di′a·thet′ic
di′a·tom′
di′a·to·ma′ceous
di′a·tom′ic
di·az′e·pam′
di′a·zine′
di·az′i·non
di·az′o
di·az′o·tize′
·tized′ ·tiz′ing
di·bas′ic
di·ceph′a·lous adj.
di·ceph′a·lus n.
di·chlo′ride

di·chlo′ro·phe·nox′y·a·ce′tic acid
di·chot′o·mi·za′tion
di·chot′o·mize′
·mized ·miz′ing
di·chot′o·mous
di·chot′o·my
·mies
di·chro′ic or di′chro·it′ic
di′chro·ism
di·chro′mate
di′chro·mat′ic
di·chro′ma·tism
di·chro′mic
di′chro·scope′ or di·chro′o·scope′
di·coe′lous
di·cou′ma·rin
di·crot′ic
di′cro·tism
dic′ty·o·some′
Di·cu′ma·rol′
(trademark)
di·dac′tic or
·dac′ti·cal
di·dym′i·um
did′y·mous adj.
(twin)
did′y·mus
(testis)
die
died dy′ing
diel′drin
di′en·ceph′a·lon′
·la
di′en·ce·phal′ic

di·er′e·sis
·ses′
di′e·ret′ic
di·es′trous adj.
di·es′trum
di·es′trus n.
di′et
di′e·tar′y
·ies
di′et·er
di′e·tet′ic or
·tet′i·cal
di′e·tet′ics
di·eth′yl·bar′bi·tu′ric acid
di·eth′yl ether
di·eth′yl·stil·bes′trol
di′e·ti′tian or
·ti′cian
Die′tl's crisis
dif′fer
dif′fer·ence
dif′fer·ent
dif′fer·en′tial
dif′fer·en′ti·ate′
·at′ed ·at′ing
dif′fer·en′ti·a′tion
dif′fi·cult
dif′flu·ence
dif′flu·ent
dif·fract′
dif·frac′tion
dif·frac′tive
dif·fu′sate
dif·fuse′
·fused′ ·fus′ing
dif·fus′er or ·fu′sor

dif·fus′i·ble
dif·fu′sion
dif·fu′sive
di·gas′tric
di·gen′e·sis
di′ge·net′ic
di·gest′ v.
di·ges′tant
di·gest′er
di·gest′i·bil′i·ty
di·gest′i·ble
di·ges·tif′
di·ges′tion
di·ges′tive
dig′it
dig′i·tal
dig′i·tal·in
dig′i·tal·is
dig′i·tal·i·za′tion
dig′i·tal·ize′
 ·ized′ ·iz′ing
dig′i·tate′ or ·tat′ed
dig′i·ta′tion
dig′i·ti·form′
dig′i·tox′in
dik′ty·o′ma
 ·mas or ·ma·ta
di·lac′er·ate′
 ·at′ed ·at′ing
di·lac′er·a′tion
di·lat′a·bil′i·ty
di·lat′a·ble
di·lat′an·cy
di·lat′ant
dil′a·ta′tion
dil′a·ta′tion·al
dil′a·ta′tor

di·late′
 ·lat′ed ·lat′ing
di·la′tion
di·la′tive
di·la′tor
dil′do
 ·dos or ·does
dil′doe
dil′u·ent
di·lute′
 ·lut′ed ·lut′ing
di·lut′er or ·lu′tor
di·lu′tion
di′men·hy′dri·nate′
di·men′sion
di′mer
di·mer′ic
di·meth′yl
 sulf·ox′ide
di·mid′i·ate′
 ·at′ed ·at′ing
di·min′ish
dim′i·nu′tion
di·mor′phic or
 ·phous
di·mor′phism
dim′ple
 ·pled ·pling
di·ner′ic
din′i·cal
di·ni′tro·ben′zene
di′no·flag′el·late
di·nu′cle·o·tide′
di·oc′tyl sodium
 sul′fo·suc′ci·nate′
di·oe′cious
di·oe′cism

di·oes′trum
di·op′ter or ·tre
di′op·tom′e·ter
di′op·tom′e·try
di·op′tral
di·op′tric or ·tri·cal
di·op′trics
di·ox′ane
di·ox′ide
di·ox′in
dip
 dipped dip′ping
di·pep′ti·dase′
di·pep′tide
di·per′o·don′
di·pet′a·lo·ne′ma
di·phal′lus
di·phase′ or di·
 pha′sic
di·pheb′u·zol′
di′phen·hy′dra·
 mine
di·phen′yl
di·phen′yl·hy·dan′
 to·in
di·phos′gene
diph·the′ri·a
diph·the′ri·al
diph′the·rit′ic or
 ·diph·ther′ic
diph′the·roid
diph·thon′gi·a
di·phyl′lo·both·ri′a·
 sis
 ·ses′

di·phyl'lo·both'ri·um
di·phy'o·dont'
dip·la·cu'sis
·ses
di·ple'gi·a
dip'lo·blas'tic
dip'lo·car'di·a
dip'lo·coc'cal or ·coc'cic
dip'lo·coc'cus
·coc'ci
dip'lo·ë
dip'lo·et'ic or dip·lo'ic
dip'loid
dip·loi'dy
dip'lo·mate'
dip'lont
di·plo'pi·a
di·plo'pic
dip'lo·scope'
di·plo'sis
·ses
dip'lo·so·ma'ti·a or dip'lo·some'
di·po'lar
di'pole'
dip·se'sis
dip·set'ic
dip'so·gen
dip'so·ma'ni·a
dip'so·ma'ni·ac'
dip'so·ma·ni'a·cal
dip'stick'
dip'ter·an
dip'ter·ous

di·rect'
di·rec'tion
di·rec'tive
di·rec'tor
dis'a·bil'i·ty
·ties
dis·a'ble
·bled ·bling
dis·a'ble·ment
di·sac'cha·ri·dase'
di·sac'cha·ride'
dis'ag·gre·ga'tion
dis'ap·pear'
dis'ar·tic'u·late'
·lat'ed ·lat'ing
dis'ar·tic'u·la'tion
dis'as·sim'i·late'
·lat'ed ·lat'ing
dis'as·sim'i·la'tion
dis'as·so'ci·ate'
·at'ed ·at'ing
dis'as·so'ci·a'tion
disc
dis·cern'
dis·cern'i·ble
dis·charge' n.
dis·charge'
·charged' ·charg'ing
dis·charge'a·ble
dis·cis'sion
dis·close'
·closed' ·clos'ing
dis'co·blas'tic
dis'co·blas'tu·la
·las or ·lae'

dis'coid or dis·coi'dal
dis·col'or
dis·col'or·a'tion
dis·com'fort
dis'con·nect'
dis'con·nec'tion
dis'con·tin'ue
·ued ·u·ing
dis'con·tin'u·ous
dis'cord
dis·cord'ance or ·an·cy
dis·cord'ant
dis·crep'an·cy
·cies
dis·crete'
(separate)
dis·crim'i·nate'
·nat'ed ·nat'ing
dis·crim'i·na'tion
dis'cus
·cus·es or ·ci
dis·cuss'
dis·cus'sive or dis·cu'tient
dis·ease'
·eased' ·eas'ing
dis'en·gage'
·gaged' ·gag'ing
dis'en·gage'ment
dis'e·qui·lib'ri·um
·ri·ums or ·ri·a
dis·fea'ture
·tured ·tur·ing
dis·fig'u·ra'tion

dis·fig'ure
  ·ured ·ur·ing
dis·fig'ure·ment
dish
dis'ha·bit'u·a'tion
dis'in·fect'
dis'in·fect'ant
dis'in·fec'tion
dis'in·fest'
dis'in·fes·ta'tion
dis'in·hi·bi'tion
dis·in'te·grate'
  ·grat'ed ·grat'ing
dis·in'te·gra'tion
dis·in'te·gra'tor
dis·joint'
dis·junc'tion *or*
  ·ture
dis·junc'tive
disk
dis'ki·form'
dis·ki'tis
dis·kog'ra·phy
dis'lo·cate'
  ·cat'ed ·cat'ing
dis'lo·ca'tion
dis·mem'ber
dis·mem'ber·ment
dis·or'der
dis·or'gan·i·za'tion
dis·or'gan·ize'
  ·ized' ·iz'ing
dis·o'ri·ent'
dis·o'ri·en·tate'
  ·tat'ed ·tat'ing
dis·o'ri·en·ta'tion
dis'pa·rate

dis·par'i·ty
  ·ties
dis·pen'sa·ry
  ·ries
dis·pen'sa·to'ry
  ·ries
dis·pense'
  ·pensed' ·pens'ing
dis·pen'ser
dis·perse'
  ·persed' ·pers'ing
dis·pers'er
dis·pers'i·ble
dis·per'sion
dis·per'sive
dis·per'soid
dis·place'
  ·placed' ·plac'ing
dis·place'ment
dis·pos'a·ble
dis·pose'
  ·posed' ·pos'ing
dis'po·si'tion
dis'po·si'tion·al
dis'pro·por'tion
dis·rupt'
dis·rup'tion
dis·rup'tive
dis·sect'
dis·sec'tion
dis·sec'tor
dis·sem'i·nate'
  ·nat'ed ·nat'ing
dis·sem'i·na'tion
dis·sep'i·ment
dis·sim'i·lar

dis·sim'i·late'
  ·lat'ed ·lat'ing
dis·sim'i·la'tion
dis·sim'u·late'
  ·lat'ed ·lat'ing
dis·sim'u·la'tion
dis'si·pate'
  ·pat'ed ·pat'ing
dis'si·pa'tion
dis·so'ci·ate'
  ·at'ed ·at'ing
dis·so'ci·a'tion
dis·so'ci·a'tive
dis·sol'u·ble
dis'so·lu'tion
dis·solv'a·ble
dis·solve'
  ·solved' ·solv'ing
dis·sol'vent
dis'tad
dis'tal
dis'tance
dis·tem'per·a·ture
dis·tend'
dis·ten'si·ble
dis·ten'tion *or* ·sion
dis·ti·chi'a·sis *or*
  dis·tich'i·a
dis·till' *or* ·til'
  ·tilled' ·till'ing
dis'til·late'
dis'til·la'tion
dis·till'ment *or*
  ·til'ment
dis'to·buc'cal
dis'to·buc'co–oc·
  clu'sal

72

dis·tort'
dis·tor'tion
dis·tract'i·bil'i·ty
dis·trac'tion
dis·trac'tive
dis·tress'
dis·trib'ute
· ut·ed · ut·ing
dis'tri·bu'tion
dis·trib'u·tive
dis'trix
dis·turb'
dis·turb'ance
di·sul'fate
di·sul'fide
di·sul'fi·ram
di'u·re'sis
· ses
di'u·ret'ic
di·ur'nal
di'va·gate'
· gat'ed · gat'ing
di'va·ga'tion
di·va'lent
di·var'i·cate'
· cat'ed · cat'ing
di·var'i·ca'tion
di·verge'
· verged' · verg'ing
di·ver'gence
di·ver'gent
di'ver·tic'u·lar
di'ver·tic'u·li'tis
di'ver·tic'u·lo'sis
di'ver·tic'u·lum
· la

di·vide'
· vid·ed · vid·ing
div'i–div'i
di·vi'sion
di·vulse'
· vulsed' · vuls'ing
di·vul'sion
di·vul'sor
di'zy·got'ic
diz'zi·ness
diz'zy
· zi·er · zi·est
· zied · zy·ing
DMSO
DMT
DNA
Do·bell's' solution
Do'bie's globule
doc'tor
doc'tor·al
doc'trine
doc'u·ment'
doc'u·sate'
dod'der·ing
dog bite
dog·mat'ic
Dog'ma·tism
Dog'ma·tist
dog'tooth'
· teeth'
dol
dol'i·cho·ce·phal'ic
  or
· cho·ceph'a·lous
dol'i·cho·ceph'a·ly
dol'i·cho·cra'ni·al
  or · cho·cra'nic

dol'i·cho·cra'ny
dol'i·cho·pel'lic or
· pel'vic
do'lo·mite'
do'lor
do·lo'res
do'lo·rif'ic
do'lo·rous
do·main'
dom'i·nance or
· nan·cy
dom'i·nant
dom'i·nate'
· nat·ed · nat·ing
dom'i·na'tor
do'nate
· nat·ed · nat·ing
do·nee'
do'nor
Don'o·van bodies
do'pa
do'pa·mine'
dop'ant
dope
  doped dop'ing
Dop'pler effect
dor'man·cy
dor'mant
dor'nase
Dorn'i·er
  lithotripter
dor'sad'
dor'sal
dor·sal'gi·a
dor·sa'lis
dor'si·flex'ion
dor'si·ven'tral

dor'so·lat'er·al
dor'so·ven'tral
dor'sum
 ·sa
dos'age
dose
 dosed dos'ing
do·sim'e·ter
do·si·met'ric
do·sim'e·try
dot
 dot'ted dot'ting
dot'age
dot'ard
dou'ble
 ·bled ·bling
double bind
dou'ble–blind' test
double bond
dou'ble–joint'ed
double pneumonia
dou'blet
douche
 douched douch'ing
Doug'las's
 cul–de–sac
dove'tail'
dow'el
down n.
Down's syndrome
DPT vaccine
drachm
drac'on·ti'a·sis or
 dra·cun'cu·li'a·sis
 or dra·cun'cu·lo'-
 sis
draft

drag n.
dra·gée'
drain'age
dram
dram'a·tism
dram'a·ti·za'tion
dram'a·tize'
 ·tized' ·tiz'ing
drape
 draped drap'ing
dras'tic
draw
 drew drawn
 draw'ing
dream
 dreamed or dreamt
 dream'ing
drep'a·no·cyte'
drep'a·no·cyt'ic
dress
 dressed or drest
 dress'ing
DRG
drib'ble
 ·bled ·bling
drift
drill
drink
 drank
 drunk
 drink'ing
drip
 dripped or dript
 drip'ping
drip'per
drive n.

driv'el
 ·eled or ·elled
 ·el·ing or ·el·ling
dro'mo·trop'ic
drool
droop
drop'let
drop'per
drop'si·cal or ·sied
drop'sy
drown
drow'sy
 ·si·er ·si·est
drug
 drugged drug'ging
drug'–fast'
drug'gist
drug'store'
drum'head'
drum'stick'
drunk'ard
drunk'en
drunk'en·ness
drunk·o'me·ter
dru'sen
 (sing. druse)
dry
 dri'er dri'est
 dried dry'ing
dry ice
dry'–nurse'
 ·nursed' ·nurs'ing
dry nurse
du'al
du'al·ism
du'al·ist
du'al·is'tic

duct
duc'tal
  (of a duct)
duc'tile
  (stretchable)
duc·til'i·ty
duct'less
duct'ule
duc'tu·lus
  ·li'
duc'tus
Dukes'
  classification
dull'ness *or* dul'
  ness
dumb
dumb'bell'
dum'my
  ·mies
dump'ing
  syndrome
du'o·de'nal
du'o·de'no·cho·led'
  o·chot'o·my
du'o·de'no·je·ju'
  nos'to·my
du'o·de'num
  ·de'na *or* ·de'nums
du'pli·ca'tion
du'ral
du'ra mat'er
du'ra·plas'ty
du·ra'tion
dust
du'ty
  ·ties
dwarf

dwarf'ish
dwarf'ism
dy'ad
dy·ad'ic
dye
  dyed dye'ing
dy'ing
dy·nam'ic *or*
  ·nam'i·cal
dy·nam'ics
dy'na·mo·gen'e·sis
  ·ses'
dy'na·mo·gen'ic
dy'na·mom'e·ter
dyne
dy·ne'in
dys'a·cou'si·a *or*
  ·cous'ma
dys'a·cou'sis *or*
  ·cu'sis
dys·cra'si·a
dys·cra'sic *or*
  ·crat'ic
dys'en·ter'ic
dys'en·ter'y
dys·func'tion
dys·func'tion·al
dys·gam'ma·glob'u·
  li·ne'mi·a
dys·gam'ma·glob'u·
  li·ne'mic
dys·gen'e·sis
  ·ses'
dys·gen'ic
dys·gen'ics
dys·gna'thi·a
dys·gnath'ic

dys·gon'ic
dys·graph'i·a
dys'ki·ne'si·a
dys·la'li·a
dys·lec'tic
dys·lex'i·a
dys·lex'ic
dys·men'or·rhe'a
dys'o·rex'i·a
dys'pa·reu·ni·a
dys·pep'si·a *or*
  ·pep'sy
dys·pep'tic
dys·pha'gi·a
dys·phag'ic
dys·pha'si·a
dys·pha'sic
dys·pho'ni·a
dys·phon'ic
dys·pho'ri·a
dys·phor'ic
dys·pla'si·a
dys·plas'tic
dysp·ne'a
dysp·ne'al *or* ·ne'ic
dys·pro'si·um
dys·rhyth'mi·a
dys·rhyth'mic
dys·so'cial
dys·sta'si·a
dys·stat'ic
dys·tax'i·a
dys·to'ci·a
dys·to'ni·a
dys·ton'ic
dys·to'pi·a

dys·top′ic
dys·troph′ic
dys′tro·phy
dys·u′ri·a
dys·u′ric

E

ear′ache′
ear′drop′
ear′drum′
ear dust
ear′lap′
ear′lobe′
ear′plug′
ear trum′pet
ear′wax′
eat
    ate eat′en eat′ing
ebb
e′bo·na′tion
e·bri′e·ty
e′bur·na′tion
e·cau′date
ec·bol′ic
ec·cen′tric
ec′chy·mo′sis
    ·ses
ec′chy·mot′ic
ec′crine
ec·crit′ic
ec′dy·sis
    ·ses′
ech′i·nate′ or
    ·nat′ed
e·chi′no·coc′cus
    ·ci
ech′o
    ·oes
e′cho·car′di·o·
    gram′
e′cho·car′di·og′ra·
    phy
ech′o·gram′
e·chog′ra·phy
ech′o·la′li·a
ech′o·la′lic
ec·lamp′si·a
ec·lamp′tic
ec·lec′tic
ec·lec′ti·cism
e·clipse′
E. co′li
ec′o·log′i·cal or
    ·log′ic
e·col′o·gist
e·col′o·gy
ec′sta·sy
    ·sies
ec·stat′ic
ec′to·blast′
ec′to·car′di·a
ec′to·com·men′sal
ec′to·derm′
ec′to·der′mal or
    ·der′mic
ec′to·gen′e·sis
ec·tog′e·nous or ec′
    to·gen′ic
ec′to·mere′
ec′to·mer′ic
ec′to·morph′
ec′to·mor′phic
ec′to·mor′phy
ec′to·par′a·site′
ec′to·par′a·sit′ic
ec·to′pi·a
ec·top′ic
ec′to·plasm
ec′to·plas′mic
ec′to·sarc′
ec·tro′pi·on or
    ·pi·um
ec′ze·ma
ec·zem′a·tous
e·de′ma
    ·mas or ·ma·ta
e·dem′a·tous
e·den′tate
e·den′tu·lous
ed′e·tate′
edge′–strength′
ed′i·ble
EDTA
ed′u·ca·bil′i·ty
ed′u·ca·ble
e′duct
e·duc′tion
ef·face′ment
ef·fect′
    (*result;* SEE affect)
ef·fec′tive
    (*producing effect;*
    SEE affective)
ef·fec′tor
ef·fec′tu·al
ef·fem′i·nate
ef·fem′i·na′tion

ef′fer·ent
  (*carrying away;* SEE
  afferent)
ef′fer·vesce′
  ·vesced′ ·vesc′ing
ef′fer·ves′cence
  (*a bubbling up;*
  SEE efflorescence)
ef′fer·ves′cent
ef·fi′cien·cy
  ·cies
ef·fi′cient
ef′flo·resce′
  ·resced′ ·resc′ing
ef′flo·res′cence
  (*skin eruption;* SEE
  effervescence)
ef′flo·res′cent
ef′flu·ent
ef·flu′vi·al
ef·flu′vi·um
  ·vi·a *or* ·vi·ums
ef′fort
ef·fuse′
  ·fused′ ·fus′ing
ef·fu′sion
e·gest′
e·ges′ta
e·ges′tion
e′go
  e′gos
e′go·cen′tric
e′go·cen·tric′i·ty
  ·ties
e′go·cen′trism
e′go–dys·ton′ic
e′go·ism
e′go·ist

e′go·is′tic *or*
  ·is′ti·cal
e′go·ma′ni·a
e′go·ma′ni·ac′
e′go·ma·ni′a·cal
e·goph′o·ny
e′go·tism
e′go·tist
e′go·tis′tic *or*
  ·tis′ti·cal
Ehr′lich's reaction
ei·det′ic
ein·stein′i·um
e·jac′u·late′
  ·lat′ed ·lat′ing
e·jac′u·la′tion
e·jac′u·la′tor
e·jac′u·la·to′ry
e·ject′
e·jec′ta
e·jec′tion
e·jec′tor
e·lab′o·rate′
  ·rat′ed ·rat′ing
e·lab′o·ra′tion
el′ae·op′tene
e·las′tase
e·las′tic
e·las·tic′i·ty
  ·ties
e·las′ti·cize′
  ·cized′ ·ciz′ing
e·las′tin
e·las′to·mer
e·lat′er·in
el′a·te′ri·um
e·la′tion

el′bow
el′der
  (*shrub*)
eld′er·ly
el′drin
el′e·cam·pane′
e·lec′tive
E·lec′tra complex
e·lec′tric
e·lec′tri·cal
e·lec′tri·fy′
  ·fied′ ·fy′ing
e·lec′tri·za′tion
e·lec′tro·ac′u·scope′
e·lec′tro·a·nal′y·sis
  ·ses′
e·lec′tro·an′a·lyt′ic
  or ·lyt′i·cal
e·lec′tro·car′di·o·
  gram′
e·lec′tro·car′di·o·
  graph′
e·lec′tro·car′di·o·
  graph′ic
e·lec′tro·car′di·og′
  ra·phy
e·lec′tro·chem′i·cal
e·lec′tro·chem′is·try
e·lec′tro·con·vul′
  sive
e·lec′tro·cute′
  ·cut′ed ·cut′ing
e·lec′tro·cu′tion
e·lec′trode
e·lec′tro·di·al′y·sis
  ·ses′

77

e·lec′tro·en·ceph′a·lo·gram′
e·lec′tro·en·ceph′a·lo·graph′
e·lec′tro·en·ceph′a·lo·graph′ic
e·lec′tro·en·ceph′a·log′ra·phy
 ·phies
e·lec′tro·graph′
e·lec′trol′o·gist
e·lec′trol′y·sis
 ·ses′
e·lec′tro·lyte′
e·lec′tro·lyt′ic
e·lec′tro·lyze′
 ·lyzed′ ·lyz′ing
e·lec′tro·mag′net
e·lec′tro·mag·net′ic
e·lec′tro·mag′net·ism
e·lec·trom′e·ter
elec′tro·my′o·gram′
e·lec′tro·my′o·graph′
e·lec′tro·my·og′ra·phy
e·lec′tron
e·lec′tro·neg′a·tive
e·lec′tron′ic
e·lec′tron·i·za′tion
e·lec′tron–volt′
e·lec′tro·op′tic or ·op′ti·cal
e·lec′tro·op′tics
e·lec′tro·pho·re′sis
 ·ses

e·lec′tro·pho·ret′ic
e·lec′tro·phys′i·o·log′i·cal
e·lec′tro·phys′i·ol′o·gist
e·lec′tro·phys′i·ol′o·gy
e·lec′tro·plate′
 ·plat′ed ·plat′ing
e·lec′tro·pos′i·tive
e·lec′tro·scis′sion
e·lec′tro·scope′
e·lec′tro·shock′
e·lec′tro·sleep′
e·lec′tro·sur′ger·y
e·lec′tro·syn′the·sis
e·lec′tro·ther′a·pist
e·lec′tro·ther′a·py
e·lec′tro·ton′ic
e·lec′trot′o·nus
e·lec′tro·va′lence
e·lec′tu·ar′y
 ·ar′ies
e·le′i·din
el′e·ment
el′e·men′ta·ry
 (basic; SEE alimentary)
el′e·mi
el′e·op′tene
el′e·phan·ti′a·sis
 ·ses
el′e·vate′
 ·vat′ed ·vat′ing
el′e·va′tion
el′e·va′tor
e·lim′i·nant

e·lim′i·nate′
 ·nat′ed ·nat′ing
e·lim′i·na′tion
e′lin·gua′tion
e·lix′ir
el·lip′sis
 ·ses
el·lip′soid or el·lip′soi′dal
e·lon′gate
 ·gat·ed ·gat·ing
e·lon′ga·tion
el′u·ate
el′u·ent or ·ant
e·lute′
 ·lut′ed ·lut′ing
e·lu′tion
e·lu′tri·ate′
 ·at′ed ·at′ing
e·lu′tri·a′tion
e·ma′ci·ate′
 ·at′ed ·at′ing
e·ma′ci·a′tion
em′an
em′a·nate′
 ·nat′ed ·nat′ing
em′a·na′tion
em′a·na′tor
e·mas′cu·late′
 ·lat′ed ·lat′ing
e·mas′cu·late adj.
e·mas′cu·la′tion
em·balm′
em·bed′
 ·bed′ded ·bed′ding
em′bo·lec′to·my
 ·mies

em·bol'ic
em'bo·lism
em'bo·lus
  ·li
em'bo·ly
em·bouche·ment'
em·brace' reflex
em·bra'sure
em'bro·cate'
  ·cat'ed ·cat'ing
em·bro·ca'tion
em'bry·ec'to·my
  ·mies
em'bry·o'
  ·os'
em'bry·o·gen'ic or
  ·ge·net'ic
em'bry·og'e·ny or
  ·o·gen'e·sis
em'bry·o·log'ic or
  ·log'i·cal
em'bry·ol'o·gist
em'bry·ol'o·gy
em'bry·o·nal
em'bry·on'ic
e·mer'gence
e·mer'gen·cy
  ·cies
e·mer'gent
em'er·y
em'e·sis
  ·ses'
e·met'ic
em'e·tine'
em'i·grate'
  ·grat'ed ·grat'ing
em'i·gra'tion

em'i·nence
em'is·sar'y
  ·sar'ies
e·mis'sion
em'is·siv'i·ty
e·mit'
  ·mit'ted ·mit'ting
em·men'a·gogue'
em·me'ni·a
em·men'ic
em'me·tro'pi·a
em'me·trop'ic
e·mol'li·ent
e·mo'tion
e·mo'tion·al
e·mo'ti·o·vas'cu·lar
e·mo'tive
e'mo·tiv'i·ty
em·path'ic or em'
  pa·thet'ic
em'pa·thize'
  ·thized' ·thiz'ing
em'pa·thy
  ·thies
em'phy·se'ma
em'phy·se'ma·tous
em·pir'ic
em·pir'i·cal
em·pir'i·cism
emp'ty
  ·tied ·ty'ing
empty nest
  syndrome
em'py·e'ma
  ·ma·ta
em'py·e'ma·tous or
  ·e'mic

e·mul'si·fi'a·ble or
  ·si·ble
e·mul'si·fi·ca'tion
e·mul'si·fi'er
e·mul'si·fy'
  ·fied' ·fy'ing
e·mul'sin
  (enzyme)
e·mul'sion
  (mixture of liquids)
e·mul'sive
e·mul'soid'
e·munc'to·ry
  ·ries
en·am'el
e·nam'e·lum
en·an'ti·o·morph'
en·an'ti·o·mor'phic
en·an'ti·o·mor'
  phism
en'ar·thro'sis
  ·ses
en bloc'
en·cap'su·late'
  ·lat'ed ·lat'ing
en·cap'su·la'tion
en·cap'sule
  ·suled ·sul·ing
en·ceinte'
en·ce·phal'ic
en·ceph'a·lit'ic
en·ceph'a·li'tis
encephalitis le·thar'
  gi·ca
en·ceph'a·lo·gram'
en·ceph'a·log'ra·
  phy

en·ceph'a·lo·my'e·li'tis
en·ceph'a·lon'·la
en·ceph'a·lop'a·thy
en'chon·dro'ma
  ·ma·ta or ·mas
en'chon·drom'a·tous
en'clave
en·clit'ic
en·crust'
en'crus·ta'tion
en·cyst'
en·cyst'ment or en'cys·ta'tion
en'da·moe'ba or ·me'ba
en'da·moe'bic
en·dar·ter·ec'to·my
en·dau'ral
end'–bod'y
end'brain'
end'–bulb'
en·dem'ic or ·dem'i·cal
en'der·gon'ic
en·der'mic
en'do·bi·ot'ic
en'do·blast'
en'do·car'di·al
en'do·car·di'tis
en'do·car'di·um
  ·di·a
en'do·com·men'sal
en'do·cra'ni·um
  ·ni·a

en'do·crine
en'do·cri'no·log'i·cal
en'do·cri·nol'o·gist
en'do·cri·nol'o·gy
en'do·cy·to'sis
  ·ses
en'do·derm'
en'do·der'mal or ·der'mic
en'do·don'tic
en'do·don'tics or ·don'tia
en'do·don'tist
en'do·en'zyme
en·dog'a·mous or en'do·gam'ic
en·dog'a·my
en·dog'e·nous
en·dog'e·ny
en'do·lymph'
en'do·me'tri·al
en'do·me'tri·o'sis
  ·ses
en'do·me·tri'tis
en'do·me·tri'um
  ·tri·a
en'do·mi·to'sis
  ·ses
en'do·mix'is
en'do·morph'
en'do·mor'phic
en'do·mor'phy
en'do·par'a·site'
en'do·pep'ti·dase'
en·doph'a·gous
en'do·phyte'

en'do·phyt'ic
en'do·plasm
en'do·plas'mic
end organ
en·dor'phin
en'do·sal'pin·go'sis
en'do·scope'
en'do·scop'ic
en·dos'co·py
  ·pies
en'do·skel'e·tal
en'do·skel'e·ton
en·dos·mo'sis
  ·ses
en'dos·mot'ic
en'do·spore'
en'do·spor'ic
en'do·spo'ri·um
  ·ri·a
en·dos'te·al
en·dos'te·um
  ·te·a
en·dos·to'sis
  ·ses
en'do·the'li·al
en'do·the'li·oid' or en·doth'e·loid'
en'do·the'li·um
  ·li·a
en'do·ther'mic or ·ther'mal
en'do·tox'in
en'do·tra'che·al
end'–piece'
end plate
end product
en'drin

en·e·ma
  ·mas *or* ·ma·ta
en·er·get'ics
en'er·gid
en'er·giz'er
en'er·gy
  ·gies
en'er·vate'
  ·vat'ed ·vat'ing
  (weaken; SEE
  innervate)
en·er'vate *adj.*
en'er·va'tion
en'er·va'tor
en'flu·rane'
en·gage'ment
en'gine
en'gi·neer'ing
en·globe'
  ·globed' ·glob'ing
en·globe'ment
en·gorge'
  ·gorged' ·gorg'ing
en·gorge'ment
en'gram
en·gram'mic
en·hance'
  ·hanced' ·hanc'ing
en·hance'ment
en·keph'a·lin
en·large'
  ·larged' ·larg'ing
en·large'ment
e'nol
en·rich'ment
en·san'guine
  ·guined ·guin·ing

en·sheathe'
  ·sheathed'
  ·sheath'ing
en'si·form'
en·swathe'
  ·swathed'
  ·swath'ing
en'tad
en'tal
en'ta·moe'ba
  ·bae *or* ·bas
en·ta'si·a
en'ta·sis
  ·ses'
en·ter'ic *or* en'
  ter·al
en·ter·i'tis
en'ter·o·bi'a·sis
  ·ses'
en'ter·o·coc'cal
en'ter·o·coc'cus
  ·coc'ci
en'ter·o·coele' *or*
  ·coel'
en'ter·o·co·li'tis
en'ter·o·gas'trone
en'ter·o·ki'nase
en'ter·on'
en'ter·os'to·my
  ·mies
en·thal'py
en·tire'
en'ti·ty
  ·ties
en'to·blast' *or*
  ·derm'
en'to·mol'o·gy

en'to·phyte'
en'to·phyt'ic
en'to·zo'al *or* ·zo'ic
en'to·zo'on
  ·zo'a
en'trails
en'trance
en'tro·py
  ·pies
en'try
  ·tries
e·nu'cle·ate'
  ·at'ed ·at'ing
e·nu'cle·ate *adj.*
e·nu'cle·a'tion
en'u·re'sis
  ·ses
en'u·ret'ic
en've·lope'
en·ven'om
en·ven'om·a'tion
en·vi'ron·ment
en·vi'ron·men'tal
en'vy
en'zo·ot'ic
en'zy·mat'ic *or* en·
  zy'mic
en'zyme
en'zy·mo·log'ic
en'zy·mol'o·gist
en'zy·mol'o·gy
e'o·sin
e'o·sin'ic
e'o·sin'o·phil' *or*
  ·phile'
e'o·sin'o·phil'ic
ep·ax'i·al

ep'en·ceph'a·lon'
ep·en'dy·ma
ep·en'dy·mal
e·phe'be·at'rics
e·phe'bic
e·phed'rine
ep'i·an·dros'ter·one'
ep'i·blast'
ep'i·bol'ic
e·pib'o·ly
ep'i·can'thic
ep'i·can'thus
ep'i·car'di·a
ep'i·car'di·al
ep'i·car'di·um
·di·a
ep'i·cau'ma
ep'i·cra'ni·al
ep'i·cra'ni·um
·ni·a
ep'i·dem'ic
ep'i·dem'i·cal
ep'i·de·mi·o·log'ic or ·log'i·cal
ep'i·de·mi·ol'o·gy
·gies
ep'i·der'mal or ·der'mic
ep'i·der'mis
ep'i·der'moid or ·der·moi'dal
ep'i·di'a·scope'
ep'i·did'y·mal
ep'i·did'y·mis
·di·dym'i·des'
ep'i·gas'tric

ep'i·gas'tri·um
·tri·a
ep'i·gen'e·sis
·ses'
ep'i·ge·net'ic
ep'i·glot'tal or ·glot'tic
ep'i·glot'tis
ep'i·lep'sy
ep'i·lep'tic
ep'i·lep'toid or ·lep'ti·form'
ep'i·mas'ti·gote'
ep'i·mer
ep'i·mere
ep'i·mor'phic
ep'i·mor'pho·sis
·ses'
ep'i·my'si·um
·si·a
ep'i·neph'rine
ep'i·neu'ral
(on a neural arch)
ep'i·neu'ri·al
(of the epineurium)
ep'i·neu'ri·um
ep'i·phe·nom'e·nal
ep'i·phe·nom'e·nal·ism
ep'i·phe·nom'e·non'
·na
e·piph'o·ra
ep'i·phys'e·al or ·i·al
e·piph'y·sis
·ses'
epiphysis cer'e·bri'

e·pi'si·ot'o·my
·mies
ep'i·sode'
ep'i·sod'ic or ·sod'i·cal
ep'i·some'
e·pis'ta·sis
·ses'
ep'i·stat'ic
ep'i·stax'is
ep'i·ster'num
·na
ep'i·the'li·al
ep'i·the'li·al·i·za'tion
ep'i·the'li·al·ize'
·ized' ·iz'ing
ep'i·the'li·oid'
ep'i·the'li·o'ma
·ma·ta or ·mas
ep'i·the'li·um
·li·ums or ·li·a
ep'i·the'li·za'tion
ep'i·the'lize
·lized' ·liz'ing
ep'i·troch'le·a
ep'i·zo'an
·zo'a
ep'o·nym'
ep'o·nym'ic
ep·ox'ide
ep·ox'i·dize'
·dized' ·diz'ing
ep·ox'y
ep'si·lon'
Ep'som salts or salt

Ep′stein–Barr′ virus
ep·u′lis
  li·des
ep′u·loid′
ep′u·lo′sis
e′qual
e′qual·ize′
  ized′ ·iz′ing
e·quat′a·ble
e·quate′
  ·quat·ed ·quat′ing
e·qua′tion
e·qua′tion·al
e·qua′tor
e′qua·to′ri·al
e′qui·lat′er·al
e·quil′i·brate′
  ·brat′ed ·brat′ing
e·quil′i·bra′tion
e·quil′i·bra′tor
e′qui·lib′ri·um
  ·ri·ums or ·ri·a
e′qui·lin
e′qui·mo′lal
e′qui·mo′lar
e′qui·mo·lec′u·lar
e·quip′
  ·quipped′
  ·quip′ping
e′qui·po·ten′tial
e·quiv′a·lence or ·len·cy
e·quiv′a·lent
e·ra′sion
er′bi·um
e·rect′

e·rec′tile
e·rec′til′i·ty
  ·ties
e·rec′tion
e·rec′tor
e·rep′sin or e·rep′tase
er′e·thism
er′e·this′mic or ·this′tic
er′e·thit′ic
erg
er′go·graph′
er·gom′e·ter
er·gom′e·try
er′go·nom′i·cal
er′go·nom′ics
er·gon′o·mist
er′go·no·vine
er·gos′ter·ol′
er′got
er·got′a·mine′
er·got′ic
er′got·ism
e·rode′
  ·rod·ed ·rod′ing
e·rod′i·ble
e·rog′e·nous or er′o·gen′ic
e′ros *(libido)*
e·rose′
  (irregular)
e·ro′sion
e·ro′sion·al
e·ro′sive
e·rot′ic

e·rot′i·cize′
  ·cized′ ·ciz′ing
er′o·tism or e·rot′i·cism
er′o·tize′
  ·tized′ ·tiz′ing
e·ro′to·gen′ic
e·ro′to·ma′ni·a
er·rat′ic
er′ror
e·ruct′
e·ruc′tate
  ·tat·ed ·tat·ing
e·ruc·ta′tion
e·rupt′
e·rup′tion
e·rup′tive
er′y·sip′e·las
er′y·si·pel′a·tous
er′y·sip′e·loid
er′y·the′ma
er′y·the′mic or ·them′a·tous
e·ryth′rism
er′y·thris′mal
er′y·thris′tic
e·ryth′rite′
e·ryth′ri·tol′
e·ryth′ro·blast′
e·ryth′ro·blas′tic
e·ryth′ro·blas·to′sis
  ·ses
e·ryth′ro·cyte′
e·ryth′ro·cyt′ic
e·ryth′ro·der′ma
er′y·throid′
e·ryth′ro·me·lal′gi·a

e·ryth'ro·my'cin
er'y·thron'
e·ryth'ro·poi·e'sis
·ses
e·ryth'ro·poi·et'ic
e·ryth'ro·poi·e'tin
e·ryth'ro·sin *or*
·sine'
es·cape'
·caped' ·cap'ing
es·cap'ism
es·cap'ist
es'char
es·cha·rot'ic
Esch'e·rich'i·a co'li
es'cu·lent
es·cutch'eon
es'er·ine'
Es'march's
 bandage
e·soph'a·ge'al
e·soph'a·gus
·a·gi'
es'o·tro'pi·a
es'sence
es·sen'tial
est
es'ter
es'ter·ase'
es·ter'i·fi·ca'tion
es·ter'i·fy'
·fied' ·fy'ing
es'ter·ize'
·ized' ·iz'ing
es·the'si·a
es·the'si·om'e·ter
es·thet'ic

es·thet'ics
es'ti·val
es'ti·vo·au·tum'nal
es·tra·di'ol
es'trin
es'tri·ol'
es'tro·gen
es'tro·gen'ic
es'trone
é·tat' ma·me·lon·né'
eth'a·nal'
 (acetaldehyde)
eth'ane
eth'a·nol'
 (alcohol)
eth'ene
e'ther
e·the're·al
e·ther'i·fi·ca'tion
e·ther'i·fy'
·fied' ·fy'ing
e'ther·i·za'tion
e'ther·ize'
·ized' ·iz'ing
eth'i·cal
eth'ics
eth'moid *or*
·moi'dal
eth'nic *or* ·ni·cal
eth·nog'ra·pher
eth'no·graph'ic *or*
·graph'i·cal
eth·nog'ra·phy
eth'no·log'i·cal *or*
·log'ic
eth·nol'o·gist

eth·nol'o·gy
eth'yl
eth'yl·ate *n.*
eth'yl·ate'
 ·at'ed ·at'ing
eth'y·la'tion
eth'yl·ene'
e'ti·o·late'
 ·lat'ed ·lat'ing
e'ti·o·la'tion
e'ti·o·log'ic *or*
·log'i·cal
e'ti·ol'o·gy
·gies
et'y·mol'o·gy
·gies
eu·caine'
eu'ca·lypt'
eu'ca·lyp'tol *or*
·tole
eu'ca·lyp'tus
·tus·es *or* ·ti
eu·car'y·ote'
eu'car·y·ot'ic
eu·cho'li·a
eu·chro'ma·tin
eu·cil'i·ate
eu'di·om'e·ter
eu'di·o·met'ric *or*
·met'ri·cal
eu·gen'ic *or*
·gen'i·cal
eu·gen'i·cist
eu·gen'ics
eu'gen·ist
eu·gen·ol'
eu·gle'na

eu·gle'noid
eu·kar'y·ote'
eu·kar'y·ot'ic
eu'nuch
eu·pep'si·a
eu·pep'tic
eu·phen'ics
eu·phor'bi·a
eu·pho'ri·a
eu·pho'ri·ant
eu·phor'ic
eu'phra·sy
·sies
eu·plas'tic
eu'ploid
eu'ploi·dy
·dies
eup'ne·a or eup'noe·a
eu·rhyth'mi·a
eu·rhyth'mic
eu·ro'pi·um
eu'ry·ther'mal or ·mous or ·mic
Eu·sta'chi·an tube
eu·tec'tic
eu·tec'toid
eu·tha·na'si·a
eu·then'ics
eu·thy'roid
eu·to'ci·a
e·vac'u·ant
e·vac'u·ate'
·at·ed ·at·ing
e·vac'u·a'tion
e·vac'u·a'tive
e·vac'u·a'tor

e·vag'i·nate'
·nat'ed ·nat'ing
e·vag'i·na'tion
e·val'u·a'tion
ev'a·nesce'
·nesced' ·nesc'ing
ev'a·nes'cent
Ev'ans blue
e·vap'o·ra·bil'i·ty
e·vap'o·ra·ble
e·vap'o·rate'
·rat'ed ·rat'ing
e·vap'o·ra'tion
e·vap'o·ra'tive
e·vap'o·ra'tor
e·va'sion
e·vent'
e'ven·tra'tion
e·ver'sion
e·vert'
e·ver'tor
é'vide·mon'
e'vil, St. John's
e·vis'cer·ate'
·at'ed ·at'ing
e·vis'cer·a'tion
ev'o·ca'tion
ev'o·ca'tor
ev'o·lu'tion
e·volve'
·volved' ·volv'ing
e·vul'sion
ex·ac'er·bate'
·bat·ed ·bat'ing
ex·ac'er·ba'tion
ex·am'i·na'tion

ex·am'ine
·ined ·in·ing
ex·am'in·er
ex·an'them
ex'an·the'ma
·mas or ·ma·ta
ex'ca·vate'
·vat'ed ·vat'ing
ex'ca·va'tion
ex'ca·va'tor
ex·cep'tion·al
ex·cess'
ex·ces'sive
ex·change'
·changed'
·chang'ing
ex·chang'er
ex·cide'
·cid'ed ·cid'ing
ex·cip'i·ent
ex·cis'a·ble
ex·cise'
·cised' ·cis'ing
ex·ci'sion
ex·cit'a·bil'i·ty
·ties
ex·cit'a·ble
ex·cit'ant
ex·ci·ta'tion
ex·cit'a·to'ry
ex·cite'
·cit'ed ·cit'ing
ex·cit'er
(generator)
ex·ci'tor
(nerve)
ex'clave

85

ex·clu'sion
ex·co'ri·ate'
　·at'ed ·at'ing
ex·co'ri·a'tion
ex'cre·ment
ex'cre·men'tal or
　·men·ti'tious
ex·cres'cence
ex·cres'cen·cy
　·cies
ex·cres'cent
ex·cre'ta
ex·cre'tal
ex·crete'
　·cret'ed ·cret'ing
ex·cre'tion
ex'cre·to'ry
　·ries
ex·cur'sion
ex·cur'sive
ex'cur·va'tion
ex·e'mi·a
ex·en'ter·ate'
　·at'ed ·at'ing
ex·en'ter·a'tion
ex'er·cis·a·ble
ex'er·cise'
　·cised' ·cis'ing
ex'er·cis'er
ex·er'tion
ex·fo'li·ate'
　·at'ed ·at'ing
ex·fo'li·a'tion
ex·fo'li·a'tive
ex·hal'ant
ex'ha·la'tion

ex·hale'
　·haled' ·hal'ing
ex·haust'
ex·haust'i·bil'i·ty
ex·haus'tion
ex·haus'tive
ex·hib'it
ex'hi·bi'tion
ex'hi·bi'tion·ism
ex'hi·bi'tion·ist
ex'hi·bi'tion·is'tic
ex·hil'a·rant
ex·hil'a·rate'
　·rat'ed ·rat'ing
ex·hil'a·ra'tion
ex·hu·ma'tion
ex·hume'
　·humed' ·hum'ing
ex'is·ten'tial
ex'i·tus
ex'o·crine
ex'o·cy·to'sis
　·ses
ex'o·don'tia
ex'o·don'tics
ex'o·don'tist
ex·og'a·mous or ex'·
　o·gam'ic
ex·og'a·my
　·mies
ex·og'e·nous
　*(originating
　externally;* SEE
　axogenous)
ex'on
ex'o·pep'ti·dase'
ex'oph·thal'mic

ex'oph·thal'mos or
　·thal'mus or
　·thal'mi·a
ex'o·skel'e·tal
ex'o·skel'e·ton
ex·os·mo'sis
　·ses
ex'os·mot'ic
ex·os·to'sis
　·ses
ex·os·tot'ic
ex'o·ter'ic
ex'o·ther'mic or
　·ther'mal
ex·ot'ic
ex'o·tox'in
ex'o·tro'pi·a
ex'o·tro'pic
ex·pand'er
ex·pan'sile
ex·pan'sion
ex·pan'sive
ex·pect'ance
ex·pect'an·cy
　·cies
ex·pect'ant
ex'pec·ta'tion
ex·pec'to·rant
ex·pec'to·rate'
　·rat'ed ·rat'ing
ex·pec'to·ra'tion
ex·pel'
　·pelled' ·pel'ling
ex·pel'lant or ·lent
ex·pe'ri·ence
　·enced ·enc·ing
ex·pe'ri·en'tial

ex·per′i·ment
ex·per′i·men·tal
ex′pi·ra′tion
ex·pir′a·to′ry
ex·pire′
　·pired′　·pir′ing
ex·plant′
ex′plan·ta′tion
ex·plode′
　·plod′ed　·plod′ing
ex′plo·ra′tion
ex·plor′a·to′ry
ex·plore′
　·plored′　·plor′ing
ex·plor′er
ex·plo′sion
ex·plo′sive
ex·po′nent
ex·pose′
　·posed′　·pos′ing
ex·po′sure
ex·press′
ex·press′er
ex·press′i·ble
ex·pres′sion
ex·pres′sive
ex′pres·siv′i·ty
ex·pul′sion
ex·pul′sive
ex′qui·site
ex·san′gui·nate′
　·nat′ed　·nat′ing
ex·san′gui·na′tion
ex·san′guine
ex·sect′
ex·sec′tion

ex′sic·cate′
　·cat′ed　·cat′ing
ex′sic·ca′tion
ex′stro·phy
　·phies
ex·tend′
ex·tend′er
ex·ten′sion
ex·ten′si·ty
ex·ten′sor
ex·te′ri·or
ex·te′ri·or·i·za′tion
ex·te′ri·or·ize′
　·ized′　·iz′ing
ex′tern
ex·ter′nal
ex·ter′nal·ize′
　·ized′　·iz′ing
ex′ter·o·cep′tive
ex′ter·o·cep′tor
ex·tinc′tion
ex′tir·pate′
　·pat′ed　·pat′ing
ex′tir·pa′tion
ex′tir·pa′tive
ex·tor′sion
ex·tor′tor
ex′tra·cel′lu·lar
ex′tra·cor·po′re·al
ex′tract n.
ex·tract′ v.
ex·tract′a·ble or
　·i·ble
ex·tract′ant
ex·trac′tion
ex·trac′tive
ex·trac′tor

ex′tra·mu′ral
ex·tra′ne·ous
ex′tra·nu′cle·ar
ex·trap′o·late′
　·lat′ed　·lat′ing
ex·trap′o·la′tion
ex′tra·sen′so·ry
ex′tra·sys′to·le′
ex′tra·u′ter·ine
ex·trav′a·sate′
　·sat′ed　·sat′ing
ex·trav′a·sa′tion
ex′tra·vas′cu·lar
ex′tra·ver′sion
ex′tra·vert′
ex·treme′
ex·trem′i·ty
　·ties
ex·trin′sic
ex′tro·ver′sion
ex′tro·vert′
ex·trude′
　·trud′ed　·trud′ing
ex·tru′sion
ex·tu′bate
　·bat·ed　·bat·ing
ex′tu·ba′tion
ex·u′ber·ance
ex·u′ber·ant
ex′u·date′
ex·u′da·tive
ex·ude′
　·ud′ed　·ud′ing
ex·u′vi·ae′
　(sing. ·vi·a)
ex·u′vi·al

87

ex·u′vi·ate′
  ·at′ed ·at′ing
ex·u′vi·a′tion
ex vi′vo
eye′ball′
eye bank
eye′bright′
eye′brow′
eye′cup′
eye′drop′per
eye′glass′
eye′glass′es
eye′hole′
eye′lash′
eye′less
eye′lid′
eye′piece′
eye′sight′
eye′strain′
eye′tooth′
  ·teeth′
eye′wash′

F

fa·bel′la
  ·lae
fab′ri·cate′
  ·cat′ed ·cat′ing
fab′ri·ca′tion
face′–bow′
face′–lift′ *v.*
face lift′ing *or* face lift
fac′et

fa′ci·a
fa′cial
  (of the face; SEE
  fascial*)*
fa′ci·es′
fa·cil′i·tate′
  ·tat′ed ·tat′ing
fa·cil′i·ta′tion
fa·cil′i·ta′tive
fa·cil′i·ty
  ·ties
fac′ing
fa′ci·o·ceph′a·lal′gi·a
F-ac′tin
fac·ti′tious
fac′tor V
fac·to′ri·al
fac′ul·ta′tive
fac′ul·ty
  ·ties
fade
  fad′ed fad′ing
fae′cal
fae′ces
fag′o·py′rism *or*
  fag′o·py·ris′mus
Fahr′en·heit′
fail′ure
faint
faith cure *or* faith healing
fal′cate
fal′ces
  (sing. falx*)*
fal′cial
fal′ci·form′

fal·cip′a·rum malaria
fall
  fell fall′en fall′ing
Fal·lo′pi·an tube
fall′out′
false′–neg′a·tive
false′–pos′i·tive
false ribs
fal′si·fi·ca′tion
fal′si·fy′
  ·fied′ ·fy′ing
falx
  fal′ces
fa·mil′ial
fam′i·ly
  ·lies
family plan′ning
fang
fan′go
fan′ni·a
fan′ning
fan′ta·sist
fan′ta·size′
  ·sized′ ·siz′ing
fan′tasm
fan′ta·sy
  ·sies
fan′ta·sy
  ·sied ·sy·ing
fan′tod
fan′tom
far′ad
far′a·day′
fa·rad′ic
far′a·di·za′tion

far′a·dize′
·dized′ ·diz′ing
far′a·do·mus′cu·lar
fa·ri′na
far′i·na′ceous
far point
Farres' tubercles
far′sight′ed
fas′ci·a
·ae′ or ·as
fas′ci·al
(of connective
tissue; SEE facial)
fas′ci·cle
fas·cic′u·late or
·lat′ed or ·lar
fas·cic′u·lus
·li′
fas′ci·i′tis or fas·ci′
tis
fast
fas·tid′i·ous
fas·tid′i·um
fas·tig′i·um
fast′ness
fat
fat′ter fat′test
fa′tal
fa·tal′i·ty
·ties
fate
fat′i·ga·bil′i·ty
fat′i·ga·ble
fa·tigue′
·tigued′ ·tigu′ing
fat′-sol′u·ble

fat′ty
·ti·er ·ti·est
fau′ces
fau′cial
fa·ve′o·late′
fa·ve′o·lus
·li′
fa′vid
fa′vism
fa′vus
fear
fea′ture
·tured ·tur·ing
fe·brif′ic
feb′ri·fuge′
fe′brile
fe′cal
fe′ces
fec′u·lence
fec′u·lent
fe′cund
fe·cun′date′
·dat′ed ·dat′ing
fe′cun·da′tion
fe·cun′di·ty
·ties
fee′ble
·bler ·blest
fee′ble·mind′ed
feed
fed feed′ing
feed′back′
feed′-for′ward
feel
felt feel′ing
Feh′ling's solution
fel·la′ti·o′

fel′on
felt′work′
fe′male
fem′i·nine
fem′i·nism
fem′i·ni·za′tion
fem′i·nize′
·nized′ ·niz′ing
femme
femmes
fem′o·ral
fe′mur
fe′murs or fem′o·ra
fe·nes′tra
·trae
fe·nes′tral
fe·nes′trat·ed or
·nes′trate
fen′es·tra′tion
fen′nel
fer′ment n.
fer·ment′ v.
fer·ment′a·ble
fer′men·ta′tion
fer·ment′a·tive
fer′mi·um
fern′ing
fer′re·dox′in
fer′ric
fer′ri·cy′a·nide′
fer′ri·tin
fer′ro·cy′a·nide′
fer′ro·sil′i·con
fer′rous
fer·ru′gi·nous
fer′rule
fer′tile

89

fer·til′i·ty
fer′til·iz′a·ble
fer′til·i·za′tion
fer′til·ize′
  ·ized′ ·iz′ing
fer·til′i·zin
fer′u·la
  ·lae
fer·ves′cence
fes′ter
fes′ti·nant
fes′ti·nate′
  ·nat′ed ·nat′ing
fes′ti·na′tion
fes·toon′
fe′tal
fe·ta′tion
fe·ti·ci′dal
fe′ti·cide′
fet′id
fet′ish *or* ·ich
fet′ish·ism *or*
  ·ich·ism
fet′ish·ist
fet′ish·is′tic
fet′ish·ize′
  ·ized′ ·iz′ing
fe·tol′o·gist
fe·tol′o·gy
fe·tom′e·try
fe′tor
fe′to·scope′
fe·tos′co·py
fe′tus
  ·tus·es
fe′ver
fe′vered

fe′ver·ish *or* ·ous
fi′at
fi′ber *or* ·bre
fi′ber–op′tic
fiber optics
fi′ber·scope′
fi′bril
fi·bril′la
  ·lae
fi′bril·lar *or* ·lar′y
fi′bril·late′
  ·lat′ed ·lat′ing
fi′bril·la′tion
fi′bril·lose′
fi′brin
fi·brin′ase′
fi·brin′o·gen
fi′brin·o·gen′ic
fi′brin·oid′
fi′bri·nol′y·sin
fi′bri·nol′y·sis
  ·ses′
fi′bri·no·lyt′ic
fi′brin·ous
fi′bro·blast′
fi′bro·blas′tic
fi′broid
fi·bro′ma
  ·mas *or* ·ma·ta
fi·bro′ma·tous
fi′bro·pla′si·a
fi′bro·plas′tic
fi·bro′sis
fi·bro·si′tis
fi·brot′ic
fi′brous
fi′bro·vas′cu·lar

fib′u·la
  ·lae′ *or* ·las
fib′u·lar
fib′u·lo·cal·ca′ne·al
fi′cin
field
fig′ure
fi′la
  (sing. fi′lum*)*
fi·la′ceous
fil′a·ment
fil′a·men′ta·ry
fil′a·men′tous
fi′lar
fi·lar′i·a
  ·ae
fi·lar′i·al *or* ·i·an
fil′a·ri′a·sis
file
  filed fil′ing
fil′i·al
fil′i·form′
fil′let
fill′ing
film
fi′lo·po′di·um
  ·di·a
fi′lo·pres′sure
fi′lose
fil′ter
  (separating device;
  SEE philter*)*
fil′ter·a·bil′i·ty
fil′ter·a·ble
fil′tra·bil′i·ty
fil′tra·ble

fil′trate
·trat·ed ·trat·ing
fil·tra′tion
fi′lum
·la
fim′bri·a
·bri·ae
fim′bri·ate′
fim′bri·a′tion
fin′ger·nail′
fin′ger·print′
fin′ger·stall′
finger tip
fire′damp′
first aid
first′–de·gree′
first′–gen′er·a′tion
fish′skin′ disease
fis′sile
fis′sion
fis′sion·a·ble
fis·sip′a·rous
fis′sure
·sured ·sur·ing
fis′tu·la
·las or ·lae′
fis′tu·li·za′tion
fis′tu·lize′
·lized′ ·liz′ing
fis′tu·lous or ·lar
fit
fit′ted or fit,
fit′ted, fit′ting
fit′ful
fit′ness
fix
fixed fix′ing

fix′ate
·at·ed ·at·ing
fix·a′tion
fix′a·tive
flab′by
·bi·er ·bi·est
fla·bel′lum
·la
flac′cid
flac·cid′i·ty
·ties
flag′el·lant
flag′el·late′
·lat′ed ·lat′ing
flag′el·late or
·lat′ed
flag′el·la′tion
flag′el·la′tor
flag′el·la·to′ry
fla·gel′li·form′
fla·gel′lin
fla·gel′lum
·la or ·lums
flail
flam′ma·bil′i·ty
flam′ma·ble
flange
flank
flan′nel
flap
flare
flared flar′ing
flash
flask
flat
flat′ter flat′test
flat′foot′

flat′–foot′ed
flat′u·lence or
·len·cy
flat′u·lent
fla′tus
flat′worm′
fla′va·none′
fla′vin
fla′vine
fla′vo·bac·te′ri·um
fla′vone
fla′vo·nol′
fla′vo·pro′tein
fla′vor
flax′seed′
flea′bite′
flea′wort′
fleck
flec′tion
fleece
flesh
flesh′y
Fletch′er·ism
flex
flex′i·bil′i·ty
flex′i·ble
flex′ile
flex′ion
flex′or
flex′ur·al
flex′ure
flick′er
float·a′tion
float′ers
floc
floc′cose

floc′cu·lant
floc′cu·late′
　·lat′ed　·lat′ing
floc′cu·la′tion
floc′cule
floc′cu·lence
floc′cu·lent
floc′cu·lus
　·li′
flo′ra
　·ras *or* ·rae
flo′rid
floss
flo·ta′tion
flow′er
flow′me·ter
flu
fluc′tu·ant
fluc′tu·ate′
　·at′ed　·at′ing
fluc′tu·a′tion
flu′id
fluid dram *(or* drachm)
flu′id·ex′tract
flu·id′ic
flu·id′i·ty
fluid ounce *or* flu′id·ounce′
fluid pressure
fluke
flu′or
flu·o·resce′
　·resced′　·resc′ing
flu′o·res′ce·in
flu′o·res′cence
flu′o·res′cent

fluor′i·date′
　·dat′ed　·dat′ing
fluor′i·da′tion
flu′o·ride
fluor′i·nate′
　·nat′ed　·nat′ing
fluor′i·na′tion
flu′o·rine
flu′o·ro·graph′ic
flu·o·rog′ra·phy
flu·o·rom′e·ter
flu·o·ro·met′ric
flu·o·rom′e·try
fluor′o·scope′
　·scoped′　·scop′ing
fluor′o·scop′ic
flu·o·ros′co·pist
flu·o·ros′co·py
　·pies
flu·o·ro′sis
　·ses
flush
flut′ter–fi′bril·la′tion
flux
flux′ion
fly
　flies
foam′y
　·i·er　·i·est
fo′cal
fo′cal·i·za′tion
fo′cal·ize′
　·ized′　·iz′ing
fo·cim′e·ter
fo′cus
　·cus·es *or* ·ci

fo′cus
　·cused *or* ·cussed
　·cus·ing *or* ·cus·sing
fo′cus·er
For words beginning foet–
see FET–
fog
　fogged fog′ging
foil
fo·la′cin
fo′li·a
　(sing. fo′li·um)
fo·li·a′ceous
fo′li·ar
fo′li·ate *adj.*
fo′lic acid
fo·lie à deux′
folie de gran·deur′
fo·lin′ic acid
fo′li·um
　·li·ums *or* ·li·a
fol′li·cle
fol′li·cle–stim′u·lat·ing hormone
fol·li·c′u·lar
fol·lic′u·late *or* ·lat·ed
fol·lic′u·lin
fol·lic′u·lus
　·li′
fol′low-up′
fo·ment′
fo·men·ta′tion
fo′mes
　·mi·tes

fon'ta·nel' *or* ·nelle'
food poi'son·ing
foot'–and–mouth' disease
foot'–can'dle
foot'–drop'
foot'mark'
foot'–pound'
foot'–pound'al
foot'print'
fo·rage'
fo·ra'men
  ·ram'i·na *or* ·ra'mens
foramen mag'num
fo·ram'i·nal *or* ·i·nate
force
  forced forc'ing
force'–feed'
  –fed' –feed'ing
for'ceps
for'ci·ble
fore'arm'
fore'brain'
fore'fin'ger
fore'gut'
fore'head
for'eign
fo·ren'sic
fore'play'
fore'skin'
fore'tooth'
  ·teeth'
form·al'de·hyde'

for'mal·ize'
  ·ized' ·iz'ing
for'mate
for·ma'tion
form'a·tive
for'mic
for'mi·ca'tion
for'mu·la
  ·las *or* ·lae'
for'mu·lar'y
  ·lar'ies
for'mu·late'
  ·lat'ed ·lat'ing
for'mu·la'tion
for'mu·la'tor
for'mu·li·za'tion
for'mu·lize'
  ·lized' ·liz'ing
for'myl
for'ni·cate'
  ·cat'ed ·cat'ing
for'ni·cate *adj.*
for'ni·ca'tion
for'ni·ca'tor
for'nix
  for'ni·ces'
for'ti·fi·ca'tion
for'ti·fy'
  ·fied' ·fy'ing
fos'sa
  ·sae
fos'sate
fos·sette'
fos'ter
fou·droy'ant
fou·lage'
foun·da'tion

four·chette'
fo've·a
  ·ae' *or* ·e·as
fovea cen·tra'lis
fo've·al *or* ·ate
fo've·i·form'
fo·ve'o·la
  ·lae' *or* ·las
fo've·ole
fox'glove'
frac'tion
frac'tion·al
frac'tion·ate'
  ·at'ed ·at'ing
frac'tion·a'tion
frac'tur·al
frac'ture
frae'num
  ·nums *or* ·na
frag'ile
fra·gil'i·ty
  ·ties
frag'ment
frag'men·tate'
  ·tat'ed ·tat'ing
frag'men·ta'tion
frail
frail'ty
fram·be'si·a *or* ·boe'si·a
frame
  framed fram'ing
frame'work'
fran'ci·um
fra·ter'nal
freck'le
  ·led ·ling

free′–liv′ing
freez′a·ble
freeze
  froze fro′zen
  freez′ing
freeze′–dry′
  –dried′ –dry′ing
frem′i·tus
fre′nal
fren′u·lum
  ·lums or ·la
fre′num
  ·nums or ·na
fren′zy
  ·zies
fre′quen·cy
  ·cies
Fres·nel′ mirrors
fre′tum
  ·ta
Freud′i·an
Freud′i·an·ism
fri′a·bil′i·ty
  ·ties
fri′a·ble
fric′a·tive
fric′tion
fric′tion·al
frig′id
fri·gid′i·ty
frig′o·la′bile
frig′o·rif′ic
frig′o·rism
frit
fron′tal
fron′to·tem′po·ral
fron′to·tem′po·ra′le

frost′bite′
  ·bit′ ·bit′ten
  bit′ing
froth
fruc′tose
fruc′to·su′ri·a
fruit sugar
frus′trate
  ·trat·ed ·trat·ing
frus·tra′tion
fuch′sin or ·sine
fu′gi·tive
fugue
fugu′ist
ful′crum
  ·crums or ·cra
ful·fill′ment or
  ·fil′ment
ful′gu·rant
ful′gu·rate′
  ·rat′ed ·rat′ing
ful′gu·ra′tion
fu·lig′i·nous
full′–term′
ful′mi·nant
ful′mi·nate′
  ·nat′ed ·nat′ing
ful′mi·na′tion
ful′mi·na′tor
ful·min′ic acid
fu·mar′ic acid
fume
  fumed fum′ing
fu′mi·gant
fu′mi·gate′
  ·gat′ed ·gat′ing
fu′mi·ga′tion

fu′mi·ga′tor
fu′mi·to′ry
  ·ries
func′tion
func′tion·al
fun′da·ment
fun′dic
fun′dus
  ·di
fun′gal
fun′gi
  (sing. fun′gus)
fun′gi·ci′dal
fun′gi·cide′
fun′gi·form′
fun′goid
fun′gous adj.
fun′gus n.
  ·gi or ·gus·es
fu′nic
fu′ni·cle
fu·nic′u·lar
fu·nic′u·late
fu·nic′u·lus
  ·li′
fun′nel
fun′ny bone
fur
  furred fur′ring
fur′cate
  ·cat·ed ·cat·ing
fur·ca′tion
fur′cu·la
  ·lae
fur′cu·lar
fur′cu·lum
  ·la

fur′fur
   fur′fur·es′
fur′fu·ra′ceous
fur′fu·ral′
fur′nace
fu′ror
fur′row
fu′run·cle
fu·run′cu·lar or
   ·cu·lous
fu·run′cu·lo′sis
   ·ses
fu·sa′ri·um
fuse
   fused fus′ing
fu′si·ble
fu′si·form′
fu′sion
fu′so·spi′ro·che′tal
fus′tic
fus′ti·gate′
   ·gat′ed ·gat′ing
fus′ti·ga′tion
fuze
   fuzed fuz′ing
fuzz′y
   ·i·er ·i·est

G

gad′o·lin′i·um
gag
   gagged gag′ging
gage
gain
gait
ga·lac′ta·gogue′
ga·lac′tan
ga·lac′tase
ga·lac′tic
gal′ac·tom′e·ter
ga·lac′tor·rhe′a
ga·lac·tos′a·mine′
ga·lac′tose
ga·lac′to·se′mi·a
ga·lac′to·side′
ga·lan′gal
gal′ba·num
ga′le·a
   ·le·ae′
ga′le·ate′ or ·at·ed
ga′le·at′o·my
ga·len′ic
ga·len′i·cal
ga′len·ism
gal′in·gale′
gall′blad′der
gal′le·in
gal′lic acid
Gal′lie transplant
gal′li·pot′
gal′li·um
gal′lon
gal′lop
gall′stone′
gal·van′ic or
   ·van′i·cal
gal′van·ism
gal′va·ni·za′tion
gal′va·nize′
   ·nized′ ·niz′ing
gal′va·no·con′trac·
   til′i·ty
gal′va·nom′e·ter
gal′va·no·met′ric
gal′va·nom′e·try
gam′bier or ·bir
gam·boge′
gam′ete
ga·met′ic
ga·met′o·cyte′
ga·me′to·gen′e·sis
   ·ses′
ga·me′to·gen′ic or
   gam′e·tog′e·nous
gam′e·tog′e·ny
   ·nies
   (gamete formation)
gam′e·tog′o·ny
   ·nies
   (gamete
   reproduction)
ga·me′to·phyte′
ga·me′to·phyt′ic
gam′ic
gam′ma
gamma globulin
gam·mop′a·thy
Gam′na's disease
gam′o·gen′e·sis
   ·ses′
gam′o·ge·net′ic
gam′ont
gan′gli·at′ed or ·ate
gan′gli·form′ or
   ·gli·o·form′
gan′gli·o·blast′

gan′gli·on
 ·gli·a or ·gli·ons
gan′gli·o·neu·ro′ma
 ·mas or ·ma·ta
gan′gli·on′ic
gan′gli·o·side′
gan′gli·o·si·do′sis
 ·ses
gan·go′sa
gan′grene
 grened gren·ing
gan′gre·nous
gan′ja or ·jah
gan′o·blast′
gan′try
 ·tries
gap′toothed′
Gard′ner's syndrome
gar′gle
 ·gled ·gling
gar′goyl·ism
gar′lic
Gart′ner's duct
gas
 gassed gas′sing
gas′e·ous
gash
gas′i·form′
gas·om′e·ter
gas′o·met′ric
gas·om′e·try
gasp
gas′ser·ec′to·my
gas·se′ri·an
gas′sy
 ·si·er ·si·est

gas′ter
gas·tral′gi·a or ·ter·al′gi·a
gas′trec·ta′si·a or gas·trec′ta·sis
gas·trec′to·my
 ·mies
gas′tric
gas′trin
gas·tri′tis
gas′tro·col′ic
gas′tro·co·li′tis
gas′tro·derm′
gas′tro·en·ter·i′tis
gas′tro·en·ter·ol′o·gist
gas′tro·en·ter·ol′o·gy
 ·gies
gas′tro·in·tes′ti·nal
gas′tro·lith
gas·trol′o·gist
gas′trop·to′sis or ·to′si·a
gas′tro·scope′
gas′tro·scop′ic
gas·tros′co·pist
gas·tros′co·py
gas·tros′to·my
gas·trot′o·my
 ·mies
gas′tro·trich
gas′tro·trop′ic
gas′tru·la
 ·lae′ or ·las
gas′tru·la′tion
gauge

gauge′a·ble
gaunt′let
gauss
Gauss′i·an curve
gauze
gauz′y
 ·i·er ·i·est
ga·vage′
Gay–Lus·sac's′ law
gaze
 gazed gaz′ing
Gei·ger–Mül′ler counter
Geiss′ler tube
gel
 gelled gel′ling
gel′a·tin or ·tine
ge·lat′in·ase′
gel′a·tin·if′er·ous
ge·lat′i·ni·za′tion
ge·lat′i·nize′
 ·nized′ ·niz′ing
ge·lat′i·noid′
ge·lat′i·nous
ge·la′tion
gel′ose
ge·lo′sis
gel′o·ther′a·py or ·o·to·ther′a·py
gel′o·trip′sy
gel·se′mi·um
gem′i·nate′
 ·nat·ed ·nat·ing
gem′i·na′tion
gem′ma
 ·mae
gem·ma′ceous

gem′mate
  ·mat·ed ·mat·ing
gem·ma′tion
gem′mu·la′tion
gem′mule
ge′nal
gen′der
gene
ge′ne·a·log′i·cal
ge′ne·al′o·gist
ge′ne·al′o·gy
  ·gies
ge′ne·og′e·nous
gen′er·a
  (*sing.* ge′nus)
gen′er·a·ble
gen′er·al
gen′er·al·ist
gen′er·al·i·za′tion
gen′er·al·ize′
  ·ized′ ·iz′ing
gen′er·ate′
  ·at′ed ·at′ing
gen′er·a′tion
gen′er·a·tive
gen′er·a′tor
ge·ner′ic
ge·ne·si·ol′o·gy
gen′e·sis
  ·ses′
ge·net′ic *or*
  ·net′i·cal
ge·net′i·cist
ge·net′ics
ge·net′o·troph′ic
ge·ni′al
gen′ic

ge·nic′u·late *or*
  ·lat′ed
ge·nic′u·lum
  ·la
ge·ni′o·plas′ty
gen′i·tal
gen′i·ta′li·a
gen′i·tals
gen′i·to·plas′ty
gen′i·to·u′ri·nar′y
ge′nius
  ·nius·es
gen′o·der·ma·to′sis
  ·ses
ge′nome
ge·nom′ic
gen′o·type′
gen′o·typ′ic *or*
  ·typ′i·cal
gen·ta·mi′cin
gen′tian·o·phil *or*
  ·phile′
gen′tian vi′o·let
gen·tis′ic acid
ge′nu
  gen′u·a
gen′u·clast′
gen′u·pec′to·ral
ge′nus
  gen′er·a *or* ge′
  nus·es
gen′y·an′trum
gen′y·plas′ty
ge′ode
ge·od′ic
ge′o·med′i·cine
ge·oph′a·gy

ge′o·tac′tic
ge′o·tax′is
ge·ot′ri·cho′sis
ge·ra′ni·ol
ge·rat′ic
ger′i·at′ric
ger′i·a·tri′cian *or*
  ·i·at′rist
ger′i·at′rics
Ger′lach's valve
germ
ger·ma′ni·um
Ger′man measles
germ cell
ger′men
  ·mens *or* ·mi·na
ger′mi·ci′dal
ger′mi·cide′
ger′mi·nal
ger′mi·nant
ger′mi·nate
  ·nat′ ·nat′ing
ger′mi·na′tion
ger′mi·na′tive
germ plasm
ger′o·der′ma *or*
  ·der′mi·a
ger′o·don′tics
ger′o·don·tol′o·gy
ger′o·mor′phism
ge·ron′tal
ge·ron′to·log′i·cal
ger′on·tol′o·gist
ger′on·tol′o·gy
ge·ron′to·phil′i·a
ge·ron′to·pho′bi·a

ge·stalt' *or* Ge·stalt'
·stalt'en *or* ·stalts'

ge·stalt'ism

ges'tate
·tat·ed ·tat·ing

ges·ta'tion

ges·to'sis
·ses

ges'ture
·tured ·tur·ing

Gian·nuz'zi's cells

gi'ant·ism

gi'ar·di'a·sis
·ses'

gib·bos'i·ty

gib'bous
(protuberant)

gib'bus
(a hump)

gid'dy
·di·er ·di·est

gift'ed

gi·gan'tism

gi·gan'to·cyte'

gin'ger

gin·gi'va
·vae

gin·gi'val

gin'gi·vec'to·my

gin'gi·vi'tis

gin'gi·vo·glos·si'tis

gin'gly·moid

gin'gly·mus
·mi'

gin'seng

gir'dle

git'a·lin

give
gave giv'en giv'ing

gla·bel'la
·lae

gla·bel'lar

gla'brate

gla'brous

gla'cial

glad'i·ate'

glad'i·o·lus

glair'y

gland

glan'ders

glan'di·lem'ma

glan'du·la
·lae

glan'du·lar

glan'dule

glans
glan'des

glare
glared glar'ing

glass'es

glass'y
·i·er ·i·est

Glau'ber's *(or* Glau'ber) salt *or* salts

glau·co'ma

glau·co'ma·tous

glaze

gleet

gle'noid

gli'a

gli'a·cyte'

gli'a·din

gli'al

glide
glid'ed glid'ing

gli·o'ma
·ma·ta *or* ·mas

gli·o'ma·tous

gli'o·sar·co'ma
·mas *or* ·ma·ta

glis·sade'

glis·sad'ic

glo'bal

globe

glo'bin

glo'boid

glo'bose *or* ·bous

glo·bos'i·ty

glob'u·lar

glob'ule

glob'u·lin

glo'bus
·bi

glo·man'gi·o'ma
·mas *or* ·ma·ta

glom'er·ate

glo·mer'u·lar

glom'er·ule'

glo·mer'u·lo'ne·phri'tis

glo·mer'u·lus
·li'

glo'mus
glom'er·a *or* glo'mi

glos'sa
·sae *or* ·sas

glos'sal

glos·sal'gi·a

glos·sit'ic

glos·si'tis
glos'so·cele'
glos·sor'rha·phy
gloss'y
 ·i·er ·i·est
glot'tal or ·tic
glot'tis
 ·tis·es or ·ti·des'
glu'ca·gon'
glu·ca·to'ni·a
glu'ci·phore' or
 ·co·phore'
glu'co·cor'ti·coid'
glu'co·nate'
glu·con'ic acid
glu'cose
glu·co'si·dase'
glu'co·side'
glu'co·sid'ic
glu'co·su'ri·a
glue
glue'–sniff'ing
glu'ta·mate'
glu·tam'ic acid
glu'ta·mine'
glu·ta·thi'one
glu·te'al
glu'ten
 (wheat protein; SEE
 glutin)
glu'ten·ous
glu·te'us
 ·te'i
glu'tin
 (gelatin; SEE
 gluten)
glu'ti·nous

glut'ton·y
 ·ies
glyc'er·al'de·hyde'
gly·cer'ic acid
glyc'er·ide'
glyc'er·in
glyc'er·in·ate'
 ·at'ed ·at'ing
glyc'er·i·na'tion
glyc'er·ol'
 (sweet, oily fluid)
glyc'er·o·lize'
 ·lized' ·liz'ing
glyc'er·yl'
 (glycerol radical)
gly'cine
gly'co·gen
gly·co·gen'e·sis
 (glycogen
 formation)
gly'co·gen'ic
gly'co·ge·no'sis
 (metabolic disorder)
gly'col
gly·col'ic acid
gly·col'y·sis
 ·ses'
gly'co·lyt'ic
gly'co·ne·o·gen'e·sis
gly'co·pro'tein
gly'co·side'
gly'co·sid'ic
gly'co·su'ri·a
gly'co·su'ric
gnash
gnat
gnath'ic

gna'thi·on
gna·thi'tis
gna'tho·dy·na·
 mom'e·ter
gna·thos'to·mi'a·sis
 ·ses'
gno'si·a
gnos'tic
gno'to·bi·ol'o·gy
gno'to·bi·o'ta
gno'to·bi·ote
gno'to·bi·ot'ic
gno'to·bi·ot'ics
Go'a powder
gob'let
gog'gle–eyed'
goi'ter or ·tre
goi'tro·gen
goi'tro·gen'ic
gold'en·seal'
gold'–filled'
Gol'gi apparatus
 (or body)
go·mit'o·li'
gom·pho'sis
 ·ses
go'nad
go·nad'al
go·nad'o·trope
go'na·do·trop'ic or
 ·troph'ic
go·na·do·trop'in or
 go·na·do·tro'phin
go·nid'i·al
go·nid'i·um
 ·nid'i·a

go′ni·on′
  ·ni·a
go′ni·o·scope′
go′ni·os′co·py
gon′o·coc′cal
gon′o·coc′cus
  ·coc′ci
gon′o·phore′
gon′o·phor′ic *or*
go·noph′o·rous
gon′or·rhe′a *or*
  ·rhoe′a
gon′or·rhe′al *or*
  ·rhoe′al
good will *or*
  good′will′
goose flesh (*or*
  bumps *or* pimples
  or skin)
gore
  gored gor′ing
gor′get
gos′sy·pol′
gouge
gout′y
  ·i·er ·i·est
Graaf′i·an follicle
  (*or* vesicle)
grac′ile
gra·cil′i·ty
grade
  grad′ed grad′ing
gra′di·ent
grad′u·ate′
  ·at′ed ·at′ing
graft
grain

gram
gram atom *or*
  gram′–a·tom′ic
  weight
gram calorie
gram′–e·quiv′a·lent
gram′i·ci′din
gram molecule *or*
  gram′–mo·lec′u·
  lar weight
Gram′–neg′a·tive
  or gram′–neg′a·tive
Gram′–pos′i·tive
  or gram′–pos′i·tive
Gram's method
Gran·cher's′
  system
gran′di·ose′
gran′di·os′i·ty
grand mal
gran′u·lar
gran′u·lar′i·ty
gran′u·late′
  ·lat′ed ·lat′ing
gran′u·la′tion
gran′u·la′tive
gran′u·la′tor *or*
  ·lat′er
gran′ule
gran′u·lo·cyte′
gran′u·lo·cyt′ic
gran′u·lo′ma
  ·mas *or* ·ma·ta
gran′u·lo′ma·tous
gran′u·lo·poi·e′sis
  ·ses
gran′u·lo·poi·et′ic

gran′u·lose′
gran′u·lo′sis
  ·ses
graph
graph′ic
graph′ite
gra·phit′ic
graph′o·a·nal′y·sis
graph·ol′o·gy
graph′or·rhe′a
grasp
grat′i·fi·ca′tion
grat·tage′
grave
grav′el
grav′el–blind′
Graves' disease
grav′id
grav′i·da
gra·vid′i·ty
  ·ties
grav′i·do·car′di·ac
gra·vim′e·ter
grav′i·met′ric *or*
  ·met′ri·cal
gra·vim′e·try
  ·tries
grav′i·tate′
  ·tat′ed ·tat′ing
grav′i·ta′tion
grav′i·ta′tion·al
grav′i·ta′tive
grav′i·ty
  ·ties
gray matter
great calorie
green′ie

green'sick·ness
green soap
green'stick' fracture
grid
grief
grin·de·li·a
grippe *or* grip
gris'e·o·ful'vin
gris'e·ous
gris'tle
groan
groin
groove
  grooved groov'ing
gross
group'–spe·cif'ic
grow
  grew grown grow'ing
growth
gru'el
gruffs
grume
gru'mose *or* ·mous
gry·po'sis
  ·ses
G spot
G'–suit'
guai'ac
guai'a·col'
guai'a·cum
guan·eth'i·dine
guan'i·dine'
gua'nine
guard'ed

gu'ber·nac'u·lum
  ·la
guid'ance
guide
  guid'ed guid'ing
guide'line'
Guil·lain'–Bar·ré' syndrome
guil'lo·tine'
guin'ea pig
Guin'ea worm
gu'lar
gul'let
gulp
gum
  gummed gum'ming
gum ar'a·bic
gum'boil'
gum'ma
  ·mas *or* ·ma·ta
gum·ma·tous
gum'mous
gum'my
  ·mi·er ·mi·est
gur'gle
  ·gled ·gling
gur'ney
  ·neys
gus·ta'tion
gus'ta·to'ry *or* gus'ta·tive *or* gus'ta·to'ri·al
gut
gut'ta
  ·tae
gut'ta–per'cha
gut'tate

gut·ta'tim
gut'tur
gut'tur·al
gym·nas'tics
gym'no·spore'
gyn·an'dro·morph'
gyn·an'dro·mor'phic *or* ·mor'phous
gyn·an'dro·mor'phism *or* ·an'dro·mor'phy
gy·ne'cic
gyn'e·coid'
gyn'e·co·log'ic *or* ·log'i·cal
gyn'e·col'o·gist
gyn'e·col'o·gy
gyn'e·co·mas'ti·a
gyn'e·cop'a·thy
gyn'e·pho'bi·a
gyn'i·at'rics
gyn'o·path'ic
gyp'sum
gy'ral
gy'rate
  ·rat·ed ·rat·ing
gy·ra'tion
gyre
gy·rec'to·my
  ·mies
gy'ren·ce·phal'ic
gy'rose *or* ·rous
gy'ro·spasm'
gy'rus
  ·ri

H

ha·be′na
 ·nae
ha·be′nal or
 ·be′nar
ha·ben′u·la
 ·lae
ha·ben′u·lar
hab′it
hab′it-form′ing
ha·bit′u·al
ha·bit′u·ate′
 ·at′ed ·at′ing
ha·bit′u·a′tion
hab′i·tude′
hab′i·tus
 ·tus′
hache·ment′
hack
*For words
 beginning* haem—
 see also HEM—
hae′ma·tox′y·lon′
haf′ni·um
hah′ni·um
hair′line′
ha·kim′
ha·la′tion
hal′a·zone′
hale
half
 halves
half′-life′ or half
 life
half′-val′ue layer
half′way′ house
hal′ide
hal′i·to′sis
 ·ses
hal′i·tus
hal′lex
 ·li·ces′
hal·lu′ci·nate′
 ·nat′ed ·nat′ing
hal·lu′ci·na′tion
hal·lu′ci·na′tive
hal·lu′ci·na·to′ry
hal·lu′ci·no·gen
hal·lu′ci·no·gen′ic
hal·lu′ci·no′sis
 ·ses
hal′lux
 ·lu·ces′
ha′lo
 ·los or ·loes
hal′o·gen
hal′o·gen·ate′
 ·at′ed ·at′ing
hal′o·gen·a′tion
ha·log′e·nous
hal′o·ge·ton′
hal′oid
hal′o·per′i·dol′
hal′o·phile′ or hal′o·phil
hal′o·phil′ic or ha·loph′i·lous
hal′o·thane′
halve
 halved halv′ing
ha·mar′ti·a
ham′ate bone
ham′mer
ham′mer·toe′
ham′ster
ham′string′
ham′u·lar
ham′u·lus
 ·li′
hand′ed·ness
hand′i·cap′
 ·capped′ ·cap′ping
hand′piece′
hang
 hung hang′ing
hang′nail′
hang′o′ver
Han′sen's disease
hap′loid
hap′loi′dy
 ·dies
hap′lont
hap·lo′sis
hap′ten or ·tene′
hap·ten′ic
hap′tic
hard drug
hard′ened
har′den·ing
hare′lip′
har′le·quin
har′mine
har·mo′ni·a
har·mon′ic
har′mo·ny
 ·nies
har·poon′
harts′horn′

Hash·i·mo'to's disease
hash'ish or ·eesh
hatch'et
haunch bone
haus'tral
haus'trum
 ·tra
haus'tus
Ha·ver'sian canal
hay fever
haz'ard·ous
head'ache'
head cold
head'gear'
heal'er
health food
health'ful
health'y
 ·i·er ·i·est
hear
 heard hear'ing
heart attack
heart'beat'
heart block
heart'burn'
heart'throb'
heart'y
 ·i·er ·i·est
heat'ing pad
heat'stroke'
heave
 heaved heav'ing
heav'y
 ·i·er ·i·est
he·be·phre'ni·a
he·be·phren'ic

he·bet'ic
heb'e·tude'
he·bos'te·ot'o·my or he·bot'o·my
 ·mies
hec'tic
hec'to·gram'
hec'to·li'ter
he·do'ni·a
he·don'ic
he·don'ics
he'do·nism
he'do·nist
he'do·nis'tic
heel
height
Heim'lich maneuver
Heinz bodies
For words beginning hekto–
 see HECTO–
He'La cell
hel'coid
 (like an ulcer)
hel'i·cal
hel'i·ces'
 (sing. helix*)*
hel'i·cin
 (a glycoside)
hel'i·cine
 (of a helix)
hel'i·coid' or hel'i·coi'dal
 (spiral-shaped)
he'li·o·ther'a·py
 ·pies

he'li·um
he'lix
 ·lix·es or ·li·ces'
hel'le·bore'
hel'minth
hel'min·thi'a·sis
 ·ses
hel·min'thic
hel'min·thol'o·gy
he·lo'ma
 ·mas or ·ma·ta
he·lo'sis
help'er T cells
help'less
he'ma·cy·tom'e·ter
he'mad
hem'a·den
he·mag'glu'ti·nate'
 ·nat'ed ·nat'ing
he·mag'glu·ti·na'tion
he·mag'glu·ti·nin
he'mal or he'ma·tal
he·man'gi·o'ma
 ·mas or ·ma·ta
hem·ar·thro'sis
 ·ses
he·ma·te'in
he'ma·ther'mal
he·mat'ic
hem'a·tin
hem'a·tin'ic
hem'a·to·blast'
hem'a·to·blas'tic
he·mat'o·crit'
hem'a·to·cyst'

hem'a·to·gen'e·sis
·ses'
hem'a·to·gen'ic *or*
·ge·net'ic
hem'a·tog'e·nous
he'ma·to·log'ic *or*
·log'i·cal
he'ma·tol'o·gist
he'ma·tol'o·gy
·gies
he'ma·to'ma
·mas *or* ·ma·ta
hem'a·to·pha'gi·a
hem'a·toph'a·gous
hem'a·to·poi·e'sis
·ses
hem'a·to·poi·et'ic
he'ma·to·ther'mal
he'ma·tox'y·lin
hem'a·to·zo'ic *or*
·zo'al
hem'a·to·zo'on
·zo'a
he'ma·tu'ri·a
heme
hem'er·a·lo'pi·a
hem'er·a·lop'ic
he'mic
hem'i·cel'lu·lose'
hem'i·chor'date
hem'i·cra'ni·a
hem'i·fa'cial
he'min
hem'i·pa·re'sis
·ses
hem'i·ple'gi·a
hem'i·ple'gic

he·mip'ter·an
he·mip'ter·oid
he·mip'ter·ous
hem'i·sphere'
hem'i·spher'i·cal *or*
·spher'ic
hem'i·sphe'roid
hem'i·ter'pene
hem'lock
he'mo·chro'ma·to'-
sis
·ses
he·mo·cy'a·nin
he'mo·cyte'
he'mo·cy·tom'e·ter
he'mo·di·al'y·sis
·ses'
he'mo·fil·tra'tion
he'mo·flag'el·late'
he'mo·glo'bin
he'mo·glo·bin'ic
he'mo·glo'bin·ous
he'mo·glo'bin·u'ri·a
he'mo·glo'bin·u'ric
he'moid
he'mo·lymph'
he·mol'y·sin
he·mol'y·sis
·ses
he'mo·lyt'ic
he'mo·lyze'
·lyzed' ·lyz'ing
he'mo·phile' *or*
·phil
he'mo·phil'i·a
he'mo·phil'i·ac
he'mo·phil'ic

he·moph'i·lus
he·mop'ty·sis
·ses'
hem'or·rhage
·rhaged ·rhag·ing
hem'or·rhag'ic
hem'or·rhoid'
hem'or·rhoid'al
hem'or·rhoid·ec'-
to·my
·mies
he·mos'ta·sis
·ses'
he'mo·stat'
he'mo·stat'ic
he'mo·tho'rax
he'mo·tox'ic
he'mo·tox'in
hemp
hen'bane'
Hen'le's loop
hen'na
hen'ry
·rys *or* ·ries
hep'a·rin
hep'a·rin·ize'
·ized' ·iz'ing
hep'a·tec'to·my
·mies
he·pat'ic
hep'a·ti'tis
hep'a·ti·za'tion
hep'a·to'ma
·mas *or* ·ma·ta
hep'ta·chlor'
hep'tad
hep'ta·va'lent

hep′tose
her′ald patch
herb′al
herb′al·ist
her·biv′o·rous
he·red′i·ta·ble
he·red′i·tar′i·an
he·red′i·tar′i·an·ism
he·red′i·tar′y
he·red′i·ty
·ties
her′it·a·bil′i·ty
her′it·a·ble
her′i·tage
her·maph′ro·dite′
her·maph′ro·dit′ic
or ·dit′i·cal
her·maph′ro·dit′ism
or ·ro·dism
her·met′ic or
·met′i·cal
her′ni·a
·ni·as or ·ni·ae′
her′ni·al
her′ni·ate′
·at′ed ·at′ing
her′ni·a′tion
her′ni·o·plas′ty
her′o·in
her′pes sim′plex
herpes zos′ter
her·pet′ic
her·sage′
hertz
hes·per′i·din
het′er·o·chro·mat′ic
het′er·o·chro′ma·tin
het′er·o·chro′mo·some′
het′er·o·cy′clic
het′er·oe′cious
het′er·o·ga′mete
het′er·o·ga·met′ic
het′er·og′a·mous
het′er·og′a·my
·mies
het′er·o·ge·ne′i·ty
·ties
het′er·o·ge′ne·ous
het′er·o·gen′e·sis
·ses′
het′er·o·ge·net′ic
het′er·og′e·nous
(of different origin)
het′er·og′o·nous
*(of the alternation
of generations)*
het′er·og′o·ny
·nies
het′er·o·graft′
het′er·ol′o·gous
het′er·ol′o·gy
het′er·ol′y·sis
·ses′
het′er·o·lyt′ic
het′er·o·mor′phic
or ·mor′phous
het′er·o·mor′phism
het′er·on′o·mous
(differentiated)
het′er·on′o·my
het′er·on′y·mous
*(of crossed optical
images)*
het′er·op′a·thy
het′er·o·phil′
het′er·o·phil′ic
het′er·o·plas′tic
het′er·o·plas′ty
·ties
het′er·o·ploid′
het′er·o·ploi′dy
·dies
het′er·op′ter·ous
het′er·o·sex′u·al
het′er·o·sex′u·al′i·ty
·ties
het′er·o·tax′i·a or
·tax′y
het′er·o·tax′is
·tax′es
het′er·o·tac′tic or
·tac′tous or ·tax′ic
het′er·o·thal′lic
het′er·o·thal′lism
het′er·o·to′pi·a
het′er·o·top′ic
het′er·ot′o·py
·pies
het′er·o·troph′
het′er·o·troph′ic
het′er·o·typ′ic or
·typ′i·cal
het′er·o·zy·go′sis
·ses
het′er·o·zy′gote
het′er·o·zy′gous
heu·ris′tic
hex′a·ba′sic

hex′a·chlo′ro·eth′ane *or* ·chlor·eth′ane
hex′a·chlo′ro·phene′
hex′ad
hex·ad′ic
hex·a·hy′dric
hex′a·mer
hex′a·me·tho′ni·um
hex′a·meth′yl·ene·tet′ra·mine′
hex′ane
hex′a·va′lent
hex′one
hex′os·a·mine′
hex′o·san′
hex′ose
hex′yl
hex′yl·res·or′cin·ol′
hi·a′tal
hi·a′tus
 ·tus·es *or* ·tus
hic′cough
hic′cup
 ·cuped *or* ·cupped
 ·cup·ing *or* ·cup·ping
hide′bound′
hi·dro′sis
 ·ses
hi·drot′ic
hi′e·mal
hi′er·ar′chy
 ·chies
hi′er·o·pho′bi·a
high′–strung′

hi′la
 (*sing.* hi′lum)
hi′lar
hil′lock
hi′lum
 ·la
hi′lus
 ·li
hind′brain′
hind′gut′ *or* –gut′
hinge′–bow′
hinge joint
hip′bone′
hip joint
hipped
hip′po·cam′pal
hip′po·cam′pus
 ·pi
Hip·poc′ra·tes′
Hip′po·crat′ic oath
hip pointer
hip·pu′ric acid
hip′shot′
hir′sute
hir′sut·ism
hir′u·din
hir′u·di·ni′a·sis
 ·ses′
hi·ru′di·ni·za′tion
hi·ru′di·nize′
 ·nized′ ·niz′ing
His′s bundle
his·tam′i·nase′
his′ta·mine′
his′ta·min′ic
his′ti·dase′
his′ti·dine′

his′ti·o·cyte′
his′ti·o·cyt′ic
his′ti·on′ic
his′to·blast′
his′to·chem′i·cal
his′to·chem′is·try
his′to·com·pat′i·bil′i·ty
 ·ties
his′to·gen′e·sis
 ·ses′
his′to·ge·net′ic
his′to·gram′
his′toid
his′to·ki·ne′sis
his′to·log′ic *or* ·log′i·cal
his·tol′o·gist
his·tol′o·gy
his·tol′y·sis
 ·ses′
his′to·lyt′ic
his·to′ma
 ·mas *or* ·ma·ta
his′tone
his′to·pa·thol′o·gy
his′to·phys′i·ol′o·gy
his′to·plas·mo′sis
 ·ses
his′to·ry
 ·ries
his′tri·on′ic
hives
hoarse
hoars′en
hob′ble
 ·bled ·bling

hob′nail′
hof
Hoff′mann's duct
Hof′mann's bacillus
hol′er·ga′si·a
hol′er·gas′tic
ho′lism
ho·lis′tic
hol′low
hol′mi·um
hol′o·blas′tic
ho′lo·crine
hol′o·en′zyme
ho·log′a·mous
ho·log′a·my
  ·mies
hol′o·gram′
hol′o·graph′
hol′o·graph′ic
ho·log′ra·phy
  ·phies
hol′o·me·tab′o·lous
hol′o·thu′rin
hol′o·zo′ic
hom′a·lu′ri·a
hom·at′ro·pine′
ho′me·o·mor′phism
ho′me·o·mor′phous
ho′me·o·path′
ho′me·o·path′ic
ho′me·op′a·thist
ho′me·op′a·thy
ho′me·o·pla′si·a
ho′me·o·plas′tic
ho′me·o·sta′sis
  ·ses

ho′me·o·stat′ic
ho′me·o·ther′mal
ho′me·o·typ′ic
hom′i·ci′dal
hom′i·cide′
hom′i·nid
hom′i·noid′
ho′mo
  hom′i·nes′
ho′mo·cy′to·trop′ic
ho′mo·e·rot′ic
ho′mo·e·rot′i·cism
ho·mog′a·mous
ho·mog′a·my
ho′mo·ge′nate′
ho′mo·ge·ne′i·ty
ho′mo·ge′ne·ous
ho·mog′e·ni·za′tion
ho·mog′e·nize′
  ·nized′ ·niz′ing
ho·mog′e·nous
ho·mog′e·ny
ho′mo·graft′
ho·moi′o·ther′mal
  or ·ther′mic
ho·moi′o·ther′my
ho·mo·lat′er·al
ho′mo·log′i·cal
ho·mol′o·gize′
  ·gized′ ·giz′ing
ho·mol′o·gous
hom′o·logue′ or ·log′
ho·mol′o·gy
  ·gies
ho·mol′y·sin

ho′mo·mor′phic or ·mor′phous
ho′mo·mor′phism or ho′mo·mor′phy
ho·mon′y·mous
ho·mon′y·my
ho′mo·phil
  (antibody)
ho′mo·phile′
  (homosexual)
ho′mo·phobe
ho′mo·pho′bi·a
ho′mo·pho′bic
ho′mo·plas′tic
ho′mo·plas′ty
Ho′mo sa′pi·ens
ho′mo·sex′u·al
ho′mo·sex′u·al′i·ty
  ·ties
ho′mo·thal′lic
ho′mo·thal′lism
ho′mo·trans′plant
ho′mo·trans′plan·ta′tion
ho′mo·zy·go′sis
  ·ses
ho′mo·zy′gote
ho′mo·zy·got′ic
ho′mo·zy′gous
ho·mun′cu·lus
  ·li
hon′ey
  ·eys
hook′nose′
hook′worm′
hor·de′o·lum
  ·la

107

hore′hound′
ho·ri′zon
hor′i·zon′tal
hor·mo′nal *or* ·mo′nic
hor′mone
hor′mo·no·poi·e′sis *or* hor′mo·poi·e′sis
 ·ses
hor′mo·no·ther′a·py
hor′ni·fi·ca′tion
horn′y
 ·i·er ·i·est
hor·rop′ter
hor·rip′i·late′
 ·lat′ed ·lat′ing
hor·rip′i·la′tion
hor′ror au′to·tox′i·cus
horse′shoe′ kidney
hos′pice
hos′pi·tal
hos′pi·tal·ism
hos′pi·tal·i·za′tion
hos′pi·tal·ize′
 ·ized′ ·iz′ing
host
hos·til′i·ty
 ·ties
hot flash
hot line
hot′ten·tot′ism
hour′glass′ contraction
house′maid′s′ knee

house physician
Hous′ton′s muscle
How′ship′s lacunae
H′–tet′a·nase′
huck′le·bone′
hue
hum
 hummed hum′ming
hu′man
hu·mane′
hu·mec′tant
hu′mec·ta′tion
hu′mer·al
hu′mer·o·scap′u·lar
hu′mer·us
 ·mer·i′
hu′mid
hu·mid′i·fi·ca′tion
hu·mid′i·fi′er
hu·mid′i·fy′
 ·fied′ ·fy′ing
hu·mid′i·ty
 ·ties
hu′mor
hu′mor·al
hump′back′
hump′backed′
humped
hu′mus
hunch′back′
hunch′backed′
hun′ger
hun′gry
 ·gri·er ·gri·est
hunt′ing reaction
Hun′ting·ton′s chorea

hurt
 hurt hurt′ing
Hux′ley′s layer
hy′a·lin *n.*
hy′a·line *adj.*
hy′a·lin′i·za′tion
hy·al′o·gen
hy′a·loid′
hy′a·lo·plasm
hy·al′o·some
hy′al·u·ron′ic acid
hy′al·u·ron′i·dase′
hy′brid
hy′brid·ism *or* hy·brid′i·ty
hy′brid·i·za′tion
hy′brid·ize′
 ·ized′ ·iz′ing
hy′brid·iz′er
hy′da·tid
hy′da·tid′i·form′
hy·drac′id
hy·drar′gy·rum
hy·dras′tine
hy·dras′tis
hy′drate
 ·drat·ed ·drat·ing
hy·dra′tion
hy′dra·tor
hy·drau′lics
hy′dra·zide′
hy′dra·zine′
hy′dra·zo′ate
hy′dra·zo′ic acid
hy·dre′mi·a
hy′dric
hy′dride

hy′dri·od′ic acid
hy·dro′a
hy′dro·bro′mic acid
hy′dro·car′bon
hy′dro·cele′
hy′dro·ce·phal′ic
hy′dro·ceph′a·loid′
hy′dro·ceph′a·lous *adj.*
hy′dro·ceph′a·lus or ·a·ly *n.*
hy′dro·chlo′ric acid
hy′dro·chlo′ride
hy′dro·col′loid
hy′dro·cor′ti·sone′
hy′dro·cy·an′ic acid
hy′dro·dy·nam′ics
hy′dro·e·lec′tric
hy′dro·fluor′ic acid
hy′dro·gel′
hy′dro·gen
hy′dro·gen·ate′ ·at′ed ·at′ing
hy′dro·gen·a′tion
hy′dro·gen·ize′ ·ized′ ·iz′ing
hy·drog′e·nous
hy′dro·ki·net′ic
hy′dro·ki·net′ics
hy′dro·log′ic or ·log′i·cal
hy·drol′o·gy
hy·drol′y·sate′ or ·zate′

hy·drol′y·sis ·ses′
hy′dro·lyte′
hy′dro·lyt′ic
hy′dro·lyz′a·ble
hy′dro·lyze′ ·lyzed′ ·lyz′ing
hy·dro′ma
hy·drom′e·ter
hy′dro·met′ric or ·met′ri·cal
hy·drom′e·try
hy·dro′ni·um
hy′dro·path′ic
hy′drop·a·thist
hy·drop′a·thy ·thies
hy′dro·phil′ic or hy′dro·phile′
hy·droph′i·lous
hy′dro·pho′bi·a
hy′dro·pho′bic or hy′dro·phobe′
hy·drop′ic
hy′drops′
hy′dro·qui·none′ or ·quin′ol
hy′dro·sol′
hy′dro·stat′ic or ·stat′i·cal
hy′dro·stat′ics
hy′dro·sul·fu′rous acid
hy′dro·tac′tic
hy′dro·tax′is ·es

hy′dro·ther′a·peu′tic
hy′dro·ther′a·peu′tics
hy′dro·ther′a·py ·pies
hy′dro·ther′mal
hy′dro·tho′rax
hy′drous
hy·drox′ide
hy·drox′y
hy·drox′y·bu·tyr′ic acid
hy·drox′yl
hy·drox′yl·a·mine′
hy·drox′yl·ate′ ·at′ed ·at′ing
hy·drox′yl·a′tion
hy·drox·yl′ic
hy·drox′y·zine′
hy′dro·zo′an
hy·dru′ri·a
hy·dru′ric
hy′giene
hy′gi·en′ic
hy′gi·en′ics
hy′gi·en·ist
hy·gro′ma ·mas or ·ma·ta
hy·gro′ma·tous
hy·grom′e·ter
hy′gro·met′ric
hy·grom′e·try
hy′gro·scope′
hy′gro·scop′ic
hy′gro·sco·pic′i·ty ·ties

hy′la
hy′lic
hy·lo′ma
· ·mas *or* ·ma·ta
hy′men
hy′men·al
hy′me·nop′ter·an
hy′me·nop′ter·ous
hy′oid
hy′o·scine′
hy′os·cy′a·mine′
hy·pa·cu′sis *or*
· ·cu′si·a *or*
· ·cou′si·a
hy·pax′i·al
hy′per·ac′id
hy′per·a·cid′i·ty
· ·ties
hy′per·ac′tive
hy′per·ac·tiv′i·ty
hy′per·bar′ic
hy′per·cal·ce′mi·a
hy′per·cap′ni·a
hy′per·cap′nic
hy′per·e′mi·a
hy′per·es·the′sia
hy′per·es·thet′ic
hy′per·eu·tec′tic
hy′per·eu·tec′toid
hy′per·gly·ce′mi·a
hy′per·gly·ce′mic
hy′per·in′su·lin·ism
hy′per·ker′a·to′sis
· ·ses
hy′per·ker′a·tot′ic
hy′per·ki·ne′sis *or*
· ·ne′sia

hy′per·ki·net′ic
hy′per·me·tro′pi·a
hy′per·me·trop′ic
hy′perm·ne′sia
hy′perm·ne′sic
hy′per·o′pi·a
hy′per·op′ic
hy′per·os·to′sis
· ·ses
hy′per·os·tot′ic
hy′per·pi·tu′i·ta·
rism
hy′per·pi·tu′i·tar′y
hy′per·pla′si·a
hy′per·plas′tic
hy′per·pne′a
hy′per·pne′ic
hy′per·py·ret′ic
hy′per·py·rex′i·a
hy′per·sen′si·tive
hy′per·sen′si·tiv′i·ty
· ·ties
hy′per·sex′u·al
hy′per·sex′u·al′i·ty
hy′per·sthe′ni·a
hy′per·sthen′ic
hy′per·ten′sion
hy′per·ten′sive
hy′per·thy′roid
hy′per·thy′roid·ism
hy′per·to′ni·a
hy′per·ton′ic
hy′per·troph′ic
hy·per′tro·phy
· ·phies
hy·per′tro·phy
· ·phied ·phy·ing

hy′per·ven′ti·late′
· ·lat′ed ·lat′ing
hy′per·ven′ti·la′tion
hy′per·vi′ta·min·o′
· sis
hy′per·vo·le′mi·a
hy′per·vo·le′mic
hyp·es·the′si·a
hyp′es·the′sic *or*
· ·es·thet′ic
hy′pha
· ·phae
hy′phal
hyp′he·do′ni·a
hyp′na·gog′ic
hyp′nic
hyp·no′a·nal′y·sis
· ·ses′
hyp′no·gen′e·sis
· ·ses′
hyp′no·gen′ic *or*
· ·ge·net′ic
hyp′noid *or* hyp·
noid′al
hyp·nol′o·gy
hyp′no·pom′pic
hyp·no′sis
· ·ses
hyp′no·ther′a·py
· ·pies
hyp·not′ic
hyp′no·tism
hyp′no·tist
hyp′no·tiz′a·ble
hyp′no·tize′
· ·tized′ ·tiz′ing

hy′po
·pos
hy′po
·poed ·po·ing
hy′po·ac′tive
hy′po·ac·tiv′i·ty
hy′po·al′ler·gen′ic
hy′po·al′ler·gen·ize′
·ized′ ·iz′ing
hy′po·blast′
hy′po·blas′tic
hy′po·chlo′rite
hy′po·chlo′rous acid
hy′po·chon′dri·a
hy′po·chon′dri·ac′
hy′po·chon·dri′a·cal
hy′po·chon·dri′a·sis
·ses′
hy′po·chon′dri·um
·dri·a
hy′po·derm′ or hy′po·der′ma
hy′po·der′mal
hy′po·der′mic
hy′po·der′mis
hy′po·eu·tec′tic
hy′po·eu·tec′toid
hy′po·gas′tric
hy′po·gas′tri·um
·tri·a
hy′po·geu′si·a
hy′po·glos′sal
hy′po·gly·ce′mi·a
hy′po·gly·ce′mic
hy·pog′na·thous
hy′po·ki·ne′si·a
hy′po·ki·ne′sis
·ses
hy′po·ki·net′ic
hy′po·ma′ni·a
hy′po·man′ic
hy′po·ni′trite
hy′po·ni′trous acid
hy′po·phos′phate
hy′po·phos′phite
hy′po·phos·phor′ic acid
hy′po·phos′pho·rous acid
hy′po·phys′e·al
hy·poph′y·sis
·ses′
hy′po·pi·tu′i·ta·rism
hy′po·pi·tu′i·tar′y
hy′po·pla′si·a
hy′po·plas′tic
hy·po′py·on
hy′po·sen′si·tive
hy′po·sen′si·ti·za′tion
hy′po·sen′si·tize′
·tized′ ·tiz′ing
hy·pos′ta·sis
·ses′
hy′po·stat′ic
hy′po·sul′fite
hy′po·ten′sion
hy′po·ten′sive
hy′po·tha·lam′ic
hy′po·thal′a·mus
hy′po·ther′mal
hy′po·ther′mi·a
hy·poth′e·sis
·ses′
hy·poth′e·size′
·sized′ ·siz′ing
hy′po·thy′roid
hy′po·thy′roid·ism
hy′po·ton′ic
hy′po·to·nic′i·ty
·ties
hy′po·xan′thine
hy·pox·e′mi·a
hy·pox′i·a
hy·pox′ic
hyp′sar·rhyth′mi·a
hyp′si·loid′
hy·pur′gi·a
hys′ter·ec′to·my
·mies
hys′ter·e′sis
·ses
hys′ter·et′ic
hys′ter·eu·ryn′ter
hys·te′ri·a
hys·ter′ic
hys·ter′i·cal
hys′ter·o·gen′ic
hys′ter·oid′
hys′ter·o·scope′
hys′ter·o·scop′ic
hys′ter·os′co·py
·pies
hys′ter·ot′o·my
·mies
hys′tri·ci′a·sis or hys′tri·cism

I

i·an'thi·nop'si·a
i'a·tra·lip'tics
i·at'ric *or* ·at'ri·cal
i·at'ro·gen'ic
i'a·trol'o·gy
i·at'ro·phys'i·cal
i·bo'ga·ine
i·bu'pro·fen
ice bag
Ice'land disease
i'chor
i'chor·ous
ich'thy·oid'
ich'thy·oph'a·gous
ich'thy·oph'a·gy
ich'thy·o'sis
·ses
ich'thy·ot'ic
ic'tal
ic·ter'ic
ic'ter·o·gen'ic
ic'ter·oid'
ic'ter·us
ic·tom'e·ter
ic'tus
·tus·es *or* ·tus
id
i·de'a
i·de'al
i·de'al·i·za'tion
i·de'al·ize'
·ized' ·iz'ing
i'de·a'tion
i'de·a'tion·al

i·dée fixe'
i·den'ti·cal
i·den'ti·fi'a·ble
i·den'ti·fi·ca'tion
i·den'ti·fi·ca·to'ry
i·den'ti·fi'er
i·den'ti·fy'
·fied' ·fy'ing
i·den'ti·ty
·ties
i'de·o·ge·net'ic *or*
·de·og'e·nous
i'de·ol'o·gy
·gies
i'de·o·mo'tor
id'i·o·cy
·cies
id'i·o·path'ic
id'i·op'a·thy
·thies
id'i·o·plasm
id'i·o·syn'cra·sy
·sies
id'i·o·syn·crat'ic
id'i·ot sa·vant'
id'i·ots sa·vants' *or*
id·i·ots sa·vants'
id'i·ot sa·van'tism
i'dose
ig·na'ti·a
ig'ni·pe·di'tes
ig'nis in·fer·na'lis
il·e·ac' *or* ·e·al
(of the ileum; SEE
ileac)
il'e·ec'to·my
·mies

il'e·i'tis
il'e·o·col'ic
il'e·os'to·my
·mies
il'e·ot'o·my
·mies
il'e·um
·a *(small intestine;*
SEE ilium)
il'e·us
il'i·ac'
(of the ilium; SEE
ileac)
il'i·o·fem'o·ral
il'i·om'e·ter
il'i·o·spi'nal
il'i·o·tib'i·al
il'i·um
·a *(hip bone;* SEE
ileum)
ill
worse worst
il·laq'ue·a'tion
il·le·git'i·ma·cy
·cies
il·le·git'i·mate
il·lin'i·tion
il·lin'i·um
il·lu'mi·nance
il·lu'mi·nant
il·lu'mi·nate'
·nat'ed ·nat'ing
il·lu'mi·na'tion
il·lu'mi·na'tor
il·lu'min·ism
il·lu'sion

il·lu′sion·al *or* ·sion·ar′y
i′ma
im′age
 ·aged ·ag·ing
im′age·ry
 ·ries
i·mag′i·nar′y
i·mag′i·na′tion
i·mag′ine
 ·ined ·in·ing
i·ma′go
 i·ma′go *or* i·ma′gos *or*
 i·mag′i·nes′
im·bal′ance
im′be·cile
im′be·cil′ic
im′be·cil′i·ty
 ·ties
im·bed′
 ·bed′ded ·bed′ding
im·bibe′
 ·bibed′ ·bib′ing
im′bi·bi′tion
im′bri·cate *adj.*
im′bri·cate′
 ·cat′ed ·cat′ing
im′bri·ca′tion
im·brue′
 ·brued′ ·bru′ing
im·bue′
 ·bued′ ·bu′ing
im′id·az′ole
im′ide
im′i·do′
im′ine
im′i·no′
i·mip′ra·mine′
im′i·tate′
 ·tat′ed ·tat′ing
im′i·ta′tion
im′i·ta·tive
im′i·ta′tor
im·ma·ture′
im′ma·tu′ri·ty
 ·ties
im·me′di·a·cy
im·me′di·ate
im·med′i·ca·ble
im·merse′
 ·mersed′ ·mers′ing
im·mers′i·ble
im·mer′sion
im′mis·ci·bil′i·ty
 ·ties
im·mis′ci·ble
im·mo′bile
im′mo·bil′i·ty
 ·ties
im·mo′bi·li·za′tion
im·mo′bi·lize′
 ·lized′ ·liz′ing
im·mune′
im′mu·ni·fa′cient
im·mu′ni·ty
 ·ties
im′mu·ni·za′tion
im′mu·nize′
 ·nized′ ·niz′ing
im′mu·no·as′say
im′mu·no·chem′is·try
im′mu·no·flu′o·res′cence
im′mu·no·flu′o·res′cent
im′mu·no·gen
im′mu·no·ge·net′ics
im′mu·no·gen′ic
im′mu·no·glob′u·lin
im′mu·no·log′i·cal *or* ·log′ic
im′mu·nol′o·gist
im′mu·nol′o·gy
im′mu·no·pro′tein
im′mu·no′re·ac′tion
im′mu·no′sup·pres′sion
im′mu·no·ther′a·py
im′mu·no·tox′in
im·mu′ta·ble
im·pact′ed
im·pac′tion
im·pair′ment
im·pal′pa·bil′i·ty
im·pal′pa·ble
im′par
im·par′i·dig′i·tate′
im′passe
im·pa′ten·cy
im·pa′tent
im·ped′ance
im·ped′i·ment
im·per′a·tive
im′per·cep′tion
im·per′fect
im·per′fo·rate *or* ·rat′ed
im·per′fo·ra′tion
im·pe′ri·ous

im·per′me·a·bil′i·ty
im·per′me·a·ble
im·per′me·ant
im·per′vi·ous
im·pe·tig′i·nous
im·pe·ti′go
im·plant′ v.
im′plant′ n.
im·plan·ta′tion
im·ple′tion
im·plode′
  ·plod′ed ·plod′ing
im·plo′sion
im·pon′der·a·ble
im·pos′tume or
  ·thume
im′po·tence
im′po·ten·cy
  ·cies
im′po·tent
im·preg′nate
  ·nat·ed ·nat·ing
im′preg·na′tion
im·pres′si·o
  im·pres′si·o′nes
im·pres′sion
im′print
im·print′ing
im·prov′a·ble
im·prove′
  ·proved′ ·prov′ing
im·prove′ment
im′pulse
im·pul′sion
im·pul′sive
im·pure′

im·pu′ri·ty
  ·ties
i′mus
in′a·bil′i·ty
  ·ties
in·ac′tion
in·ac′ti·vate′
  ·vat′ed ·vat′ing
in·ac′ti·va′tion
in·ac′tive
in·ac·tiv′i·ty
  ·ties
in·ad′e·qua·cy
  ·cies
in·ad′e·quate
in·ad·vis′a·ble
in·al′ter·a·ble
in·an′i·mate
in′a·ni′tion
in·ap·par′ent
in·ap′pe·tence or
  ·ten·cy
in·ap′pe·tent
in·ar·tic′u·late
in·as·sim′i·la·ble
in·at·ten′tive
in·au′di·ble
in′born′
in′breathe′
  ·breathed′
  ·breath′ing
in′breed′
  ·bred′ ·breed′ing
in·can·des′cent
in′ca·pa·bil′i·ty
in·ca′pa·ble
in·ca·pa′cious

in′ca·pac′i·tate′
  ·tat′ed ·tat′ing
in′ca·pac′i·ta′tion
in′ca·pac′i·ty
  ·ties
in·cap·a·ri′na
in·cap′su·late′
  ·lat′ed ·lat′ing
in·car′cer·ate′
  ·at′ed ·at′ing
in·car′cer·a′tion
in·car′na·dine′
in·car′nant or
  ·car′na·tive
in·case′
  ·cased′ ·cas′ing
in·cen′tive
in·cep′tion
in′cest
in·ces′tu·ous
in′ci·dence
in′ci·dent
in·cip′i·ence or
  ·en·cy
in·cip′i·ent
in·ci′sal
in·cise′
  ·cised′ ·cis′ing
in·ci′sion
in·ci′sive
in·ci′sor
in·ci·su′ra
  ·rae
in·ci′sure
in′cli·na′tion
in·cline′
  ·clined′ ·clin′ing

in'cline n.
in·clu'sion
in'co·er'ci·ble
in'co·her'ence or ·her'en·cy
in'co·her'ent
in'com·bus'ti·ble
in'com·pat'i·bil'i·ty
   ·ties
in'com·pat'i·ble
in·com'pe·tence or ·ten·cy
in·com'pe·tent
in'com·plete'
in'com·pres'si·ble
in·con'stant
in·con'ti·nence
in·con'ti·nent
in·con·trol'la·ble
in·co·or'di·nate
in·co·or'di·na'tion
in·cor'po·rate'
   ·rat·ed ·rat·ing
in·cor'po·ra'tion
in'crease n.
in·crease'
   ·creased' ·creas'ing
in'cre·ment
in'cre·men'tal
in·cre'tion
in'cross'
in·crust'
in'crus·ta'tion
in'cu·bate'
   ·bat·ed ·bat·ing
in'cu·ba'tion
in'cu·ba'tor

in'cu·bus
   ·bus·es or ·bi'
in'cu·dal
in·cu'des
   (sing. in'cus)
in·cu·do·mal'le·al
in·cur'a·bil'i·ty
in·cur'a·ble
in·cur'vate
   ·vat·ed ·vat·ing
in'cur·va'tion or in·cur'va·ture
in'cus
   in·cu'des
in'da·ga'tion
in'dent n.
in·dent' v.
in'den·ta'tion
in·de·pend'ent
in·de·ter'mi·nate
in'dex
   ·dex·es or ·di·ces'
Index Med'i·cus
in'di·can'
in'di·cant'
in'di·cate'
   ·cat·ed ·cat·ing
in'di·ca'tion
in·dic'a·tive or
   ·dic'a·to'ry
in'di·ca'tor
in'di·ces'
   (sing. in'dex)
in·dif'fer·ent
in·dig'e·nous
in'di·gest'ed
in'di·gest'i·bil'i·ty

in'di·gest'i·ble
in'di·ges'tion
in'di·ges'tive
in·dig'i·ta'tion
in'di·go'
   ·gos' or ·goes'
in·dig'o·tin
in·di·rect'
in'dis·crete'
in·dis·crim'i·nate
in'dis·posed'
in'dis·po·si'tion
in'dis·sol'u·bil'i·ty
in'dis·sol'u·ble
in'di·um
in'di·vid'u·al
in'di·vid'u·al'i·ty
   ·ties
in'di·vid'u·al·i·za'tion
in'di·vid'u·al·ize'
   ·ized' ·iz'ing
in'di·vid'u·ate'
   ·at·ed ·at·ing
in'di·vid'u·a'tion
in'dole
in'dole·a·ce'tic acid
in'do·lence
in'do·lent
in'do·phe'nol
in·dox'yl
in·duce'
   ·duced' ·duc'ing
in·duc'er
in·duc'i·ble
in·duct'
in·duct'ance

115

in·duc′tion
in·duc′tor
in·duc′to·therm′
in′du·line
in′du·rate′
  ·rat′ed ·rat′ing
in′du·ra′tion
in′du·ra′tive
in·du′si·al
in·du′si·um
  ·si·a
in·dwell′
  ·dwelt′ ·dwell′ing
in·e′bri·ant
in·e′bri·ate′
  ·at′ed ·at′ing
in·e′bri·ate *adj., n.*
in·e′bri·a′tion
in′e·bri′e·ty
in·ed′i·ble
in′ef·fec′tive
in′ef·fec′tu·al
in′ef·fi·ca′cious
in·ef′fi·ca·cy
  ·cies
in′e·las′tic
in′e·las·tic′i·ty
in·ert′
in·er′tia
in′ex·er′tion
in′ ex·tre′mis
in′fan·cy
  ·cies
in′fant
in·fan′ti·cide′
in′fan·tile′
in·fan′ti·lism

in·fan′ti·lize′
  ·lized′ ·liz′ing
in·farct′
in·farc′tion
in·fect′
in·fec′tion
in·fec′tious
in·fec′tive
in′fec·tiv′i·ty
in·fec′tor
in·fe′cund
in′fe·cun′di·ty
in·fe′ri·or
in·fe′ri·or′i·ty
in·fe′ro·an·te′ri·or
in·fe′ro·pos·te′ri·or
in·fer′tile
in′fer·til′i·ty
in·fest′
in′fes·ta′tion
in·fest′er
in·fib′u·la′tion
in·fil′trate
  ·trat·ed ·trat·ing
in′fil·tra′tion
in·fil′tra·tive
in·fil′tra·tor
in′fi·nite
in·firm′
in·fir′ma·ry
  ·ries
in·fir′mi·ty
  ·ties
in·flame′
  ·flamed′ ·flam′ing
in·flam′ma·ble
in′flam·ma′tion

in·flam′ma·to′ry
in·flat′a·ble
in·flate′
  ·flat′ed ·flat′ing
in·flat′er *or* ·fla′tor
in·fla′tion
in·flec′tion
in·flexed′
in·flex′i·bil′i·ty
in·flex′i·ble
in′flu·ence
in′flu·en′za
in′flu·en′zal
in·fold′
in′fra–ax′il·lar′y
in·frac′tion
in′fra·red′
in′fra·son′ic
in·fun·dib′u·lar
in·fun·dib′u·li·form′
in·fun·dib′u·lum
  ·la
in·fuse′
  ·fused′ ·fus′ing
in′fu·si·bil′i·ty
in·fu′si·ble
in·fu′sion
in′fu·so·de·coc′tion
in′fu·so′ri·al
in′fu·so′ri·an
in·gen′er·ate *adj.*
in·gest′
in·ges′ta
in·ges′tant
in ges′tion
in·ges′tive
in′gra·ves′cent

in·gre'di·ent
in·grow'ing
in'grown'
in'growth'
in'guen
  ·gui·na
in'gui·nal
in'gui·no·dyn'i·a
in·hal'ant
in·ha·la'tion
in'ha·la'tor
in·hale'
  ·haled' ·hal'ing
in·hal'er
in·her'ent
in·her'it
in·her'it·a·bil'i·ty
  ·ties
in·her'it·a·ble
in·her'it·ance
in·hib'it
in·hi·bi'tion
in·hib'i·tive *or*
  ·to'ry
in·hib'i·tor *or*
  ·hib'it·er
in·ho'mo·ge·ne'i·ty
  ·ties
in'ho·mo·ge'ne·ous
in'i·ac *or* ·i·al
in'i·on
in·i'tial
in·ject'a·ble
in·jec'tion
in·jec'tor
in'jure
  ·jured ·jur·ing

in·ju'ri·ous
in'ju·ry
  ·ries
ink'blot'
in'lay'
  ·lays'
in'let *n.*
in'mate'
in·nate'
in'ner
in·ner'vate
  ·vat·ed ·vat·ing
  (stimulate; SEE
  enervate)
in'ner·va'tion
in·nerve'
  ·nerved' ·nerv'ing
in·nid'i·a'tion
in'no·cent
in·noc'u·ous
in·nom'i·nate
in·nox'ious
in·nu·tri'tion
in·oc'u·la·bil'i·ty
in·oc'u·la·ble
in·oc'u·late'
  ·lat·ed ·lat·ing
in·oc'u·la'tion
in·oc'u·la'tive
in·oc'u·la'tor
in·oc'u·lum *or* ·lant
in·o'dor·ous
in·op'er·a·ble
in·or·gan'ic
in·os'cu·late'
  ·lat·ed ·lat·ing
in·os'cu·la'tion

in'o·sine
in·o'si·tol' *or* in'o·
  site'
in'o·trop'ic
in·pa'tient
in'quest
in·qui'e·tude'
in·sal'i·vate'
  ·vat·ed ·vat·ing
in·sal'i·va'tion
in'sa·lu'bri·ous
in'sa·lu'bri·ty
in·sane'
in·san'i·tar'y
in·san'i·ty
  ·ties
in·sa'ti·a·ble
in·sa'ti·ate
in·scrip'ti·o
  in·scrip'ti·o'nes
in·scrip'tion
in'sect
in·sec·tar'i·um
  ·i·a
in·sec'tar·y
  ·tar'ies
in·sec'ti·cid'al
in·sec'ti·cide'
in·sec'ti·fuge'
in·sec'tile *or* in'sec·
  ti'val
in·se·cure'
in·se·cu'ri·ty
  ·ties
in·sem'i·nate'
  ·nat·ed ·nat·ing
in·sem'i·na'tion

in·sen′sate
in·sen′si·bil′i·ty
in·sen′si·ble
in·sen′si·tive
in·sen′si·tiv′i·ty
·ties
in·sert′ v.
in′sert n.
in·sert′ed
in·ser′tion
in·sheathe′
·sheathed′
·sheath′ing
in·sid′i·ous
in′sight′
in·sip′id
in si′tu
in′so·bri′e·ty
·ties
in′so·la′tion
in·sol′u·bil′i·ty
·ties
in·sol′u·ble
in·som′ni·a
in·som′ni·ac′
in·spec′tion
in·sper′sion
in′spi·ra′tion
in·spir′a·to·ry
in·spire′
·spired′ ·spir′ing
in·spis′sate
·sat·ed ·sat·ing
in′spis·sa′tion
in′spis·sa′tor
in′sta·bil′i·ty
in·sta′ble

in′stance
in′step′
in·still′ or ·stil′
·stilled′ ·still′ing
in′stil·la′tion
in·still′er
in·still′ment or
·stil′ment
in′stinct
in·stinc′tive
in′sti·tute′
·tut·ed ·tut′ing
in′sti·tu′tion
in′sti·tu′tion·al·i·za′tion
in′sti·tu′tion·al·ize′
·ized′ ·iz′ing
in′stru·ment
in′stru·men′tal
in′stru·men·ta′tion
in′suf·fi′cience
in′suf·fi′cien·cy
·cies
in′suf·fi′cient
in·suf′flate
·flat·ed ·flat·ing
in′suf·fla′tion
in′suf·fla′tor
in′su·lar
in′su·lar′i·ty or in′su·lar·ism
in′su·late′
·lat·ed ·lat′ing
in′su·la′tion
in′su·la′tor
in′su·lin

in′su·lin·o′ma
·mas or ·ma·ta
in′sult n.
in′sus·cep′ti·bil′i·ty
·ties
in′sus·cep′ti·ble
in·tact′
in′take′
in′te·grate′
·grat′ed ·grat′ing
in′te·gra′tion
in′te·gra′tor
in·teg′ri·ty
in·teg′u·ment
in·teg′u·men′ta·ry
in′tel·lect′
in′tel·lec′tion
in′tel·lec′tu·al
in′tel·lec′tu·al·i·za′tion
in′tel·lec′tu·al·ize′
·ized′ ·iz′ing
in·tel′li·gence
in·tel′li·gent
in·tel′li·gen′tial
in·tem′per·ance
in·tem′per·ate
in·tense′
in·ten′si·fi·ca′tion
in·ten′si·fy′
·fied′ ·fy′ing
in·ten′si·ty
·ties
in·ten′sive
in·ten′tion
in′ter·act′
in′ter·act′ant

in′ter·ac′tion
in′ter·ac′tion·ist
in′ter·ac′tive
in′ter·a′tri·al
in′ter·brain′
in·ter·ca·lar′y
in·ter·ca·late′
  ·lat′ed ·lat′ing
in′ter·cel′lu·lar
in′ter·clav′i·cle
in′ter·cla·vic′u·lar
in′ter·cos′tal
in′ter·course′
in′ter·cur′rent
in′ter·den′tal
in′ter·dig′i·tate′
  ·tat′ed ·tat′ing
in′ter·face′
  (boundary; SEE
  interphase)
in′ter·fa′cial
in′ter·fere′
  ·fered′ ·fer′ing
in′ter·fer′ence
in′ter·fer·om′e·ter
in′ter·fer′o·met′ric
in′ter·fer·om′e·try
in′ter·fer′on
in′ter·grade′
in·te′ri·or
in′ter·me′di·ar′y
  ·ar′ies
in′ter·me′di·ate
in′ter·me′din
in′ter·mis′sion
in′ter·mit′
  ·mit′ted ·mit′ting

in′ter·mit′tence
in′ter·mit′tent
in′ter·mo·lec′u·lar
in′ter·mu′ral
in′ter·mus′cu·lar
in′tern
in·ter′nal
in·ter′nal·i·za′tion
in·ter′nal·ize′
  ·ized′ ·iz′ing
in′ter·na′tion·al
  unit
in′terne′
in′ter·neu′ron
in′ter·nist
in′ter·nod′al
in′ter·node′
in′tern·ship′
in′ter·nun′ci·al
in′ter·o·cep′tive
in′ter·o·cep′tor
in′ter·phase′
  (cell stage; SEE
  interface)
in′ter·plant′
in·ter′po·late′
  ·lat′ed ·lat′ing
in·ter′po·la′tion
in′ter·po·si′tion
in·ter′pret
in·ter′pre·ta′tion
in′ter·prox′i·mal
in′ter·ra′di·al
in′ter·rupt′ed
in′ter·sect′
in′ter·sec′ti·o
  ·sec′ti·o′nes

in′ter·sec′tion
in′ter·sex′
in′ter·sex′u·al
in′ter·space′
in·ter′stice
  ·stic·es
in′ter·sti′tial
in′ter·val
in′ter·vene′
  ·vened′ ·ven′ing
in′ter·ven′tion
in′ter·ver′te·bral
in·tes′ti·nal
in·tes′tine
in′ti·ma
  ·mae′
in′ti·mal
in′tine
in·tol′er·ance
in·tox′i·cant
in·tox′i·cate′
  ·cat′ed ·cat′ing
in·tox′i·cate *adj.*
in·tox′i·ca′tion
in′tra–a′tri·al
in′tra·cel′lu·lar
in·trac′ta·bil′i·ty
in·trac′ta·ble
in′tra·cu·ta′ne·ous
in′tra·der′mal
in′tra·mo·lec′u·lar
in′tra·mu′ral
in′tra·mus′cu·lar
in′tra·per′son·al
in′tra·psy′chic *or*
  ·psy′chi·cal
in′tra·u′ter·ine

in·trav′a·sa′tion
in′tra·ve′nous
in′tra·ven·tric′u·lar
in·trin′sic
in′tro·duc′er
in′tro·gres′sion
in′tro·gres′sive
in·troi′tus
in′tro·ject′
in′tro·jec′tion
in′tro·mis′sion
in′tro·mit′
 ·mit′ted ·mit′ting
in′tro·mit′tent
in′tron
in′tro·spect′
in′tro·spec′tion
in′tro·spec′tive
in′tro·ver′sion
in′tro·vert′
in′tro·vert′ed
in·trude′
 ·trud′ed ·trud′ing
in′tu·bate′
 ·bat′ed ·bat′ing
in′tu·ba′tion
in′tu·i′tion
in′tu·mesce′
 ·mesced′ ·mesc′ing
in′tu·mes′cence
in′tu·mes′cent
in′tus·sus·cept′
in′tus·sus·cep′tion
in′tus·sus·cep′tive
in·u′lase′
in·u′lin
in·unc′tion

in u′ter·o
in va·cu·o′
in·vade′
 ·vad′ed ·vad′ing
in·vad′er
in·vag′i·nate′
 ·nat′ed ·nat′ing
in·vag′i·na′tion
in′va·lid
in·va·lid·ism
in·va′sion
in·va′sive
in′ven·to′ry
 ·ries
in·verse′
in·ver′sion
in·ver′sive
in·vert′ v.
in′vert′ n.
in·vert′ase
in·ver′tor
in·vest′ment
in·vet′er·ate
in·vi′a·bil′i·ty
in·vi′a·ble
in vi′tro
in vi′vo
in′vo·lu′cral
in′vo·lu′crate
in′vo·lu′cre
in′vo·lu′crum
 ·cra
in·vol′un·tar′y
in′vo·lute′
 ·lut′ed ·lut′ing
in′vo·lu′tion
in′vo·lu′tion·al

in′vo·lu′tion·ar′y
in′ward
i′o·date′
 ·dat′ed ·dat′ing
i′o·da′tion
i·od′ic
i′o·dide′
i′o·di·nate′
 ·nat′ed ·nat′ing
i′o·di·na′tion
i′o·dine′
i′o·dism
i′o·dize′
 ·dized′ ·diz′ing
i·o′do·form′
i′o·do·met′ric
i′o·dom′e·try
i′o·do·phil′i·a
i·o′do·phor′
i′o·do·pro′tein
i′o·dop′sin
i·o′dous
i′on
i·on′ic
i·o′ni·um
i′on·i·za′tion
i′on·ize′
 ·ized′ ·iz′ing
i′on·iz′er
i·on′o·gen
i·on′o·gen′ic
i·on′o·phore′
i on′o·pho·re′sis
 ·ses
i·on′o·pho·ret′ic
i·on′to·pho·re′sis
 ·ses

i·on'to·pho·ret'ic
i·o'ta·cism
ip'e·cac'
ip'e·cac'u·an'ha
ip'o·moe'a
i'pro·ni'a·zid
ip'si·lat'er·al
i·ras'ci·bil'i·ty
i·ras'ci·ble
i'ri·dal'gi·a
ir'i·dec'to·mize'
  ·mized' ·miz'ing
ir'i·dec'to·my
  ·mies
ir'i·des'
  (sing. i'ris)
ir'i·des'cence
ir'i·des'cent
i·rid'e·sis
  ·ses'
i·rid'ic
i·rid'i·um
i'ris
  i'ris·es or ir'i·des'
i·rit'ic
i·ri'tis
i'ron
iron lung
ir·ra'di·ate'
  ·at'ed ·at'ing
ir·ra'di·a'tion
ir·ra'di·a'tive
ir·ra'di·a'tor
ir·ra'tion·al
ir·ra'tion·al'i·ty
  ·ties

ir're·duc'i·bil'i·ty
  ·ties
ir're·duc'i·ble
ir·reg'u·lar
ir·reg'u·lar'i·ty
  ·ties
ir're·me'di·a·ble
ir·rep'a·ra·ble
ir're·spir'a·ble
ir're·vers'i·bil'i·ty
ir're·vers'i·ble
ir'ri·ga·ble
ir'ri·gate'
  ·gat'ed ·gat'ing
ir'ri·ga'tion
ir'ri·ga'tive
ir'ri·ga'tor
ir'ri·ta·bil'i·ty
ir'ri·ta·ble
ir'ri·tan·cy
ir'ri·tant
ir'ri·tate'
  ·tat'ed ·tat'ing
ir'ri·ta'tion
ir'ri·ta'tive
is·che'mi·a
is·che'mic
is·che'sis
is'chi·al or is'chi·
  ad'ic or is'chi·
  at'ic
is'chi·um
  ·chi·a
i'sin·glass'
is'land
is'let

i'so·ag·glu'ti·na'
  tion
i'so·ag·glu'ti·nin
i'so·al·lox'a·zine'
i'so·an'ti·bod'y
  ·bod·ies
i'so·an'ti·gen
i'so·bar'
i'so·bar'ic
i'so·chro·mat'ic
i·soch'ro·nal or
  ·soch'ro·nous
i·soch'ro·nism
i·soch'ro·nize'
  ·nized' ·niz'ing
i·soch'ro·ous
i'so·cy'a·nate
i'so·cy'clic
i'so·di'a·met'ric
i'so·dose'
i'so·dy·nam'ic
i'so·e·lec'tric
i'so·en'zyme
i'so·gam'ete
i'so·ga·met'ic
i·sog'a·mous
i·sog'a·my
i·sog'e·nous
i·sog'e·ny
i·sog'o·ny
i'so·graft'
i'so·late'
  ·lat'ed ·lat'ing
i'so·late n.
i'so·la'tion
i'so·la'tor
i'so·leu'cine

i·sol′o·gous
i·so·logue′ or ·log′
i·sol′y·sin
i·sol′y·sis
  ·ses′
i·so·lyt′ic
i′so·mer
i′so·mer′ic
i·som′er·ism
i·som′er·ous
i′so·met′ric or
  ·met′ri·cal
i′so·met′rics
i′so·me·tro′pi·a
i·som′e·try
i′so·morph′
i′so·mor′phic or
  ·phous
i′so·mor′phism
i′so·ni′a·zid
i′so·prene′
i′so·pro′pan·ol
i′so·pro′pyl
i′so·pro·ter′e·nol
is′os·mot′ic
i′sos·the·nu′ri·a
i′so·therm′
i′so·ther′mal
i′so·tone′
i′so·ton′ic
i′so·to·nic′i·ty
i′so·tope′
i′so·top′ic
i·sot′o·py
i′so·trop′ic or i·sot′
  ro·pous

i·sot′ro·py
i′so·zyme′
is′sue
isth′mi·an
isth·mi′tus
isth′mus
  ·mus·es or ·mi
i·tai′–i·tai′
itch
itch′y
  ·i·er ·i·est
i′ter
i′ter·al
ith′y·lor·do′sis
ix′o·di′a·sis
ix′o·my′e·li′tis

J

jack′et
jack′knife′ position
jack′screw′
Jack·so′ni·an
  epilepsy
Ja′cob·son's organ
jac′ti·tate′
  ·tat·ed ·tat·ing
jac′ti·ta′tion or jac·
  ta′tion
Ja′kob–Creutz′feldt
  disease
jal′ap
jal′a·pin
jar′gon

jaun′dice
  ·diced ·dic·ing
Ja·velle′ (or Ja·
  vel′) water
jaw′bone′
je·ju′nal
je′ju·nec′to·my
  ·mies
je′ju·ni′tis
je′ju·no·il′e·i′tis
je′ju·nos′to·my
  ·mies
je′ju·not′o·my
  ·mies
je·ju′num
  ·na
jel′li·fi·ca′tion
jel′li·fy′
  ·fied′ ·fy′ing
Jen′ner's stain
jerk
jet lag
jig′ger
jim′son (or jimp′
  son) weed
Jo·cas′ta complex
jock itch
jock′strap′
Jof·froy's′ reflex
jog
  jogged jog′ging
jog′ger
john′ny
  ·nies
joint
joint′ed
joule

jowl
ju'gal
ju·ga'le
ju'gate
ju'go·max·il·lar'y
jug'u·lar
ju'gu·late'
  ·lat'ed ·lat'ing
ju'gum
  ·ga or ·gums
juice
ju·men'tous
junc'tion
junc·tu'ra
  ·rae
junc'ture
Jung'i·an
jun'gle fever
ju'ni·per
junk food
Ju·nod's' boot
ju'ris·pru'dence,
  dental
ju'ry-mast'
jus'cu·lum
jus'to ma'jor
ju·van'ti·a
ju'ven·ile
jux'ta-ar·tic'u·lar
jux·tan'gi·na
jux'ta·pose'
  ·posed' ·pos'ing
jux'ta·po·si'tion

K

Kaes' line
Kahn test
kaif
Kai'ser·ling's
  fixative
kak'i·dro'sis
  ·ses
ka'la a·zar'
ka·le'mi·a
ka'li
kal'li·din
ka·lim'e·ter
kal'li·din
kal'li·kre'in
kal'li·krei'no·gen
kal·u·re'sis or ka'li·
  u·re'sis
kal·u·ret'ic
ka·ma'la
kan'a·my'cin
ka'o·lin
ka'o·lin·o'sis
  ·ses
Kap'o·si's sarcoma
kap'pa
kar'y·og'a·my
  ·mies
kar'y·o·gen'e·sis
  ·ses'
kar'y·o·gen'ic
kar'y·o·ki·ne'sis
  ·ses
kar'y·o·ki·net'ic
kar'y·o·lymph'

kar'y·o·plasm
kar'y·o·plas'mic
kar'y·o·some'
kar'y·o·tin
kar'y·o·type'
kar'y·o·typ'ic or
  ·typ'i·cal
ka·sai'
kat
For words
  beginning kat–
  and kata– *see*
  CAT–, CATA–
ka·thar'sis
kath'i·so·pho'bi·a
kath'ode
kat'i·on
kat'o·tro'pi·a or
  ·phor'i·a
ka'va or ka'va·
  ka'va
Ka'wa·sa'ki disease
ke'bu·zone'
keck
kef or ke·ef'
Ke'gel's exercises
ke'lis
Kell blood group
ke'loid
ke·loi'dal
kelp
Kel'vin scale
Ken'ny method
ke·ram'ics
ker'a·sin
ker'a·tal'gi·a

ker′a·tec·ta′si·a
ker′a·tec′to·my
·mies
ke·rat′ic
ker′a·tin
ker′a·tin′i·za′tion
ker′a·tin·ize′
·ized′ ·iz′ing
ke·rat′i·nous or ·noid′
ker′a·ti′tis
ker′a·to·ac′an·tho′ma
·mas or ·ma·ta
ker′a·to·cele′
ker′a·to·co′nus
ker′a·tog′e·nous
ker′a·to·glo′bus
ker′a·toid′
ker′a·tom′e·ter
ker′a·to·met′ric
ker′a·tom′e·try
ker′a·to·plas′ty
·ties
ker′a·tose′
ker′a·to′sis
·ses
ker′a·tot′ic
ker′a·tot′o·my
·mies
ke′ri·on
ker′i·ther′a·py or ker′o·ther′a·py
ker·nic′ter·us
Ker′nig's sign
ker′o·sene′ or ·sine
ke′tal

ket′a·mine′ hydrochloride
ke′tene
ke′to·gen′e·sis
·ses′
ke′to·gen′ic
ke′tol
ke·tol′y·sis
·ses′
ke′to·lyt′ic
ke′tone
ke′to·ne′mi·a
ke′to·nu′ri·a
ke′tose
ke·to′sis
·ses
ke′to·ster′oid
ke·tot′ic
kev or Kev
khat
khel′lin
ki
kibe
kid′ney
·neys
Kien′böck disease
kif or kief
kil′o·cal′o·rie
kil′o·cy′cle
kil′o·gram′
kil′o·gram′-me′ter
kil′o·hertz′
kil′o·joule′
kil′o·li′ter
kil·lo′me·ter
ki′lo·met′ric
kil′o·nem′

kil′o·volt′
kil′o·volt′–am′pere
kil′o·watt′
Kim′mel·steil′–Wil′son syndrome
kin′aes·the′si·a or ·the′sis
kin′aes·thet′ic
ki′nase
kin′e·mat′ic or ·mat′i·cal
kin′e·mat′ics
ki·ne′sic
ki·ne′sics
ki·ne′si·ol′o·gy
ki·ne′sis
·ses
kin′es·the′si·a or ·the′sis
kin′es·the′si·om′e·ter
kin′es·thet′ic
ki·net′ic
ki·net′ics
ki′ne·tin
ki·ne′to·car′di·og′ra·phy
ki·ne′to·chore′
ki·net′o·plast′
ki·ne·to′sis
·ses
ki·ne′to·some′
king′dom
king's e′vil
ki′nin
ki·nin′o·gen′

kink
ki′no
ki′no·hapt′
kin′o·mom′e·ter
kin′ship′
ki′o·tome′
kit′a·sa·my′cin
kleb′si·el′la
klep′to·ma′ni·a
klep′to·ma′ni·ac′
Kline′fel′ter's syndrome
Kline test
knead
knee′cap′ or ·pan′
knee jerk
knife
  knives
knis′mo·gen′ic
knob
knock′–kneed′
knot
  knot′ted knot′ting
knuck′le·bone′
knuckle joint
Koch's bacillus
koi′lo·nych′i·a
koi′lo·ster′ni·a
ko′jic acid
ko′la nut
Kol′mer test
ko·lyt′ic
ko′ni·o·cor′tex
ko·phe′mi·a
Kop′lik's spots
ko′ro

ko·ro′ni·on
  ·ni·a
Kor′sa·koff′s' psychosis
krait
krau·ro′sis
  ·ses
Krebs cycle
kryp′ton
ku·bis′a·ga′ri
Kupf′fer's cells
ku′ru
Kuss′maul's aphasia
kwa′shi·or′kor
ky′mo·gram′
ky′mo·graph′
ky′mo·graph′ic
ky·mog′ra·phy
ky′mo·scope′
kyn·u·ren′ine or ·in
ky′phos
ky·pho′sis
  ·ses
ky·phot′ic
kyr′tor·rhach′ic

L

la′bel
  ·beled or ·belled
  ·bel·ing or ·bel·ling
la′bi·a
  (sing. la′bi·um)
la′bi·al

la′bi·al·ism
labia ma·jo′ra
labia mi·no′ra
la′bile
la·bil′i·ty
  ·ties
la′bi·o·al·ve′o·lar
la′bi·o·cer′i·cal
la′bi·o·den′tal
la′bi·o·glos′so·la·ryn′ge·al
la′bi·o·graph′
la′bi·o·na′sal
la′bi·um
  ·bi·a
la′bor
lab′o·ra·to′ry
  ·ries
la′bra
  (sing. la′brum)
lab′y·rinth′
lab′y·rin·thec′to·my
lab′y·rin′thine or
  ·thi·an or ·thic
lab′y·rin·thi′tis
lab′y·rin·thot′o·my
lab′y·rin′thus
  ·thi
lac
  lac′ta
lac′er·a·ble
lac′er·ate′
  ·at′ed ·at′ing
lac′er·a′tion
la·cer′tus
  la·cer′ti
lach·ry′mal

125

lach′ry·ma′tor
lach′ry·ma·to′ry
lac′ri·mal
lac′ri·ma′tion
lac′ri·ma′tor
lac′ri·ma·to′ry
lac′ri·mot′o·my
lac·tac′i·de′mi·a
lac·tac′i·du′ri·a
lac′ta·gogue′
lac′tam
lac·tam′ide
lac′tase
lac′tate
 ·tat·ed ·tat·ing
lac·ta′tion
lac′te·al
lac′te·nin
lac·tes′cence
lac·tes′cent
lac′tic
lac·tif′er·ous
lac′ti·fuge
lac·tig′e·nous
lac·tig′er·ous
lac′to·ba·cil′lus
 ·cil′li
lac′to·cele′
lac′to·fla′vin
lac′to·gen′ic
lac·tom′e·ter
lac′tone
lac′to·pro′te·in
lac′tor·rhe′a
lac′tose
lac′to·veg′e·tar′i·an

lac′tu·lose′
la·cu′na
 ·nas or ·nae
la·cu′nar or ·nal
la·cu′nose
la·cu′nule
la′cus
For words beginning lae– *see* LE–
La·en′nec's′ cirrhosis
la′e·trile′
la·gen′i·form
lag′oph·thal′mos or ·mus
lag phase
lake
 laked lak′ing
lak′y
 ·i·er ·i·est
la·li′a·try
lal·la′tion
lal′og·no′sis
la·lop′a·thy
La·marck′ism
La·maze′
lamb′da
lamb′da·cism
lamb′doid
lam′bert
lam·bli′a·sis
lame
 lamed lam′ing
la·mel′la
 ·lae or ·las
la·mel′lar

lam′el·late′ or ·lat′ed
lam′el·la′tion
la·mel′lose
lam′el·los′i·ty
lam′i·na
 ·nae or ·nas
lam′i·na·ble
lam′i·nar or ·nal
lam′i·nate′
 ·nat·ed ·nat·ing
lam′i·nate n.
lam′i·na′tion
lam′i·nec′to·my
 ·mies
lam′i·not′o·my
lan·at′o·side′ C
lance
 lanced lanc′ing
lan′ce·o·late′
lan′cet
lan′ci·nate′
 ·nat·ed ·nat·ing
lan′ci·na′tion
land′mark′
Lane's kink
Lang′er·hans′ islets *or* islands
lan′guage
lan′guor
la′ni·ar′y
lan′o·lin or ·line
lan′tha·num
la·nu′gi·nous or ·nose′
la·nu′go

lap
   lapped lap′ping
la·pac′tic
lap′a·ro·scope′
lap′a·ro·scop′ic
lap′a·ros′co·py
lap′a·rot′o·my
   ·mies
lap′ar·o·trach·e·lot′
lap′is
lapse
   lapsed laps′ing
lap′sus
lar·da′ceous
lar′va
   ·vae *or* ·vas
lar′val
lar′vate
lar′vi·cid′al
lar′vi·cide′
la·ryn′ge·al
lar′yn·gec′to·my
lar′yn·gis′mal
lar′yn·gis′mus
lar′yn·git′ic
lar′yn·gi′tis
lar·yn′go·cele
lar·yn′go·cen·te′sis
lar·yn′gol′o·gist
lar·yn′gol′o·gy
lar·yn′go·phar′yn·gi′tis
la·ryn′go·scope′
la·ryn′go·scop′ic *or* ·scop′i·cal
lar′yn·gos′co·py
   ·pies

lar′yn·got′o·my
lar·yn′go·xe·ro′sis
lar′ynx
   lar′ynx·es *or* la·ryn′ges
lase
   lased las′ing
la′ser
lash
L′–as′par·a′gin·ase′
Las′sa fever
las′si·tude′
la′tah
la′ten·cy
   ·cies
la′tent
la·ten′ti·a′tion
lat′er·ad′
lat′er·al
lat′er·al′is
lat′er·al′i·ty
lat′er·i′tious *or* ·ceous
lat′er·o·duc′tion
lat′er·o·flex′ion *or* ·i·flex′ion
lat′er·o·ver′sion
la′tex
   lat′i·ces′ *or* la′tex·es
lathe
lath′er
lath′y·rism
lath′y·ro·gen
lath′y·ro·gen′ic
la·tis′si·mus
lat′ro·dec′tus

lat′tice
   ·ticed ·tic·ing
   (openwork structure)
la′tus
   lat′er·a
   (the flank)
laud′a·ble
lau′da·nine
laud′a·num
laugh′ing gas
lau′ric acid
la·vage′ *or* la·va′tion
la·ven′du·lin
law·ren′ci·um
lax
lax·a′tion
lax′a·tive
lax′a·tor
lax′i·ty
lay′er
lay·ette′
la′zar
laz′a·ret′to
   ·tos
L′–do′pa
leach
   (to extract; SEE leech*)*
lead arsenate
lead, bipolar
lead′er
leaf′let
leak′age
lean

learn
  learned or learnt
  learn′ing
learn′a·ble
learning disability
learning–disabled
leash
leath′er·y
Le′ber's disease
Le·boy·er′ method
Le·cat's′ gulf
lech′er·y
  ·er·ies
lec′i·thal
lec′i·thin
lec′i·thin·ase′
lec′i·tho·blast′
lec′tin
ledge, dental
leech
  (bloodsucker; SEE
  leach)
lees
left′–hand′ed
Legg′–Cal·vé′–Per′
  thes disease
leg′gings
le′gion·el·la
Le′gion·naires′'
  disease
le′gion·el·o′sis
le·git′i·ma·cy
lei′o·der′mi·a
lei′o·my′o·fi·bro′ma
  ·mas or ·ma·ta
lei′o·my·o′ma
  ·mas or ·ma·ta

leish·ma′ni·a
leish′man·i′a·sis
  ·ses′
le′ma
le′mic
lem′mo·cyte′
lem·nis′cus
  ·ci
lem′o·ste·no′sis
length′en
len′i·tive
lens
len·tic′u·lar
len′ti·form′
len·tig′i·no′sis
len·tig′i·nous or
  ·nose′
len·ti′go
  len·tig′i·nes′
len′toid
le′on·ti′a·sis
lep′er
le·pid′ic
lep′i·do′sis
lep·ro·sa′ri·um
  ·ri·ums or ·ri·a
lep′rose
lep′ro·sy
  ·sies
lep′rous
lep·to·ceph′a·lus
lep·to·dac′ty·lous
lep·to·spi′ral
lep′to·spire′
lep·to·spi·ro′sis
  ·ses
le·re′sis

les′bi·an·ism
le′sion
let′–down′ reflex
le′thal
le·thal′i·ty
  ·ties
le·thar′gic
leth′ar·gize′
  ·gized′ ·giz′ing
leth′ar·gy
  ·gies
leu′cine
*For words
  beginning leuc–
  and leuco– see
  also* LEUK– *and*
  LEUKO–
leu′co·plast′
leu·cot′o·my
leu·ka·phe·re′sis
leu·ke′mi·a or
  ·kae′mi·a
leu·ke′mic
leu·ke′moid
leu′ko·cyte′
leu·ko·cyt′ic
leu′ko·cy·to·blast′
leu′ko·cy·to·blas′tic
leu′ko·cyt′oid
leu′ko·cy·to′sis
  ·ses
leu′ko·cy·tot′ic
leu·ko·der′ma
leu·ko′ma
  ·mas or ·ma·ta
leu·ko·pe′ni·a
leu′ko·pe′nic

leu′ko·pla′ki·a
leu′ko·poi·e′sis
·ses
leu′ko·poi·et′ic
leu′kor·rhe′a
leu′kor·rhe′al
leu·kot′o·my
·mies
leu′ko·tri′ene
lev′ar·te·re′nol
le·va′tor
lev′a·to′res or le·va′tors
lev′el
lev′er
lev′er·age n.
lev′i·gate′
·gat′ed ·gat′ing
lev′i·ga′tion
lev′i·tate′
·tat′ed ·tat′ing
lev′i·ta′tion
le′vo·do′pa
le′vo·duc′tion
le′vo·gy′rate or ·gy′rous
le′vo·ro·ta′tion
le′vo·ro·ta·to′ry
lev′u·lin
lev′u·lose′
lev′u·lo·se′mi·a
lew′is·ite′
Ley′den jar
lib′er·ate′
·at′ed ·at′ing
lib′er·a′tion
lib′er·a′tor

lib′er·o·mo′tor
li·bid′i·nal
li·bid′i·nize′
·nized′ ·niz′ing
li·bid′i·nous
li·bi′do
li′bra
·brae
lice
(sing. louse)
li′cense
·censed ·cens·ing
li·cen′ti·ate
li′chen
li·chen′i·fi·ca′tion
li′chen·oid′
li′chen·ous or ·ose′
lic′o·rice
lid′ded
li′do·caine′
lie
lie detector
li′en
li·e′nal
life
lives
life expectancy
life′–giv′ing
life′long′
life′sav′ing
life span
life table
life′time′
lift
lig′a·ment
lig′a·men′tum
·ta

lig′and
li′gase
li′gate
·gat·ed ·gat·ing
li·ga′tion
lig′a·ture
·tured ·tur·ing
light
light′ed or lit
light′ing
light′en·ing
light′head′ed
lig′ne·ous
lig′nin
limb
lim′ber
lim′bic
lim′bus
·bi
lime
li′men
li′mens or lim′i·na
li′mes
lim′i·nal
lim′it
lim′it·a·ble
lim′i·ta′tion
lim′i·ta′tive
lim′o·nene′
li′mo·ther′a·py
limp
lin′co·my′cin
lin′dane′
line
lined lin′ing
lin′e·a
·ae

lin′e·age
lin′e·al
lin′e·ar
lin′e·ate
lin′e·a′tion
line breed′ing
lin′er
lin′ger
lin′gua
  ·guae
lin′gual
lin′gui·form′
lin′gu·la
  ·lae
lin′gu·lar
lin′gu·late
lin′i·ment
li′nin
lin′ing
li·ni′tis
link′age
li·no′le·ate′
lin′o·le′ic acid
lin′o·le′nate
lin′o·le′nic acid
lin′seed′
lint
lip
li′pase
lip′id *or* ·ide
lip′o·fus′cin
li′poid *or* li·poi′dal
li·pol′y·sis
  ·ses′
lip′o·lyt′ic
li·po′ma
  ·ma·ta *or* ·mas

li·pom′a·tous
lip′o·phil′ic
lip′o·pro′tein
lip′o·some′
lip′o·trop′ic
lip′o·tro′pin
li·pot′ro·pism
lipped
Lip′pes loop
lip′ping
lip′–read′
  –read′ –read′ing
liq′ue·fa′cient
liq′ue·fac′tion
liq′ue·fi′a·ble
liq′ue·fi′er
liq′ue·fy′
  ·fied′ ·fy′ing
li·ques′cence
li·ques′cent
li·queur′
liq′uid
liq·uid′i·ty
liq′uid·ize′
  ·ized′ ·iz′ing
liq′uor
lisp
lis′sen·ceph′a·ly
lis′sive
lis′ten
list′less
li′ter
lit′er·a·cy
lit′er·ate
lit′er·a·ture
lith′a·gogue
lithe

lith′i·a
lith·i′a·sis
  ·ses′
lith′ic
lith′i·um
lith′oid *or* li·thoi′dal
li·thol′y·sis
lith′o·tom′ic
li·thot′o·my
  ·mies
lith′o·trip′sy
lith′o·trip′ter
lith′o·trip′tor
li·thot′ri·ty
  ·ties
lit′mus
lit′ter
lit′tle
  lit′tler *or* less *or* less′er
  lit′tlest *or* least
live
  lived liv′ing
li·ve′do
liv′er
lives
  (*sing.* life)
liv′id
li·vid′i·ty
li′vor
lix·iv′i·ate′
  ·at′ed ·at′ing
lix·iv′i·a′tion
lix·iv′i·um
  ·i·ums *or* ·i·a
load′ing

lo′bar
lo′bate
lo·ba′tion
lobe
lo·bec′to·my
  ·mies
lo·be′li·a
lo′be·line′
lo·bot′o·mize′
  ·mized′ ·miz′ing
lo·bot′o·my
  ·mies
lob′u·lar *or* ·late′
lob′ule
lo′cal
lo′cal·iz′a·ble
lo′cal·i·za′tion
lo′cal·ize′
  ·ized′ ·iz′ing
lo′cal·iz′er
lo′cate
  ·cat·ed ·cat·ing
lo′cat·er *or* ·ca·tor
lo·ca′tion
lo′chi·a
lo′ci
  (*sing.* locus)
lock′jaw′
lo′co·mo′tion
lo′co·mo′tive
lo′co·mo′tor
lo′co·mo′to·ry
loc′u·lar *or* ·late
loc′ule
loc′u·lus
  ·li
lo′cum ten′ens

lo′cus
  lo′ci
log′a·dec′to·my
lo·get′ro·nog′ra·phy
log′o·ple′gi·a
log′or·rhe′a
log′or·rhe′ic
log′wood′
lo·i′a·sis
loin
lon′ga·nim′i·ty
lon·gev′i·ty
lon·ge′vous
long′–head′ed *or*
  long′head′ed
lon′gi·tu′di·nal
long′–lived′
long′sight′ed
long′–suf′fer·ing *or*
  ·fer·ance
loop
  (*circular form;* SEE
  loupe)
loose
  loosed loos′ing
loose′–joint′ed
loose′–limbed′
loos′en·ing
lop′–eared′
lo′pho·trich′e·a
lo·phot′ri·chous
lo·quac′i·ty
lor·do′sis
lor·dot′ic
lor·gnon′
lose
  lost los′ing

loss
lo′tion
Lou Geh′rig's
  disease
loupe
  (*magnifying lens;*
  SEE loop)
louse
  lice
lous′y
love
  loved lov′ing
love′–hate′
low′–cal′
low′–grade′
lox
lox′i·a
lox·ot′ic
loz′enge
lubb′–dupp′
lu′bri·cant
lu′bri·cate′
  ·cat′ed ·cat′ing
lu′bri·ca′tion
lu′bri·ca′tive
lu′bri·ca′tor
lu′cen·cy
lu′cent
lu′cid
lu·cid′i·ty
lu·cif′er·ase′
lu·cif′u·gal
lu′co·ther′a·py
lu′es
lu·et′ic
Lu·gol's′ solution
luke′warm′

lum·ba′go
lum′bar
lum′bar·i·za′tion
lum′bri·cal
lum′bri·ca′lis
  ·ca′les
lum′bri·coid′
lu′men
  ·mi·na *or* ·mens
lu′mi·nal
lu′mi·nance
lu′mi·nesce′
  ·nesced′ ·nesc′ing
lu′mi·nes′cence
lu′mi·nes′cent
lu′mi·nif′er·ous
lu′mi·nos′i·ty
lu′mi·nous
lump
lum·pec′to·my
  ·mies
lu′na·cy
  ·cies
lu′nar
lu′nate *or* ·nat·ed
lu′na·tic
lunes
lung′wort′
lu′nu·la
  ·lae′
lu′nu·lar
lu′nule
lu′pous
(pertaining to lupus)
lu′pu·lin

lu′pus
(skin disease)
lu′pus er′y·the′ma·to′sus
lupus vul·gar′is
lu′sus na·tu′rae
lu′te·al
lu·te′ci·um
lu′te·in
lu′te·in·i·za′tion
lu′te·in·ize′
  ·ized′ ·iz′ing
lu′te·o·lin
lu′te·o′ma
  ·mas *or* ·ma·ta
lu′te·o·trop′ic *or* ·troph′ic
lu·te′ti·um
lux
  lux *or* lux′es
lux′ate
  ·at·ed ·at·ing
lux·a′tion
L–xy′lu·lo·su′ri·a
ly′ase
ly′can·thrope′
ly′can·throp′ic
ly·can′thro·py
ly′co·per′do·no′sis
  ·ses
ly′co·po′di·um
lye
ly′ing–in′
Lyme arthritis
lymph
lym·phad′e·ni′tis
lym·phad′e·noid′

lym·phan′gi·al
lym′phan·gi′tis
lym·phat′ic
lym′pho·blast′
lym′pho·cyte′
lym′pho·cyt′ic
lym′pho·cy·to′sis
  ·ses
lym′pho·cy·tot′ic
lym′pho·gran′u·lo′ma
  ·mas *or* ·ma·ta
lymph′oid
lym′pho·kine′
lym·pho′ma
  ·mas *or* ·ma·ta
lym′pho·poi·e′sis
  ·ses
ly′o·phil′ic *or* ly′o·phile′
ly·oph′i·li·za′tion
ly·oph′i·lize′
  ·lized′ ·liz′ing
ly′o·pho′bic
ly′o·trope′
ly′ra
lyse
  lysed lys′ing
ly·ser′gic acid
ly·sim′e·ter
ly′sin
ly′sine
ly′sis
  ·ses
ly′so·gen′ic
ly′so·ge·nic′i·ty
ly·sog′e·ny

ly′so·so′mal
ly′so·some′
ly′so·zyme′
lys′sa
lys′sin
ly·te′ri·an
lyt′ic
lyx′ose

M

ma·caque′
mac′er·ate′
  ·at′ed ·at′ing
mac′er·a′tion
Ma′che unit
ma·chine′
Mach number
ma′ci·es
mac′ro·bi·ot′ic
mac′ro·bi·ot′ics
mac′ro·ceph′a·lous
  or ·ce·phal′ic
mac′ro·ceph′a·ly
  ·lies
mac′ro·cyst′
mac′ro·cyte′
mac′ro·cyt′ic
mac′ro·dont′ or
  mac′ro·don′tic
mac′ro·don′ti·a
ma′cro·gam′ete
mac′ro·mo·lec′u·lar
mac′ro·mol′e·cule′
  or mac′ro·mole′

mac′ro·nu′cle·ar
mac′ro·nu′cle·us
  ·cle·i or ·cle·us·es
mac′ro·nu′tri·ent
mac′ro·phage′
mac′ro·phag′ic
mac′ro·scop′ic or
  ·scop′i·cal
mac′ro·spo·ran′
  gi·um
  ·gi·a
mac′ro·spore′
mac′ro·struc′tur·al
mac′ro·struc′ture
mac′u·la
  ·lae′ or ·las
macula lu′te·a
  ·te·ae
mac′u·lar
mac′u·late adj.
mac′u·late′
  ·lat′ed ·lat′ing
mac′u·la′tion
mac′ule
mad
  mad′der mad′dest
mad′a·ro′sis
  ·ses
mad′den
mad′der
ma·des′cent
mad′house′
mad′i·dans′
ma·du′ro·my·co′sis
  ·ses
ma·gen′ta
mag′got

mag′got·y
mag′is·ter′y
  ·ies
mag′is·tral
mag′ma
mag·mat′ic
mag·ne′sia
mag·ne′sian or
  ·ne′sic
mag′ne·site′
mag·ne′si·um
mag′net
mag·net′ic
mag′net·ism
mag′net·iz′a·ble
mag′net·i·za′tion
mag′net·ize′
  ·ized′ ·iz′ing
mag′net·iz′er
mag′ne·tom′e·ter
mag·ne′to·met′ric
mag′ne·tom′e·try
mag′ne·ton′
mag·ne′to·ther′a·py
mag′ni·fi·ca′tion
mag′ni·fi′er
mag′ni·fy′
  ·fied′ ·fy′ing
mag′ni·tude′
mag′num
mag′nus
maid′en·head′
maim
main
main′line′
  ·lined′ ·lin′ing
main′lin′er

main·tain′
main·tain′er
main′te·nance
mai′zen·ate′
ma′jor
make
　made　mak′ing
ma′la
　(sing. ma′lum)
mal′ab·sorp′tion
mal′a·chite′
ma·la′ci·a
ma·la′cic
mal′ad·ap·ta′tion
mal′a·dapt·ed
mal′a·dap′tive
mal′ad·just′ed
mal′ad·jus′tive
mal′ad·just′ment
mal′a·dy
　·dies
ma·laise′
ma′lar
ma·lar′i·a
ma·lar′i·al or
　·lar′i·an or
　·lar′i·ous
mal′ate
mal′a·thi′on
mal′ax·ate′
　·at′ed　·at′ing
mal′ax·a′tion
mal de mer′
male
male fern
ma·le′ic acid
mal′e·rup′tion

mal′for·ma′tion
mal′formed′
mal·func′tion
mal′ic acid
ma·lign′
ma·lig′nance
ma·lig′nan·cy
　·cies
ma·lig′nant
ma·lig′ni·ty
　·ties
ma·lin′ger
ma·lin′ger·er
mal′le·a·bil′i·ty
　·ties
mal′le·a·ble
mal·le′o·lar
mal·le′o·lus
　·li′
mal′let
mal′le·us
　·le·i′
mal′low
mal·nour′ished
mal′nu·tri′tion
mal′oc·clu′sion
ma·lon′ic acid
Mal·pigh′i·an body
mal·po·si′tion
mal·prac′tice
mal·prac′ti·tion·er
mal′pre·sen·ta′tion
malt
Mal′ta fever
malt′ase
Mal·thu′sian
mal′tose

mal·treat′
ma′lum
mal·un′ion
mam′ba
mam′e·lon
mam′e·lo·na′tion
mam′ma
　·mae
mam′mal
mam·ma′li·an
mam′ma·plas′ty
　·ties
mam′ma·ry
mam·mif′er·ous
mam·mil′la
　·lae
mam′mil·lar′y
mam′mil·late′ or
　·lat′ed
mam′mil·la′tion
mam·mil′li·plas′ty
　·ties
mam′mo·gram′
mam·mog′ra·phy
　·phies
mam′mose
man·chette′
man′chi·neel′
man′di·ble
man·dib′u·la
　·lae′
man·dib′u·lar
man′drake
man′drel or ·dril
man′drin
man′du·cate
man′du·ca′tion

man′du·ca·to′ry
ma·neu′ver
ma·neu′ver·a·bil′i·ty
ma·neu′ver·a·ble
man′ga·nate′
man′ga·nese′
man·gan′ic
man′ga·nite′
man′gan·ous
mange
man′gy
  ·gi·er ·gi·est
man′hood′
ma′ni·a
ma′ni·ac′
ma·ni′a·cal
man′ic
man′ic–de·pres′sive
man′i·fest′
man′i·fes·ta′tion
man′i·kin
ma·nip′u·la·ble or
  ·lat′a·ble
ma·nip′u·lar
ma·nip′u·late′
  ·lat′ed ·lat′ing
ma·nip′u·la′tion
ma·nip′u·la·tive or
  ·la·to′ry
ma·nip′u·la′tor
man′na
man′ner·ism
man′ni·kin
man′nish
man′nite
man·nit′ic

man′ni·tol′
man′nose
man·nom′e·ter
man′o·met′ric or
  ·met′ri·cal
ma·nom′e·try
ma·nos′co·py
man′so·nel·li′a·sis
man′tle
man′u·al
ma·nu′bri·um
  ·bri·a or ·bri·ums
man′u·dy′na·mom′
  e·ter
ma′nus
ma·ras′mic
ma·ras′mus
mar′ble
  ·bled ·bling
mar′ble·i·za′tion
mar′ble·ize′
  ·ized′ ·iz′ing
marc
mar′cid
mar·gar′ic acid
mar′ga·rine or ·rin
mar′gin
mar′gin·al
mar′gin·ate′ or
  ·at′ed
mar′gin·ate′
  ·at′ed ·at′ing
mar′gin·a′tion
mar·gin′o·plas′ty
mar′go
  mar′gi·nes′

ma′ri·jua′na or
  ·hua′na
mar′i·tal
mark′er
mar·mo·ra′tion
mar·mo′re·al or
  ·re·an
mar′row
mar′row·bone′
mar·su′pi·al·i·za′
  tion
mar·su′pi·al·ize′
  ·ized′ ·iz′ing
mar′tial
mas′chal·ad′e·ni′tis
mas′chal·i·a′try
mas′cu·la′tion
mas′cu·line
mas′cu·lin′i·ty
mas′cu·lin′i·za′tion
mas′cu·lin·ize′
  ·ized′ ·iz′ing
mas′cu·lin·o′vo·
  blas·to′ma
  ·mas or ·ma·ta
ma′ser
mask
mas′och·ism
mas′och·ist
mas′och·is′tic
mass
mas′sa
  ·sae
mas·sage′
  ·saged′ ·sag′ing
mas·sag′er
mas·se′ter

mas'se·ter'ic
mas'seur'
mas'seuse'
mas'sive
mas'so·ther'a·py
　·pies
mast cell
mas·tec'to·my
　·mies
Mas'ter two'–step'
　test
mas'tic
mas'ti·cate'
　·cat'ed ·cat'ing
mas'ti·ca'tion
mas'ti·ca'tor
mas'ti·ca·to'ry
mas'ti·goph'o·rous
mas·ti'tis
mas'toid
mas'toid·ec'to·my
　·mies
mas'toid·i'tis
mas·ton'cus
mas'to·pex'y
mas'tur·bate'
　·bat'ed ·bat'ing
mas'tur·ba'tion
mas'tur·ba'tor
mas'tur·ba·to'ry
ma·su'ri·um
match
mate
　mat'ed mat'ing
ma·té' *or* ma·te'
ma·te'ri·al
ma·te'ri·a med'i·ca

ma·ter'nal
ma·ter'ni·ty
　·ties
mat'rass
mat'ri·cal
mat'ri·ces'
　(*sing.* ma'trix)
mat'ri·lin'e·al
ma'trix
　·tri·ces' *or* ·trix·es
matt
mat'ter
mat'toid
mat'u·rant
mat'u·rate'
　·rat'ed ·rat'ing
mat'u·ra'tion
mat'u·ra'tion·al
ma·tur'a·tive
ma·ture'
　·tured' ·tur'ing
ma·tu'ri·ty
　·ties
ma·tu'ti·nal
max·il'la
　·lae
max'il·lar'y
　·lar'ies
max·il'lo·fa'cial
max'il·lot'o·my
max'i·mal
max'i·mum
　·mums *or* ·ma
max'well
may'hem
maz'ard
maze

ma'zic
ma'zin·dol'
ma'zo·dyn'i·a
ma'zo·pla'si·a
Mc·Bur'ney's point
Mc·Nagh'ton test
meal
mean
mea'sles
mea'sly
　·sli·er ·sli·est
meas'ur·a·bil'i·ty
　·ties
meas'ur·a·ble
meas'ure
　·ured ·ur·ing
meas'ure·ment
me·a'tal
me'a·tom'e·ter
me'a·tor'rha·phy
me·at'o·scope
me'a·tos'co·py
me·a'tus
　·tus·es *or* ·tus
me·ben'da·zole'
me·chan'i·cal
me·chan'ics
mech'a·nism
mech'a·no·car'di·
　og'ra·phy
mech'a·no·re·cep'
　tor
mech'a·no·ther'a·
　pist
mech'a·no·ther'a·py
me'cism
Meck'el's cartilage

mec′li·zine′
me·co′ni·um
me′di·a
　·di·ae′
　(coat of a blood vessel)
me′di·a
　(sing. me′di·um*)*
me′di·ad
me′di·al
me′di·an
me′di·as·ti′nal
me′di·as·ti′num
　·na
me′di·ate *adj.*
me′di·ate′
　·at′ed ·at′ing
me′di·a′tion
med′ic
med′i·ca·ble
Med′i·caid′
med′i·cal
med′i·cal·i·za′tion
med′i·cal·ize′
　·ized′ ·iz′ing
med′i·ca·ment
Med′i·care′
med′i·cate′
　·cat′ed ·cat′ing
med′i·ca′tion
med′i·ca′tive
me·dic′i·nal *or*
　·na·ble
med′i·cine
　·cined ·cin·ing
med′i·co′
　·cos′

med′i·co·chi·rur′gi·cal
med′i·co·le′gal
me′di·o·lat′er·al
me′di·o·tar′sal
me′di·sect
med′i·tate′
　·tat′ed ·tat′ing
med′i·ta′tion
me′di·um
　·di·ums *or* ·di·a
me′di·um–sized′
me·dul′la
　·dul′las *or* ·dul′lae
medulla ob′lon·ga′ta
med′ul·lar′y
med′ul·lat′ed
med′ul·la′tion
med′ul·li·za′tion
me·dul′lo·blas·to′ma
　·mas *or* ·ma·ta
mef′e·nam′ic acid
meg′a·ce·phal′ic *or* ·ceph′a·lous
meg′a·ceph′a·ly
　·lies
meg′a·cy′cle
meg′a·death′
meg′a·gam′ete
meg′a·hertz′
meg′a·lo·car′di·a
meg′a·lo·ma′ni·a
meg′a·lo·ma′ni·ac′
meg′a·lo·ma·ni′a·cal

meg′a·lo·man′ic
meg′a·scop′ic
meg′a·spo·ran′gi·um
　·gi·a
meg′a·spore′
meg′a·spor′ic
meg′a·spo′ro·phyll′
meg′a·vi′ta·min
me′grim
mei·o′sis
　·ses *(cell division;* SEE miosis*)*
mei·ot′ic
mel
mel·ag′ra
mel·al′gi·a
mel′an·cho′li·a
mel′an·cho′li·ac′
mel′an·chol′ic
mel′an·chol′y
　·chol′ies
mé·lan·geur′
me·lan′ic
mel′a·nin
mel′a·nism
mel′a·nis′tic
mel′a·nize′
　·nized′ ·niz′ing
mel′a·no·car′ci·no′ma
　·mas *or* ·ma·ta
mel′a·no·cyte′
mel′a·no·cyt′ic
mel′a·noid′
mel′a·no′ma
　·mas *or* ·ma·ta

mel′a·no·phore′
mel′a·no′sis
·ses
mel′a·no·some′
mel′a·not′ic
mel′a·nous
me·las′ma
mel′a·to′nin
me·le′na
mel′·e·nem′e·sis
·ses′
me·le′nic
mel′i·ce′ra or
·ce′ris
me′li·oi·do′sis
me·lis′sa
mel′i·tin
me·li′tis
mel′i·to·pty′a·lism
mel′i·tose or me·lit′
ri·ose
mel′i·tu′ri·a
mel′i·tu′ric
mel·li′tum
·ta
me·lom′e·lus
me·lon′cus
mel′o·plas′ty
·ties
me·lo′ti·a
melt
  melt′ed melt′ing
mem′ber
mem′brane
mem·bra′ni·form
mem′bra·no·car′ti·
lag′i·nous

mem′bra·noid
mem′bra·nous or
  mem′bra·na′ceous
mem′brum
·bra
mem′o·ry
·ries
me·nac′me
men′a·di′one
men·al′gi·a
men·ar′che
me·nar′che·al or
 ·chi·al or ·chal
Men′de·le′ev's law
men′de·le′vi·um
Men·de′li·an
Men′del·ism
Men′del·ist
Men′del's laws
Mé·nière's′
  syndrome
me·nin′ge·al
me·nin′ges
  (sing. me′ninx)
me·nin′gism
men′in·git′ic
men′in·gi′tis
me·nin′go·coc′cal
  or ·coc′cic
me·nin′go·coc′cus
·coc′ci
me·nin′go·my′e·lo·
cele′
men′in·gop′a·thy
me′ninx
  me·nin′ges
me·nis′cal

men·is·ci′tis
me·nis′co·cyte′
me·nis′co·cy·to′sis
·ses
me·nis′cus
·nis′cus·es or ·nis′ci
men′o·paus′al
men′o·pause′
men′or·rha′gi·a
men′or·rha′gic
me·nos′ta·sis or
  men′o·sta′si·a
men′o·tro′pins
men′sal
men′ses
mens sa′na in cor′
  po·re sa′no
men′stru·al
men′stru·ant
men′stru·ate′
 ·at′ed ·at′ing
men′stru·a′tion
men′stru·ous
men′stru·um
 ·stru·ums or ·stru·a
men·su′al
men′sur·a·bil′i·ty
men′sur·a·ble
men′su·ra′tion
men′su·ra′tive
men′tal
men·tal′i·ty
·ties
men·ta′tion
men′thol
men′tho·lat′ed
men′to·la′bi·al

men'ton
men'tu·lag'ra
men'tum
 ·ti
me·per'i·dine'
me·phit'ic
me·phi'tis
me·pro'ba·mate'
mer·bro'min
mer·cap'tan
mer·cap'tide
mer·cap'to
mer·cu'ri·al
mer·cu'ri·al·ism
mer·cu'ri·al·i·za·
 tion
mer·cu'ri·al·ize'
 ·ized' ·iz'ing
mer·cu'ric
mer·cu'rous
mer'cu·ry
mer'cy
me·rid'i·an
me·rid'i·o·nal
mer·is'tic
mer'o·blas'tic
mer'o·crine
mer'o·zo'ite
me'sad
mes·ar·ter·i'tis
mes·at'i·pel'lic or
 ·pel'vic
mes·ax'on
mes·cal'
mes'ca·line'
mes'en·ce·phal'ic

mes'en·ceph'a·lon'
 ·la
mes·en'chy·mal
mes'en·chyme'
mes·en·ter'ic
mes·en'ter·on'
 ·ter·a
mes·en·ter·on'ic
mes·en'ter·y
 ·ter·ies
mesh'work'
me'si·ad
me'si·al
me'si·o·dis'toc·clu'
 sal
me'si·on
me'si·o·ver'sion
mes·mer'ic
mes'mer·ism
mes'mer·ist
mes'mer·i·za'tion
mes'mer·ize'
 ·ized' ·iz'ing
mes'mer·iz'er
mes'o·blast'
mes'o·blas'tic
mes'o·ce·phal'ic or
 ·ceph'a·lous
mes'o·cra'ni·al or
 ·cra'nic
mes'o·cra'ny
mes'o·derm'
mes'o·der'mal or
 ·der'mic
mes'o·gas'tri·um
 ·tri·a
mes'o·morph'

mes'o·mor'phic
mes'o·mor'phism
mes'o·mor'phy
 ·phies
mes'on
me·son'ic
mes'o·the'li·o'ma
 ·mas or ·ma·ta
mes'o·the'li·um
 ·li·a
mes'o·tho'ri·um
mes'o·tron'
mes'o·tron'ic
mes'o·var'i·um
 ·i·a
met·a·bol'ic
me·tab'o·lism
me·tab'o·lite'
me·tab'o·liz'a·ble
me·tab'o·lize'
 ·lized' ·liz'ing
me·tab'o·lous
met'a·car'pal
met'a·car'pus
 ·pi
met'a·chro·mat'ic
met'a·chro'ma·tism
met'a·gen'e·sis
 ·ses'
met'a·ge·net'ic
met·ag·glu'ti·nin
met'a·ic·ter'ic
met'al
met'al·ize'
 ·ized' ·iz'ing
me·tal'lic
met'al·line

met′al·loid′
met′al·los′co·py
me·tal′lo·ther′a·py
·pies
met′a·mer
(chemical compound)
met′a·mere′
(body segment)
met′a·mer′ic
me·tam′er·ism
met′a·mor′phic
met′a·mor′phism
met′a·mor′phose
·phosed ·phos·ing
met′a·mor′pho·sis
·ses′
met′a·neph′ric
met′a·neph′ros
·roi
met′a·phase′
met′a·phos′phate
met′a·phos·phor′ic acid
met′a·pla′sia
met′a·plasm
met′a·plas′mic
met′a·plas′tic
met′a·pro′tein
met′a·psy′cho·log′i·cal
met′a·psy·chol′o·gy
met′a·sta′ble
me·tas′ta·sis
·ses′
me·tas′ta·size′
·sized′ ·siz′ing

met′a·stat′ic
met′a·tar′sal
met′a·tar′sus
·tar′si
me·tath′e·sis
·ses
met′a·thet′ic or ·thet′i·cal
met′a·zo′an
met′en·ce·phal′ic
met′en·ceph′a·lon′
·la
me′te·or·ism
me′te·or′o·pa·thol′o·gy
me′ter
me′ter–kil′o·gram–sec′ond
meth·ac′ry·late′
meth′a·done′
meth′am·phet′a·mine′
meth′ane
meth′a·no·gen′
meth′a·no·gen′ic
meth′a·nol′
me·than′the·line′
meth·aq′ua·lone′
met·he′mo·glo′bin
met·he′mo·glo′bi·ne′mi·a
me·the′na·mine′
me·thi′o·nine′
meth′od
meth′od·ol′o·gy
meth′o·ma′ni·a
meth′o·trex′ate

meth·ox′ide
meth·ox′y·chlor′
meth′yl
meth′yl·al′
meth′yl·a·mine′
meth′yl·ate′
·at′ed ·at′ing
meth′yl·a′tion
meth′yl·a′tor
meth′yl·ene′
me·thyl′ic
methyl i′so·bu′tyl ketone
me·top′ic
me·to′pi·on
met′o·po·dyn′i·a
met′o·pon′ hydrochloride
me′tra
·trae
me·tral′gi·a
me·tra·to′ni·a
met′re·chos′co·py
me′tri·a
me′tric
me′tri·o·ce·phal′ic
me·tri′tis
me·tro·path′ic
me·trop′a·thy
·thies
me′tro·pto′sis
·ses
me′tror·rha′gi·a
me′tro·ste·no′sis
·ses
met′ro·tome′
met′ur·e·de′pa

mev *or* Mev
mev′a·lon′ic acid
mez′ca·line′
me·ze′re·um *or*
·re·on
mho
mi·as′ma
·mas *or* ·ma·ta
mi·as′mal *or* mi′as·
mat′ic *or* mi′as′
mic
mi′ca
mi·ca′ceous
mi·ca′tion
mi′ca·to′sis
mi·cel′la
·lae
mi·cel′lar
mi·celle′
mi′cra
(*sing.* mi′cron)
mi′cra·cous′tic
mi′cren·ceph′a·lous
mi′cren·ceph′a·ly
or ·ce·pha′li·a
mi′cro·a·nal′y·sis
·ses′
mi′cro·an′a·lyst
mi′cro·bar′
mi′crobe
mi·cro′bic *or* ·bi·al
or ·bi·an
mi′cro·bi′o·log′i·cal
or ·log′ic
mi′cro·bi·ol′o·gist
mi′cro·bi·ol′o·gy
mi′cro·bi·ot′ic

mi′cro·ceph′a·lous
or ·ce·phal′ic
mi′cro·ceph′a·ly *or*
·ce·pha′li·a
mi′cro·chem′is·try
mi′cro·coc′cus
·coc′ci
mi′cro·crys′tal·line
mi′cro·cyte′
mi′cro·cyt′ic
mi′cro·dont′ *or* mi′
cro·dont′ous
mi′cro·dont′ism
mi′cro·en·cap′su·
late′
·lat′ed ·lat′ing
mi′cro·en·cap′su·la′
tion
mi′cro·far′ad
mi′cro·fiche′
mi′cro·gam′ete
mi′cro·gram′
mi′cro·graph′
mi·crog′ra·phy
·phies
mi′cro·mere′
mi·crom′e·ter
mi·crom′e·try
mi′cro·mi′cron
mi′cron
·crons *or* ·cra
mi′cron·ize′
·ized′ ·iz′ing
mi′cro·nu′cle·ar
mi′cro·nu′cle·us
·cle·i′
mi′cro·nu′tri·ent

mi′cro·or′gan·ism
mi′cro·par′a·site′
mi′cro·par′a·sit′ic
mi′cro·pho′to·
graph′
mi′cro·phyte′
mi′cro·phyt′ic
mi′cro·py′lar
mi′cro·pyle′
mi′cro·py·rom′e·ter
mi′cro·ra′di·o·
graph′
mi′cro·ra′di·o·
graph′ic
mi′cro·ra′di·og′ra·
phy
·phies
mi′cro·scope′
mi′cro·scop′ic *or*
·scop′i·cal
mi·cros′co·pist
mi·cros′co·py
mi′cro·sec′ond
mi′cro·so′mal
mi′cro·some′
mi′cro·sphere′
mi′cro·spore′
mi′cro·spor′ic
mi′cro·stam′a·tous
or mi·cros′to·
mous
mi′cro·struc′ture
mi′cro·sur′ger·y
mi′cro·sur′gi·cal
mi′cro·tome′
mi·crot′o·mist

mi·crot′o·my
 ·mies
mi′cro·volt′
mi′cro·wave′
 ·waved′ ·wav′ing
mi′cro·zo′on
 ·zo′a
mi·crur′gic
mi′cror·gy
 ·gies
mic′tion
mic′tu·rate′
 ·rat′ed ·rat′ing
mic′tu·ri′tion
mid′ax·il′la
mid′bod′y
mid′brain′
mid′dle
mid′dle–aged′
middle ear
mid′dle–sized′
midge
midg′et
mid′gut′
mid′line′
mid′rib′
mid′riff
mid′sec′tion
mid′tar′sal
mid′wife′
 ·wives′
mid′wife′ry
mi′graine
mi′grain·eur′
mi′grai·noid′
mi·grain′ous

mi′grate
 ·grat·ed ·grat·ing
mi·gra′tion
mi′gra′tion·al
mi′gra·tor
mi′gra·to′ry
For words beginning mikro–
see MICRO–
mi′kron
mild
mil′dew′
mil′i·a′ri·a
mil′i·ar′y
mi·lieu′
 ·lieus′ *or* ·lieux′
mil′i·um
 ·i·a
milk′i·ness
milk leg
milk of magnesia
milk tooth
milk′y
 ·i·er ·i·est
mil′li·am′me·ter
mil′li·am′pere
mil′li·bar
mil′li·cu′rie
mil′li·e·quiv′a·lent
mil′li·far′ad
mil′li·gram′
mil′li·hen′ry
mil′li·li·ter
mil′li·me′ter
mil′li·mi′cron
mil′li·mol *or* ·mole
mil′li·rem′

mil′li·sec′ond
mil′li·volt′
mi·me′sis
mi·met′ic
mim′ic
 ·icked ·ick·ing
mim′ic·ry
 ·ries
mim·ma′tion
Min′a·ma′ta
 disease
mind
mind′–set′
min′er·al
min′er·al·i·za′tion
min′er·al·ize′
 ·ized′ ·iz′ing
min′er·al·o·cor′ti·coid′
min′gin
min′im
min′i·ma
 (*sing.* min′i·mum)
min′i·mal brain dysfunction syndrome
min′i·mi·za′tion
min′i·mize′
 ·mized′ ·miz′ing
min′i·mum
 ·mums *or* ·ma
min′is·ter
min′i·um
mi′nor
mi·nor′i·ty
 ·ties
mi′nus

mi·nus′cu·lar
mi·nus′cule
min′ute *n.*
mi·nute′ *adj.*
min·u·the′sis
mi·o·car′di·a
mi·o·did′i·mus *or* mi·od′i·mus
mi·o·pra′gi·a
mi·o′sis
·ses *(pupil contraction;* SEE *meiosis)*
mi·ot′ic
mi·ra·cid′i·um
·i·a
mire
mir′ror
mir·yach′it
mis′-ac′tion
mis·an′dri·a
mis·an′thrope′ *or* mis·an′thro·pist
mis·an·throp′ic *or* ·throp′i·cal
mis·an′thro·py
mis·car′riage
mis·car′ry
·ried ·ry·ing
mis·ce·ge·na′tion
mis·ci·bil′i·ty
·ties
mis′ci·ble
mis·di·ag·nose′
·nosed′ ·nos′ing
mis·di·ag·no′sis
·ses

mis·match′
mi·sog′a·mist
mi·sog′a·my
(hatred of marriage)
mi·sog′y·nist
mi·sog′y·nous *or* ·y·nic
mi·sog′y·ny
(hatred of women)
mis·o′ne·ism
mis·o′ne·ist
mis·shap′en
mis·treat′
mis·tu′ra
mite
mi′ter
mith′ri·date′
mith·ra·da′tism
mi·ti·ci′dal
mi′ti·cide′
mit′i·ga·ble
mit′i·gate′
·gat′ed ·gat′ing
mit′i·ga′tion
mit′i·ga′tive
mit′i·ga′tor
mit′i·ga·to′ry
mi′tis
mi·to·chon′dri·al
mi·to·chon′dri·on
·dri·a
mi′tome
mi′to·my′cin
mit′o·plasm
mi·to′sis
·ses

mit′o·some′
mi·tot′ic
mi′tral
mi·tral·i·za′tion
mit′tel·schmerz′
mit′tor
mix
mixed *or* mixt
mix′ing
mix′ture
mne·mon′ic
mne·mon′ics
moan
mo′bile
mo·bil′i·ty
·ties
mo′bi·liz′a·ble
mo′bi·li·za′tion
mo′bi·lize′
·lized′ ·liz′ing
moc′ca·sin
mod
mod′al
mo·dal′i·ty
·ties
mode
mod′el
·eled *or* ·elled
·el·ing *or* ·el·ling
mod′er·ate′
·at′ed ·at′ing
mod′er·ate *adj.*
mod′er·a′tion
mod′er·a′tor
mod′i·fi·ca′tion
mod′i·fi′er

mod'i·fy'
  ·fied' ·fy'ing
mo·di'o·lus
  ·o·li'
mod'u·late'
  ·lat'ed ·lat'ing
mod'u·la'tion
mod'u·la'tor
mod'u·lus
  ·u·li'
mo'dus o'pe·ran'di
mog'i·ar'thri·a
mog'i·la'li·a
mog'i·pho'ni·a
Mohs'
  chemosurgery
  technique
moi'e·ty
  ·ties
moist
mois'ten
mois'ture
mois'tur·ize'
  ·ized' ·iz'ing
mois'tur·iz'er
mol
  (unit of quantity;
  SEE mole)
mo'lal
mo·lal'i·ty
  ·ties
mo'lar
mo·lar'i·ty
  ·ties
mold
mold'y
  ·i·er ·i·est

mole
  (skin spot; SEE
  mol)
mo·lec'u·lar
mo·lec'u·lar'i·ty
mol'e·cule'
mole'skin'
mol'i·la'li·a
mo·li'men
  ·lim'i·na
mol'li·fy'
  ·fied' ·fy'ing
mol'lin
mol·li'ti·es
mol·lusc'a·ci'dal
mol·lus'ci·cide or
  mol·lusc'a·cide
mol·lus'cous
mol·lus'cum
  ·ca
molt
mol'y
mo·lyb'date
mo·lyb'den'ic or
  mo·lyb·de'nous
mo·lyb'de·nize'
  ·nized' ·niz'ing
mo·lyb'de·no'sis
mo·lyb'de·num
mo·lyb'dic
mo·lyb'dous
mo·lys'mo·pho'bi·a
mo'ment
mo·men'tum
  ·tums or ·ta
mom'ism
mon·ac'id

mo'nad
mon·am'ide or
  mon'o·am'ide
mon·am'ine or
  mon'o·am'ine
mo·nar'da
mon·as'ter
mon'ath·e·to'sis
  ·ses
mon'a·tom'ic
mon·au'ral
mon·ax'i·al
mo·ne'cious
mo·nel'lin
mo'ner
mo·ner'u·la
  ·lae
mon'es·thet'ic
Mon'gol or mon'
  gol
Mon·go'li·an or
  mon·go'li·an
Mon'gol·ism or
  mon'gol·ism
Mon'gol·oid' or
  mon'gol·oid'
mo·nil'i·a
  ·i·as or ·i·a
mo·nil'i·al
mon·i·li'a·sis
  ·ses'
mo·nil'i·form'
mo·nil'i·id
mo·nil'i·o'sis
  ·ses
mon'i·tor
mon'o

mon'o·ac'id
mon'o·a·cid'ic
mon'o·am'ide
mon'o·am'ine
mon'o·am·in·er'gic
mon'o·a·tom'ic
mon'o·bas'ic
mon'o·ba·sic'i·ty
mon'o·chord'
mon'o·chro'mat
mon'o·chro·mat'ic
 or ·chro'ic
mon'o·chro'ma·
 tism
mon'o·cle
mon'o·clin'ic
mon'o·clon'al
mo·noc'u·lar
mon'o·cy'clic
mon'o·cyte'
mon'o·cyt'ic
mo·noe'cious
mo·noe'cism
mon'o·fil'a·ment or
 mon'o·fil'
mo·nog'a·my
mon'o·gen'e·sis
 ·ses
mon'o·ge·net'ic
mon'o·gen'ic
mon'o·graph'
mon'o·hy'drate
mon'o·hy'dric
mon'o·lay'er
mon'o·ma'ni·a
mon'o·ma'ni·ac'

mon'o·ma·ni'a·cal
mon'o·mer
mon'o·mer'ic
mon'o·me·tal'lic
mon'o·mo·lec'u·lar
mon'o·mor'phic or
 ·phous
mon'o·nu'cle·ar
mon'o·nu·cle·o'sis
 ·ses
mon'o·pho'bi·a
mon'o·phy·let'ic
mon'o·phy'le·tism
mon'o·ple'gi·a
mon'o·ple'gic
mon'o·ploid'
mon'o·sac'cha·ride'
mon'o·so'di·um
 glu·ta·mate'
mon'o·some'
mon'o·so'mic
mon'o·stome'
mo·nos'to·mous
mon'os·tot'ic
mon'o·ther'mi·a
mo·not'ri·chous
mon'o·typ'ic
mon'o·va'lence or
 ·va'len·cy
mon'o·va'lent
mon·ox'ide
mons
 mon'tes
mons' pu'bis
mon'ster
mon'stri·par'i·ty

mon·stros'i·ty
mon'strous
mons' ven'er·is
mon·tic'u·lus
 ·li
mood
mood'y
 ·i·er ·i·est
moon'–blind'
moon'calf'
moon'–eyed'
moon'–faced'
moon'struck' or
 ·strick'en
mor'al
mo·rale'
mor·am'ent
mor'a·men'ti·a
mor'bid
mor·bid'i·ty
 ·ties
mor·bif'ic or
 ·bif'i·cal
mor·bil'li
mor·bil'li·form'
mor'bus
 ·bi
mor·cel'
 ·celled' ·cel'ling
mor·cel·la'tion or
 ·celle·ment'
mor·da'cious
mor'dan·cy
mor'dant
mo'res
Mor·ga'gni's
 caruncle

mor'gan
morgue
mo'ri·a
mor'i·bund'
mor'i·bund'i·ty
·ties
morn'ing sickness
mo'ron
mo·ron'ic
mo·ron'i·ty *or* mo'ron·ism
mo·rose'
mor·phe'a
(skin disease; SEE morphia)
mor'pheme
mor·phe'mic
mor'phi·a
(morphine; SEE morphea)
mor'phine
mor·phin'ic
mor'phin·ism
mor'pho·gen'e·sis
·ses'
mor'pho·ge·net'ic
mor'pho·log'i·cal *or* ·log'ic
mor·phol'o·gist
mor·phol'o·gy
mor·phom'e·try
·tries
mor'phon
mor·pho'sis
·ses
mor·phot'ic

mors
(genitive mor'tis)
mor'sal
mort
mor'tal
mor·tal'i·ty
·ties
mor'tar
mor·ti'cian
mor'ti·fi·ca'tion
mor'ti·fy'
·fied' ·fy'ing
mor'tis
(death)
mor'tise
(notch)
Mor'ton's neuroma
mor'tu·ar'y
·ar'ies
mor'u·la
·lae
mor'u·lar
mor'u·la'tion
mor'u·loid'
mo·sa'ic
mo·sa'i·cism
mos·qui'to
·toes *or* ·tos
moss'y
mote
moth'er
mo'tile
mo·til'in
mo·til'i·ty
mo'tion
mo'ti·vate'
·vat'ed ·vat'ing

mo'ti·va'tion
mo'ti·va'tive
mo'ti·va'tor
mo'tive
mo·tiv'i·ty
mo'to·fa'cient
mo'to·neu'ron
mo'tor
mo·to'ri·al
mo·tor'ic
mo·tor'ic'i·ty
mo·to'ri·um
·ri·a
mo'tor·me'ter
mo·tor'pa·thy
mot'tle
·tled ·tling
mou·lage'
mound'ing
mount
mourn
mouth
mouths
mouth guard
mouth'-to-mouth'
mouth'wash'
mov'a·bil'i·ty
mov'a·ble *or* move'a·ble
move
moved mov'ing
move'ment
mox'a
mox'i·bus'tion
mu
mu'case
mu'ce·din

mu′cic acid
mu′cid
mu·cif′er·ous
mu·cig′e·nous
mu′ci·lage
mu·ci·lag′i·nous
mu′cin
mu′cin·ase′
mu·cin′o·gen
mu′cin·oid′ *or*
·cin·ous
mu′cin·o′sis
mu′cin·ous
mu·cip′a·rous
mu·ci′tis
mu′coid
mu′co·pol′y·sac′cha·ride′
mu′co·pro′tein
mu′co·pu′ru·lent
mu′co·pus′
mu′cor
mu′co·rin
mu′cor·my·co′sis
mu·co′sa
·sae *or* ·sas
mu·co′sal
mu·co′sin
mu·cos′i·ty
mu′cous
(secreting mucus;
SEE mucus)
mu′co·vis′i·do′sis
mu′cro
mu·cro′nes
mu′cro·nate *or*
·nat′ed

mu′cro·na′tion
mu′cus
(secretion; SEE mucous)
mud′pack′
mu·lat′to
·toes
mu′li·eb′ra
mu′li·eb′ri·ty
·ties
mull
mull′er
Mül′ler's duct
mul·tan′gu·lar *or*
mul′ti·an′gu·lar
mul′ti·cel′lu·lar
mul′ti·fac′tor
mul′ti·fac·to′ri·al
mul′ti·fid
mul′ti·form′
mul′ti·for′mi·ty
·ties
mul′ti·lo′bate *or*
·lo′bar *or* ·lobed
mul′ti·nu′cle·ate *or*
·cle·at′ed *or* ·cle·ar
mul·tip′a·ra
·ras *or* ·rae′
mul·tip′a·rous
mul′ti·pha′sic
mul′ti·ple
mul′ti·plex′
mul′ti·pli·ca′tion
mul′ti·pli·ca′tive
mul′ti·plic′i·ty
·ties

mul′ti·ply′
·plied′ ·ply′ing
mul′ti·va′lence
mul′ti·va′lent
mul′ti·vi′ta·min
mum′mi·fi·ca′tion
mum′mi·fy′
·fied′ ·fy′ing
mum′my
·mies
mumps
mu′mu
(filariasis)
mu mu
(micromicron)
Mun′chau′sen's syndrome
mu′ni·ty
mu′ral
mu′re·ins
mu·rex′ide
mu′ri·ate′
mu·ri·at′ic acid
mu′rid
mu′rine
mur′mur
mus′ca
·cae
mus′ca·cide′
mus′cae vo′li·tan′tes
mus′ca·rine
mus′ca·rin′ic
mus′ca·rin·ism
mus′ce·ge·net′ic *or*
·gen′ic
mus′ci·cide′

147

mus'cid
mus'cle
mus'cle–bound'
mus'cu·lar
mus·cu·la'ris
mus'cu·lar'i·ty
  ·ties
mus'cu·lar·ize'
  ·ized' ·iz'ing
mus'cu·la'tion
mus'cu·la·ture
mus'cu·lin
mus·cu·lo·ap'o·neu·rot'ic
mus·cu·lo·mem'bra·nous
mus·cu·lo·skel'e·tal
mus'cu·lus
  ·li'
mu·sen'na
mush'bite'
mush'room'
mu'si·co·gen'ic
mu'si·co·ther'a·py
mus·si·ta'tion
mus'tard
mu·ta·bil'i·ty
  ·ties
mu'ta·ble
mu'ta·cism
mu·ta·fa'cient
mu'ta·gen
mu·ta·gen'e·sis
  ·ses'
mu·ta·gen'ic
mu'tant
mu'tase

mu'tate
  ·tat·ed ·tat·ing
mu·ta'tion
mu'ta·tive
mute
mu'ti·late'
  ·lat'ed ·lat'ing
mu'ti·la'tion
mu'ti·la·tive
mut'ism
mu'tu·al·ism
my·al'gi·a
my·al'gic
my·as·the·ni'a
myasthenia gra'vis
my'as·then'ic
my'a·to'ni·a
  (lack of muscle tone; SEE *myotonia)*
my·at'o·ny
my·ce'li·al
my·ce'li·oid'
my·ce'li·um
  ·li·a
my'cete
  my·ce'tes
my'ce·the'mi·a
my'ce·tism *or* my'ce·tis'mus
my·ce·to'ma
  ·mas *or* ·ma·ta
my'ce·to·zo'an
my'cid
my'co·bac·te'ri·o'sis
  ·ses

my'co·bac·te'ri·um
  ·ri·a
my'co·log'ic *or* ·log'i·cal
my·col'o·gist
my·col'o·gy
my'co·phage'
my'co·plas'ma
  ·mas *or* ·ma·ta
my'co·plas'mal
my'co·pus'
my'cose
my·co'sis
  ·ses
my·cot'ic
my'co·tox'in
my·da'le·ine
my·dri'a·sis
  ·ses'
myd·ri·at'ic
my·ec'to·my
my·ec·to'pi·a *or* my·ec'to·py
my·el·aux'e
my·e·le'mi·a
my'e·len·ceph'a·lon'
  ·la
my·el'ic
my'e·lin
my'e·li·na'tion *or* ·li·ni·za'tion
my'e·lin'ic
my'e·lin·o·gen'e·sis
my'e·lin·o·ge·net'ic
my'e·lit'ic
my'e·li'tis

my′e·lo·blast′
my′e·lo·cele′
my′e·lo·gen′ic or
·log′e·nous
my′e·lo·gram′
my′e·log′ra·phy
·phies
my′e·loid′
my′e·lo′ma
·mas or ·ma·ta
my′e·lom′a·tous
my′e·lon′ic
my′e·lop′e·tal
my′e·lo·pore′
my′e·lot′o·my
my′e·lo·tox′ic
my′e·lo·tox′in
my′en·ter′ic
my′en·ter·on′
my·i′a·sis
·ses′
my′i·o·des·op′si·a
my·lo′dus
my′lo·hy′oid
my′o·car′di·al or
·car′di·ac
my′o·car′di·o·gram′
my′o·car′di·o·
graph′
my′o·car·di′tis
my′o·car′di·um
·di·a
my′o·cele′
(hernia)
my′o·clon′ic
my·oc′lo·nus

my′o·coele′
(cavity)
my′o·fi′bril
my′o·fi·bril′la
·lae
my′o·gen′ic
my·og′li·a
my′o·glo′bin
my′o·graph′
my′oid
my′o·log′ic or
·log′i·cal
my·ol′o·gy
my·o′ma
·mas or ·ma·ta
my·om′a·tous
my′o·neu′ral
my·op′a·thy
·thies
my′ope
my·o′pi·a
my·op′ic
my′o·por·tho′sis
my′o·sin
my′o·tac′tic
(of muscular sense;
SEE myotatic)
my·ot′a·sis
my′o·tat′ic
(of muscular
stretching; SEE
myotactic)
my′o·tome′
my′o·to′ni·a
(muscular spasm;
SEE myatonia)
my′o·ton′ic

myr·iach′it
myr′i·a·pod
myr′i·ap′o·di′a·sis
my·rin′ga
my·rin′go·dec′
to·my or myr′in·
gec′to·my
·mies
my·rin′go·plas′ty
·ties
my·rin′go·tome′
myr′in·got′o·my
·mies
myr·me′ci·a
myrrh
myr′ti·form′
my′so·phil′i·a
my′so·pho′bi·a
my′so·pho′bic
my′ta·cism
myth′o·ma′ni·a
myth′o·ma′ni·ac
myth′o·pho′bi·a
myt′i·lo·tox′ism
my·u′rous
myx·e′de′ma
·mas or ·ma·ta
myx·e′de·ma·tous
myx·id′i·o·tie or
·o·cy
myx′i·o′sis
myx′o·chon′dro·fi′
bro·sar·co′ma
·mas or ·ma·ta
myx′oid
myx·o′ma
·mas or ·ma·ta

149

myx'o·ma·to'sis
　·ses
myx·o'ma·tous
myx'o·my'cete
　·my·ce'tes
myx'o·my·ce'tous
myx'o·spore'
myx'o·vi'rus
my·ze'sis

N

na'cre·ous
nail
na'ked
nal·or'phine
nal·ox·one'
nal·trex'one
na'nism
na'noid
na'no·sec'ond
na'nous
　(dwarfish)
na'nus
　(a dwarf)
nap
　napped nap'ping
nape
naph'tha
naph'tha·lene' *or*
　·lin'
naph'tha·len'ic
naph'thol
naph'thyl
na'pi·form'

nap'kin
nap'ra·path'
na·prap'a·thy
nar'ce·ine'
nar'cis·sism *or* nar'
　cism
nar'cis·sist
nar'cis·sis'tic
nar'co·a·nal'y·sis
　·ses'
nar'co·lep'sy
　·sies
nar'co·lep'tic
nar·co'sis
　·ses
nar'co·syn'the·sis
　·ses'
nar·cot'ic
nar'co·tism
nar'co·ti·za'tion
nar'co·tize'
　·tized' ·tiz'ing
nar'es
　(sing. nar'is)
nar'i·al *or* nar'ine
nar'row
na'sal
na·sal'i·ty
nas'cence *or*
　·cen·cy
nas'cent
na'si·al
na'si·on'
na'so·fron'tal
na'so·gas'tric
na'so·pha·ryn'ge·al
na'so·phar'ynx

na'tal
na·tal'i·ty
　·ties
na'tes
　(sing. na'tis)
na'ti·mor·tal'i·ty
　·ties
na'tive
na·tiv'i·ty
na·tre'mi·a
na·trif'er·ic
na'tron
nat'u·ral
na'ture
na'tur·o·path'
na'tur·o·path'ic
na'tur·op'a·thy
nau'se·a
nau'se·ant
nau'se·ate'
　·at'ed ·at'ing
nau'se·a'tion
nau'seous
na'vel
na·vic'u·lar
near point
near'sight'ed
ne'ar·thro'sis
　·ses
neb'u·la
　·lae' *or* ·las
neb'u·lar
neb'u·li·za'tion
neb'u·lize'
　·lized' ·liz'ing
neb'u·liz'er

neb·u·los′i·ty
  ·ties
neb′u·lous *or* ·lose′
ne·ca′tor
ne·ces′si·ty
  ·ties
neck′lace
nec′ro·bi·o′sis
  ·ses
nec′ro·bi·ot′ic
nec′ro·log′i·cal
ne·crol′o·gy
  ·gies
nec′ro·pha′gi·a
ne·croph′a·gous
nec′ro·phile′
nec′ro·phil′i·a *or*
  ne·croph′i·lism
nec′ro·phil′i·ac *or*
  ne·croph′i·lous
nec′ro·pho′bi·a
nec′rop·sy
  ·sies
ne·cros′co·py
  ·pies
ne·crose′
  ·crosed′ ·cros′ing
ne·cro′sis
  ·ses
ne·crot′ic
ne·crot′o·my
  ·mies
nee′dle
  ·dled ·dling
neem
ne·gate′
  ·gat′ed ·gat′ing

neg′a·tive
neg′a·tiv·ism
neg′a·tiv·is′tic
neg′a·tron′
neg′li·gence
Ne′gri bodies
neigh′bor·wise′
neis·se′ri·a
  ·ae
nem
ne′ma
nem′a·thel′minth
nem·mat′ic
nem′a·ti·za′tion
nem′a·to·cyst′
nem′a·tode′
nem′a·to·di′a·sis
  ·ses
nem′a·toid′
nem′a·tol′o·gy
ne′o·an·throp′ic
ne′o·ars·phen′a·
  mine′
ne′o·cor′tex
ne′o·dym′i·um
ne′o·gen′e·sis
  ·ses′
ne′o·ge·net′ic
ne·ol′o·gism *or* ·gy
ne·ol′o·gis′tic *or*
  ·gis′ti·cal
ne′o·my′cin
ne′on
ne′o·na′tal
ne′o·nate′
ne′o·na·tol′o·gy
ne′o·phil′i·a

ne′o·pho′bi·a
ne′o·pla′sia
ne′o·plasm
ne′o·plas′tic
ne′o·plas′ty
ne′o·stig′mine
ne′o·stri·a′tum
ne′o·te′nic
ne·ot′e·nize′
  ·nized′ ·niz′ing
ne·ot′e·ny
ne·pen′the *or* ·thes
ne·pen′the·an
neph′e·lom′e·ter
ne·phral′gi·a
neph′ra·to′ni·a *or*
  ne·phrat′o·ny
ne·phrec′to·mize′
  ·mized′ ·miz′ing
ne·phrec′to·my
  ·mies
ne·phrid′i·al
ne·phrid′i·um
  ·i·a
ne·phrit′ic
ne·phri′tis
neph′ro·gen′ic
neph′roid
neph′ro·li·thot′
  o·my
  ·mies
ne·phrol′o·gy
neph′ron
ne·phro′sis
  ·ses
ne·phrot′ic

151

ne·phrot′o·my
·mies
neph′ro·tox′in
nep·tu′ni·um
ner′o·li oil
ner·va′tion
nerve
ner′vi
(sing. ner′vus)
ner′vi·mo′tion
ner′vine
ner′vone
ner·vos′i·ty
nerv′ous
(of the nerves; SEE nervus)
ner′vus
·vi (a nerve; SEE nervous)
ne·sid′i·o·blas·to′sis
ness′ler·i·za′tion
ness′ler·ize′
·ized′ ·iz′ing
Ness′ler's reagent
nes·ti·a′tri·a
net′tle rash
net′work′
neu·rag′mi·a
neu′ral
neu·ral′gia
neu·ral′gic
neu·ral′gi·form′
neu′ram·e·bim′e·ter
neu′ra·min′i·dase′
neu·ras·the′ni·a
neu·ras·then′ic

neu·rec′to·my
·mies
neu′ren·ter′ic
neu′rex·er′e·sis
·ses′
neu′ri·lem′ma
neu′ri·lem′mal
neu′ri·le·mo′ma or ·ri·no′ma
·mas or ·ma·ta
neu′rine
neu·rit′ic
neu·ri′tis
neu′ro·a·nat′o·my
neu′ro·bi·ol′o·gy
neu′ro·blast′
neu′ro·coel′ or ·coele′
neu′ro·ep′i·the′li·al
neu′ro·ep′i·the′li·um
·li·a
neu′ro·fi′bril
neu′ro·fi′bril·lar′y
neu′ro·gen′ic
neu·rog′li·a
neu·rog′li·al
neu′ro·hor′mone
neu′ro·hu′mor
neu′ro·hu′mor·al
neu′ro·lep′tic
neu′ro·log′i·cal
neu·rol′o·gist
neu·rol′o·gy
neu·rol′y·sis
·ses′
neu′ro·lyt′ic

neu·ro′ma
·mas or ·ma·ta
neu′ro·mast′
neu′ro·mus′cu·lar
neu′ron or ·rone
neu′ro·nal or neu·ron′ic
neu′ro·path′
neu′ro·path′ic
neu′ro·pa·thol′o·gist
neu′ro·pa·thol′o·gy
neu·rop′a·thy
neu′ro·phys′i·ol′o·gy
neu′ro·psy·chi·at′ric
neu′ro·psy·chi′a·try
neu′ro·psy·chol′o·gy
neu·ro′sis
·ses
neu′ro·sur′geon
neu′ro·sur′ger·y
neu′ro·sur′gi·cal
neu·rot′ic
neu·rot′i·cism
neu·rot′o·my
·mies
neu′ro·tox′ic
neu′ro·tox′in
neu′ro·trans·mit′ter
neu′ro·troph′ic
(of nerve tissue metabolism)
neu·rot′ro·phy

neu·ro·trop′ic
(of an affinity for nerve tissue)
neu·rot′ro·pism *or* ·ro·py
neu′ru·la
·lae
neu′ru·la′tion
neu′tral
neu′tral·i·za′tion
neu′tral·ize′
·ized′ ·iz′ing
neu′tral·iz′er
neu·tri′no
·nos
neu′tron
neu′tro·phil *or* ·phile′
ne′void
ne′vo·li·po′ma
·mas *or* ·ma·ta
ne′vose *or* ·vous
ne′vus
ne′vi
new′born′
new′ton
nex′us
nex′us·es *or* nex′us
ni′a·cin
ni′a·cin′a·mide′
niche
nick′el
Nic′ol prism
ni·co′ti·a′na
nic′o·tin′a·mide′
nic′o·tine′
nic′o·tin′ic

nic′o·tin′ism
nic′tate
·tat·ed ·tat·ing
nic·ta′tion
nic′ti·tate′
·tat′ed ·tat′ing
nic′ti·ta′tion
ni′dal
ni·da′tion
ni′dus
di *or* ·dus·es
ni′fur·ox′ime
night blind′ness
night′dress′
night′gown′
night′guard′
night light
night′mare′
night′shade′
night′shirt′
ni′gra
ni′gri·cans
ni·gri′ti·es′
ni′gro·sine′
ni′hil·ism
ni′hil·ist
ni′hil·is′tic
nim′ble
·bler ·blest
ni·mi′e·ty
ni·o′bic
ni·o′bi·um
ni·o′bous
nip′pers
nip′ple
ni′sin
Nis′sl bodies

ni′sus
nit
ni′ter
ni′ton
ni′trate
·trat·ed ·trat·ing
ni·tra′tion
ni′tric
ni′tride
ni′tri·fi·ca′tion
ni′tri·fi′er
ni′tri·fy′
·fied′ ·fy′ing
ni′trile
ni′trite
ni′tro·bac·te′ri·a
(sing. ·ri·um)
ni′tro·ben′zene
ni′tro·cel′lu·lose′
ni′tro·cel′lu·los′ic
ni′tro·fu′ran
ni′tro·gen
ni′tro·gen·ase′
ni·trog′e·nize′
·nized′ ·niz′ing
ni·trog′e·nous
ni′tro·glyc′er·in *or* ·ine
ni·trom′e·ter
ni′tros·a·mine′
ni′tro·sate′
·sat·ed ·sat·ing
ni′tro·sa′tion
ni′tro·syl
ni′trous
no·bel′i·um
no·car′di·a

no·car·di·o′sis
  ·ses
no′ci·cep′tive
no′ci·cep′tor
noc·tam′bu·lism *or* ·tam′bu·la′tion
noc·tam′bu·list
noc′ti·pho′bi·a
noc·tur′nal
noc′u·ous
nod
  nod′ded nod′ding
nod′al
no·dal′i·ty
node
no′di
  (*sing.* no′dus)
no′dose
no·dos′i·ty
nod′u·lar *or* ·u·lose′ *or* ·u·lous
nod′ule
nod′u·lus
  ·li
no′dus
  ·di
no·e′ma·tach′o·graph′
no·e′ma·ta·chom′e·ter
no·e′mat′ic
no·e′sis
no·et′ic
noise
noi′some
no′li me tan′ge·re
no′ma

no·mad′ic
no′mad·ism
no′men·cla′ture
Nom′i·na An′a·tom′i·ca
nom′i·nal
no′mo·graph′ *or* no′mo·gram′
nom′o·log′i·cal
no·mol′o·gy
nom′o·thet′ic *or* ·thet′i·cal
non′a·ge·nar′i·an
no′nan
non′ com′pos men′tis
non·con·duc′tor
non′dair′y
non′dis·junc′tion
non′e·lec′tro·lyte′
non′in·va′sive
non′o·paque′
non·po′lar
non′pro·pri′e·tar′y
non·pro′tein
non′re·straint′
non′spe·cif′ic
non·stri′at·ed muscle
non·tox′ic
non·un′ion
non·vi′a·ble
non′yl′
nor′ad·ren′a·line
nor′ad·ren·er′gic
nor′ep·i·neph′rine

norm
nor′ma
  ·mae
nor′mal
nor′mal·cy
  ·cies
nor·mal′i·ty
  ·ties
nor′mal·i·za′tion
nor′mal·ize′
  ·ized′ ·iz′ing
nor′mal·iz′er
norm′a·tive
nor′mo·cap′ni·a
nor′mo·cap′nic
nor′mo·ten′sive
nose′bleed′
nose drops
nose′piece′
nos′o·co′mi·al
no·sog′ra·pher
no·sog′ra·phy
  ·phies
nos′o·log′ic *or* ·log′i·cal
no·sol′o·gy
nos′o·ma′ni·a
nos·tal′gia
nos·tal′gic
nos′to·log′ic
nos·tol′o·gy
nos′to·ma′ni·a
nos′tril
nos′trum
no′tal
no·tal′gia
notch

note
   not′ed not′ing
no′ten·ce·phal′o·cele′
no′ti·fi′a·ble
no′ti·fy′
   ·fied′ ·fy′ing
no′to·chord′
no′to·chord′al
nou′me·nal
nou′me·non
   ·me·na
nour′ish
nour′ish·er
nour′ish·ment
nous
No′vo·cain′
   (trademark)
nox′a
   ·ae
nox′ious
nu·bec′u·la
nu′bile
nu·bil′i·ty
nu′cha
   ·chae
nu′chal
nu′cle·ar
nu′cle·ase′
nu′cle·ate *adj.*
nu′cle·ate′
   ·at′ed ·at′ing
nu′cle·a′tion
nu′cle·a′tor
nu′cle·i′
   (sing. nu′cle·us*)*
nu·cle′ic acid

nu′cle·in
nu′cle·oid
nu′cle·o·lar
nu′cle·ole
nu·cle·ol′i·form′
nu·cle′o·lus
   ·li′
nu′cle·on′
nu′cle·on′ic
nu′cle·on′ics
nu′cle·o·phile′
nu′cle·o·phil′ic
nu′cle·o·plasm
nu′cle·o·plas′mic
nu′cle·o·pro′tein
nu′cle·o·side′
nu′cle·ot′i·dase′
nu′cle·o·tide′
nu′cle·us
   ·cle·i′ *or* ·cle·us·es
nu′clide
nu·clid′ic
nude
nud′ism
nud′ist
nu′di·ty
nul·lip′a·ra
   ·a·ras *or* ·a·rae′
nul′li·par′i·ty
nul·lip′a·rous
numb
num′ber
nu′mer·al
nu·mer′i·cal
nu′min·os′i·ty
num′mi·form′
num′mu·lar

nun·na′tion
nurse
   nursed nurs′ing
nurse′maid′ *or* nurs′er·y·maid′
nurs′er
nurs′er·y
   ·er·ies
nurs′ling *or* nurse′ling
nur′tur·ance
nur′tur·ant *or* ·tur·al
nur′ture
   ·tured ·tur·ing
nur′tur·er
nu·ta′tion
nu·ta′tion·al
nu′tri·ent
nu′tri·lites′
nu′tri·ment
nu·tri′tion
nu·tri′tion·al
nu·tri′tion·ist
nu·tri′tious
nu′tri·tive
nu′tri·ture′
nux′ vom′i·ca
nyc′ta·lo′pi·a
nyc′ta·lop′ic
nyc′ter·ine′
nyc′to·pho′bi·a
ny′lon
nymph
nym′pha
   ·phae
nymph′al *or* ·e·an

nym·phec′to·my
nymph′et
nym·phet′ic
nym′pho·lep′sy
nym′pho·lept′
nym′pho·lep′tic
nym′pho·ma′ni·a
nym′pho·ma′ni·ac′
nys·tag′mic
nys·tag′mo·graph′
nys·tag′moid
nys·tag′mus
nys′ta·tin
nyx′is
·es

O

oa′kum
o·a′sis
·ses
oath
oaths
ob′ce·ca′tion
ob′dor·mi′tion
ob·duc′tion
o·be′li·ac
o·be′li·ad
o·be′li·on
·li·a
o·bese′
o·be′si·ty
·ties
o′bex

ob′fus·cate′
·cat′ed ·cat′ing
ob′fus·ca′tion
ob′ject n.
ob·jec′ti·fy′
·fied′ ·fy′ing
ob·jec′tive
ob·jec′tiv·ize′
·ized′ ·iz′ing
ob′late
ob′li·gate adj.
ob′li·gate′
·gat′ed ·gat′ing
ob·lig′a·to′ry
ob·lique′
ob·liq′ui·tous
ob·liq′ui·ty
·ties
ob·lit′er·ate′
·at′ed ·at′ing
ob·lit′er·a′tion
ob·lit′er·a′tive
ob′long
ob′lon·ga′ta
·tas or ·tae
ob′lon·ga′tal
ob·nu′bi·la′tion
ob·scu′rant or ob′
scu·ran′tic
ob′scu·ran′tism
ob′scu·ra′tion
ob·scure′
·scured′ ·scur′ing
ob·serv′a·ble
ob′ser·va′tion
ob·serve′
·served′ ·serv′ing

ob·serv′er
ob·ser′vo·scope′
ob·sess′
ob·ses′sion
ob·ses′sion·al
ob·ses′sive
ob·ses′sive–com·
pul′sive
ob′so·lesce′ ·lesc′ing
·lesced′
ob′so·les′cence
ob′so·les′cent
ob′so·lete′
ob·stet′ric or
·stet′ri·cal
ob′ste·tri′cian
ob·stet′rics
ob′sti·na·cy
·cies
ob′sti·nate
ob′sti·pate′
·pat′ed ·pat′ing
ob′sti·pa′tion
ob·struct′
ob·struct′er or
·struc′tor
ob·struc′tion
ob·struc′tive
ob·stru′ent
ob·tund′
ob·tun′dent
ob′tu·rate′
·rat′ed ·rat′ing
ob′tu·ra′tion
ob′tu·ra′tor
ob·tuse′
ob·tu′sion

ob·tu'si·ty
·ties

oc·cip'i·tal

oc'ci·put'
oc·cip'i·ta or oc'ci·puts'

oc·clude'
·clud'ed ·clud'ing

oc·clud'ent

oc·clu'der

oc·clu'sal

oc·clu'sion

oc·clu'sive

ōc·clu'so·cer'vi·cal

oc·clu'som'e·ter

oc·cult'

oc'cu·pan·cy
·cies

oc'cu·pa'tion

oc'cu·pa'tion·al

oc'cu·py'
·pied' ·py'ing

oc·cur'
·curred' ·cur'ring

oc·cur'rence

o·cel'lar

oc'el·late' or ·lat'ed

oc'el·la'tion

o·cel'lus
·li

och·le'sis

o·chrom'e·ter

o'chro·no'sis or ·sus

o'chro·not'ic

oc'tad

oc'ta·he'dron
·drons or ·dra

oc'tan

oc'tane

oc·ta'ri·us

oc'ta·va'lent

oc'to·ge·nar'i·an

oc'tose

oc'u·lar

oc'u·lar·ist

oc'u·list

oc'u·lo·car'di·ac

oc'u·lo·mo'tor

oc'u·lus
·li

OD
ODs or OD's

OD
OD'd or ODed
OD'ing or ODing

od

o'dax·es'mus

o'dax·et'ic

od·di'tis

od'ic

o'do·gen'e·sis
·ses'

o'don·tal'gi·a

o'don·tal'gic

o·don'tic

o·don'to·blast'

o·don'to·blas'tic

o·don'to·gen'e·sis
or o'don·tog'e·ny

o·don'to·graph'

o·don'toid

o·don'to·log'i·cal

o'don·tol'o·gist

o'don·tol'o·gy

o'don·to'ma
·mas or ·ma·ta

o'don·to'sis
·ses

o'dor

o'dor·ant

o'dor·if'er·ous

o'dor·ous

o'dyn·a·cu'sis

o'dyn·om'e·ter

For words beginning oe– see also E–

Oed'i·pal or oed'i·pal

Oed'i·pus

oer'sted

Oer'tel's treatment

of'fice

of·fi'cial

of·fic'i·nal

o'give

off'spring'
·spring' or ·springs'

ohm

ohm'am·me'ter

ohm'ic

ohm'me'ter

Ohm's law

o·id'i·um
·i·a

oi'ko·ma'ni·a

oi'ko·pho'bi·a

oil

oil'y
  oil'i·er oil'i·est
oint'ment
ol'a·mine'
old
  old'er *or* eld'er
  old'est *or* eld'est
o'le·ag'i·nous
o'le·an'der
o'le·ate'
o·lec'ra·nal
o·lec'ra·non'
o'le·fin *or* ·fine
o'le·fin'ic
o·le'ic acid
o'le·in
o'le·o'
o'le·o·ar·thro'sis
o'le·o·in·fu'sion
o'le·o·mar'ga·rine
  or ·rin
o'le·om'e·ter
o'le·o·res'in
o'le·um
  ·le·a
o·le'yl alcohol
ol·fac'tie *or* ·ty
ol·fac'tion
ol·fac'tive
ol·fac·tol'o·gy
ol·fac·tom'e·ter
ol·fac'to·met'ric
ol·fac·tom'e·try
  ·tries
ol·fac'to·ry
  ·ries
ol'fac·tron'ics

o·lib'a·num
ol'i·ge'mi·a
ol'i·ge'mic
ol'i·go·cho'li·a
ol'i·gop·ne'a
ol'i·go·sac'cha·ride'
ol'i·go·sper'mi·a *or*
  ol'i·go·sper'ma·
  tism *or* ol'i·go·zo'·
  o·sper'ma·tism *or*
  ol'i·go·zo'o·sper'·
  mi·a
ol'i·gu'ri·a
o·lis'thy *or* ·the
o·li'va
  ·vae
ol'i·var'y
ol'ive
ol'i·vif'u·gal
ol'i·vip'e·tal
ol'i·vo·coch'le·ar
ol'o·pho'ni·a
o·ma'gra
o·mal'gi·a
o·me'ga
o·men'tal
o'men·tec'to·my
  ·mies
o·men'to·plas'ty
  ·ties
o·men'tu·lum
  ·la
o·men'tum
  ·ta *or* ·tums
om'i·cron' *or*
  ·kron'
o·mi'tis

om·nip'o·tence
om'ni·vore'
om·niv'o·rous
o'mo·hy'oid
o'mo·pha'gi·a
o·moph'a·gist
o·moph'a·gous *or*
  o'mo·phag'ic
om·phal'ic
om'pha·los
om·phal·o·spi'nous
om·phal·o·trip'sy
o'nan·ism
o'nan·ist
o'nan·is'tic
on'cho·cer·ci'a·sis
  or ·cer·co'sis
  ·ses
on'cho·cer'cid
on'co·gene'
on'co·gen·e'sis
  ·ses'
on'co·ge·net'ic
on'co·gen'ic
on·cog'e·nous
on'co·graph'
on·coi'des
on'co·log'ic
on·col'o·gist
on·col'o·gy
on·col'y·sis
  ·ses'
on'co·lyt'ic
on·cor'na·vi'rus
on·co'sis
on·cot'ic

on·cot′o·my
·mies
on′co·vi′rus
On′dine's curse
o·nei′ric
o·nei′rism
o·nei′ro·a·nal′y·sis
·ses′
o·nei′ro·crit′ic
o·nei′ro·crit′i·cal
o·nei′ro·phre·ni·a
o′ni·o·ma′ni·a
on·kin′o·cele′
on′lay′
on′o·ma·tol′o·gy
·gies
on·to·ge·net′ic or
·gen′ic
on·tog′e·ny or on′to·gen′e·sis
on′y·al′ai
o·nych′i·a or on′y·chi′tis
on′y·cho·gry·po′sis or ·gry·pho′sis
·ses
on′y·choid′
on′y·choph′a·gy or on′y·cho·pha′gi·a
on′y·chor·rhex′is
·es
on′y·cho′sis
·ses
on′y·chot′o·my
on′yx
o·nyx′is
o′o·cyst′

o′o·cyte′
o′o·cy′tin
o·og′a·mous
o·og′a·my
o′o·gen′e·sis
·ses′
o′o·ge·net′ic
o′o·go′ni·um
·ni·a or ·ni·ums
o′o·ki·ne′sis
·ses
o′o·ki·nete′
o′oph·o·rec′to·mize′
·mized′ ·miz′ing
o′o·pho·rec′to·my
·mies
o′o·pho·ri′tis
o′o·sperm′
o′o·the′ca
·cae
o′o·the′cal
o′o·the′co·hys′ter·ec′to·my
o′o·tid
o′o·type′
ooze
oozed ooz′ing
o′o·zo′oid
o·pac′i·fi·ca′tion
o·pac′i·fy′
·fied′ ·fy′ing
o·pac′i·ty
·ties
o′pal·esce′
·esced′ ·esc′ing
o′pal·es′cence

o′pal·es′cent
o·paque′
o′pen
o′pen–heart′ surgery
op′er·a·bil′i·ty
op′er·a·ble
op′er·ant
op′er·ate′
·at′ed ·at′ing
op′er·a′tion
op′er·a′tion·al
op′er·a′tive
op′er·a′tor
op′er·a·to′ry
o·per′cu·lar
o·per′cu·late or ·lat′ed
o·per′cu·lec′to·my
o·per′cu·lum
·la or ·lums
op′er·on′
o·phi′a·sis
o′phi·di′a·sis or o′phi·dism
o′phi·din
oph·ri′tis or oph′ry·i′tis
oph′ry·on′
oph′ry·o′sis
·ses
oph·thal·mag′ra
oph·thal·mal′gi·a
oph·thal·mec′to·my
·mies
oph·thal′mi·a or oph·thal′mi·tis

159

oph·thal'mic
oph·thal'mo·di·a·
  phan'o·scope'
oph·thal'mo·dy·na·
  mom'e·ter
oph·thal'mo·dy·na·
  mom'e·try
oph·thal'mo·log'i·
  cal
oph·thal·mol'o·gist
oph·thal·mol'o·gy
oph·thal'mo·scope'
oph·thal'mo·scop'ic
oph·thal·mos'co·py
o'pi·ate n.
o'pi·ate'
  ·at'ed ·at'ing
o·pin'ion
o'pi·oid'
o·pis'thi·o·ba'si·al
o·pis'thi·on
op'is·thog'na·thism
op'is·thog'na·thous
op'is·thot'ic
o·pis'tho·ton'ic
op'is·thot'o·nos or
  ·nus
o'pi·um
o'pi·um·ism
op'o·bal'sa·mum
op'pi·late'
  ·lat'ed ·lat'ing
op'pi·la'tion
op'pi·la'tive
op·po'nens
op·po'nent

op'por·tun'ist
op'por·tun·is'tic
op·pos'a·bil'i·ty
op·pos'a·ble
op·pose'
  ·posed' ·pos'ing (to
  place opposite;
  SEE appose)
op'po·site
op'po·si'tion
op'sin
op·sin'o·gen
op'si·nog'e·nous
op·son'ic
op·son'i·fi·ca'tion
op·son'i·fy'
  ·fied' ·fy'ing
op'so·nin
op'so·ni·za'tion
op'so·nize'
  ·nized' ·niz'ing
op'so·no·cy'to·
  phag'ic
op'tic
op'ti·cal
op·ti'cian
op'ti·co·cil'i·ar'y
op'tics
op'ti·mal
op'ti·mism
op'ti·mis'tic or
  ·mis'ti·cal
op'ti·mi·za'tion
op'ti·mize'
  ·mized' ·miz'ing
op'ti·mum
  ·mums or ·ma

op'to·a·cous'tic
op·tom'e·ter
op'to·met'ric or
  ·met'ri·cal
op·tom'e·trist
op·tom'e·try
op'to·phone'
op'to·type'
o'ra
  (sing. os)
  (mouths)
o'ra
o'rae
  (edge)
o'rad
o'ral
  (of the mouth; SEE
  aural)
o·ra'le
o'ral·ism
o'ral·ist
o·ral'i·ty
  ·ties
o·ral'o·gy
or'ange
orb
or·bic'u·lar or ·late
  or ·lat'ed
or·bic'u·lar'i·ty
  ·ties
or·bic'u·lus
  ·li
or'bit
or'bit·al
or'bi·tog'ra·phy
or'bi·to·nom'e·try
or'bi·tot'o·my

or′ce·in
or·chid′ic
or′chid·ot′o·my
·mies
or′chi·ec′to·my
·mies
or′chi·o·plas′ty
or·chit′ic
or·chi′tis
or′cin·ol′
or·deal′
or′der
or′der·ly
·lies
or′di·nar′y
or′di·nate
or′dure
o·rec′tic
o·rex′i·a
o·rex′i·gen′ic
orf
or′gan
or′gan·elle′
or·gan′ic
or·gan′i·cism
or·gan′i·cist
or′gan·ism
or·gan·is′mic *or*
·mal
or′gan·iz′a·ble
or′gan·i·za′tion
or′gan·ize′
·ized′ ·iz′ing
or′gan·iz′er
or′ga·no·chlo′rine
or·gan′o·gel

or′ga·no·gen′e·sis
·ses
or′ga·no·ge·net′ic
or′ga·no·graph′ic
or′ga·nog′ra·phy
or′ga·noid′
or′ga·no·lep′tic
or′ga·no·log′ic *or*
·log′i·cal
or′ga·nol′o·gist
or′ga·nol′o·gy
or′ga·no′ma
·mas *or* ·ma·ta
or′ga·no·me·tal′lic
or′ga·non′
·na *or* ·nons′
or·ga·no·phos′phate
or′ga·no·phos′pho·rus
or·gan′o·sol′
or′ga·no·ther′a·py
or′gan·o·troph′ic
(of nutrition)
or′ga·no·trop′ic
(of tissue affinity)
or′ga·not′ro·pism
or ·not′ro·py
or′gan–spe·cif′ic
or′gan·ule
or′ga·num
·nums *or* ·na
or′gasm
or·gas′mic *or*
·gas′tic
or′gone
o′ri·ent′ *v.*
O′ri·en·tal

o′ri·en·tate′
·tat′ed ·tat′ing
o′ri·en·ta′tion
or′i·fice
or′i·fi′cial
or′i·gin
o·rig′i·nal
o·rig′i·nate′
·nat′ed ·nat′ing
o·rig′i·na′tion
o·rig′i·na′tor
o·ri′no·ther′a·py
or′nate
(describing ticks)
or′ni·thine′
or′ni·thi·ne′mi·a
or′ni·tho′sis
·ses
o′ro·pha·ryn′ge·al
o′ro·phar′ynx
or′o·tate′
o·rot′ic aciduria
O·ro′ya fever
or′phan
or′phan·age
or·phen′a·drine
hydrochloride
or·rhol′o·gy
or′rho·men′in·gi′tis
or′ris·root′
or′ther·ga′si·a
or·the′sis
·ses
or′tho·ac′id
or′tho·ce·phal′ic *or*
·ceph′a·lous

161

or′tho·ceph′a·ly
  ·lies
or′tho·chro·mat′ic
or′tho·di′a·graph′
or′tho·don′tic
or′tho·don′tics *or*
  ·ti·a
or′tho·don′tist
or′tho·dox′
or′tho·gen′e·sis
  ·ses′
or′tho·ge·net′ic
or·thog′na·thism
or·thog′na·thous
or·thog′o·nal
or′tho·grade′
or′tho·ker′a·tol′o·gy
or′tho·mo·lec′u·lar
or′tho·pe′dic *or*
  ·pae′dic
or′tho·pe′dics *or*
  ·pae′dics
or′tho·pe′dist *or*
  ·pae′dist
or′tho·phos′phate
or′tho·psy·chi·at′ric
or′tho·psy·chi′a·trist
or′tho·psy·chi′a·try
or·thop′ter·an
or·thop′tic
or·thop′tist
or′tho·scope′
or′tho·scop′ic
or′tho·stat′ic

or·thot′ic
or·thot′ics
or′tho·tist
or·thot′o·nos *or*
  ·nus
or′tho·top′ic
os
  os′sa
  (bone)
os
  o′ra
  (mouth)
o′sa·zone′
os·ce′do
os′che·al
os′che·i′tis *or* os·chi′tis
osch·el·e·phan·ti′a·sis
os′che·o·cele′
os′che·o·lith′
os′cil·late′
  ·lat′ed ·lat′ing
os′cil·la′tion
os′cil·la′tor
os·cil′la·to′ry
os·cil′lo·gram′
os·cil′lo·graph′
os·cil′lo·graph′ic
os·cil′lop′si·a
os·cil′lo·scope′
os·cil′lo·scop′ic
os′cine
os′ci·tan·cy
os′ci·tate′
  ·tat′ed ·tat′ing
os′ci·ta′tion

os′cu·lant
os′cu·lar
os′cu·late′
  ·lat′ed ·lat′ing
os′cu·la′tion
os·cu′la·to′ry
os′cu·lum
  ·la
Os′ler's disease
os′mate
os·mat′ic
os·me′sis
os′mic
os′mics
os′mi·dro′sis
  ·ses
os′mi·o·phil′ic
os′mi·um
os′mol
os·mo′lal
os′mo·lal′i·ty
  ·ties
os·mo′lar
os′mo·lar′i·ty
  ·ties
os′mole
os·mol′o·gy
os·mom′e·ter
os′mo·met′ric
os′mo·re·cep′tor
os′mo·reg′u·la·to′ry
os′mose
  ·mosed ·mos·ing
os·mo′sis
  ·ses
os·mot′ic
os′mous

os·phre′si·o·lag′ni·a
os′phre·si·ol′o·gy
os·phre′sis
os·phret′ic
os′phy·o·my·e′li·tis
os′sa
  (sing. os)
os′sa·ture
os′se·in
os′se·o·fi′brous
os′se·o·mu′cin
os′se·o·mu′coid
os′se·o·so·nom′e·ter
os′se·ous
os′si·cle
os·sic′u·lar *or* ·late
os′si·cu·lec′to·my
  ·mies
os′si·cu·lot′o·my
  ·mies
os·sic′u·lum
  ·la
os·sif′er·ous
os·sif′ic
os′si·fi·ca′tion
os·sif′lu·ence
os′si·fy′
  ·fied′ ·fy′ing
os′te·al
os′te·al′gi·a *or* os·tal′gi·a
os′te·al′gic
os′tec′to·my *or* os′te·ec′to·my
  ·mies
os′te·in
os′te·i′tis

os·tem′bry·on′
os·te′mi·a
os′tem·py·e′sis
os′te·o′ar·thri′tis
os′te·o·blast′
os′te·o·blas′tic
os′te·o·chon·dro′ma
  ·mas *or* ·ma·ta
os′te·oc′la·sis
  ·ses′
os′te·o·clast′
os′te·o·clas′tic
os′te·o·des·mo′sis
os′te·o·gen′e·sis *or* ·te·og′e·ny
os′te·o·gen′ic *or* ·ge·net′ic *or* ·te·og′e·nous
os′te·oid′
os′te·o·log′i·cal
os′te·ol′o·gist
os′te·ol′o·gy
os′te·ol′y·sis
  ·ses′
os′te·o′ma
  ·mas *or* ·ma·ta
os′te·o·ma·la′ci·a
os′te·o·my·e·li′tis
os′te·on′ *or* ·one′
os′te·o·path′
os′te·o·path′ic
os′te·o·pa·thol′o·gy
os′te·op′a·thy
os′te·o·phyte′
os′te·o·phyt′ic
os′te·o·plas′tic
os′te·o·plas′ty

os′te·o·poi′ki·lo′sis
os′te·o·poi′ki·lot′ic
os′te·o·po·ro′sis
  ·ses
os′te·o·po·rot′ic
os′te·o′sis
os′te·o·tome′
os′te·ot′o·my
  ·mies
os′ti·al
os′ti·ar′y
  ·ar′ies
os′ti·o·lar
os′ti·ole′
os′ti·um
  ·ti·a
os′to·mate′
os′to·my
  ·mies
os·to′sis
os·tra′ceous
os′tra·co′sis
os′tre·o·tox′ism *or* ·tox·is′mus
o·tal′gi·a
o·tal′gic
o′tan·tri′tis
o·tec′to·my
  ·mies
ot·he′ma·to′ma
  ·mas *or* ·ma·ta
oth′er–di·rect′ed
o′ti·at′ri·a *or* ·at′rics
o′tic
o′ti·co·din′i·a
o·tit′ic

o·ti'tis
o'to·an·tri'tis
o'to·cyst'
o'to·cys'tic
o'to·gen'ic *or* o·tog'e·nous
o'to·lar'yn·gol'o·gist
o'to·lar'yn·gol'o·gy
o'to·lith' *or* ·lite
o'to·lith'ic
o'to·log'i·cal
o·tol'o·gist
o·tol'o·gy
o'to·my·co'sis
 ·ses
o'to·rhi·no·lar'yn·gol'o·gy
o'to·scle·ro'sis
 ·ses
o'to·scle·rot'ic
o'to·scope'
o'to·scop'ic
o·tos'co·py
o·to'sis
o·tos'te·al
o·tos'te·on'
oua·ba'in
ou·li'tis
ou·lo·ni'tis
ounce
ou·ra'ri
out'break'
out'breed'ing
out'come'
out'cross'ing

out'er
out'flow'
out'grow'
 grew' grown'
 grow'ing
out'growth'
out'let'
out'line'
 lined' lin'ing
out'live'
 lived' liv'ing
out'look'
out of phase
out'pa'tient
out'put'
out'reach'
out'ward
out'weigh'
o'va
 (sing. ovum)
o'val
ov'al·bu'min
o'val·o·cyte'
o·val'o·cy·to'sis
 ·ses
o·var'i·an
o·var'i·ec'to·my
 ·mies
o·var'i·o·cen·te'sis
 ·ses
o·var'i·o·cy·e'sis
 ·ses
o·var'i·ot'o·my
 ·mies
o'va·ri'tis
o'va·ry
 ·ries

o'ver·a·chieve'
 ·chieved' ·chiev'ing
o'ver·a·chiev'er
o'ver·bite'
o'ver·clo'sure
o'ver·com'pen·sate'
 ·sat'ed ·sat'ing
o'ver·com'pen·sa'tion
o'ver·com·pen'sa·to'ry
o'ver·cor·rec'tion
o'ver·de·vel'op
o'ver·do'
 ·did' ·done'
 ·do'ing
o'ver·dose' *n.*
o'ver·dose'
 ·dosed' ·dos'ing
o'ver·due'
o'ver·eat'
 ·ate' ·eat'en
 ·eat'ing
o'ver·ex·ert'
o'ver·ex·pose'
 ·posed' ·pos'ing
o'ver·ex·po'sure
o'ver·ex·tend'
o'ver·ex·ten'sion
o'ver·flow' *n.*
o'ver·flow' *v.*
o'ver·grow'
 ·grew' ·grown'
 ·grow'ing
o'ver·growth'
o'ver·hang' *n.*

o'ver·hang'
 ·hung' ·hang'ing
o'ver·lap' *n.*
o'ver·lap'
 ·lapped' ·lap'ping
o'ver·lay' *n.*
o'ver·lay'
 ·laid' ·lay'ing
o'ver·learn'
o'ver·night' *adj.*
o'ver·night' *adv.*
o'ver·pop'u·late'
 ·lat'ed ·lat'ing
o'ver·pop'u·la'tion
o'ver·pro·duce'
 ·duced' ·duc'ing
o'ver·pro·duc'tion
o'ver·pro·tect'
o'ver·pro·tec'tive
o'ver·re·act'
o'ver·ride'
 ·rode' ·rid'den
 ·rid'ing
o'ver·strain'
o·vert'
o'ver-the–count'er
o'ver·tone'
o'ver·weight'
o'vi·ci'dal
o'vi·cide'
o'vi·duct'
o'vi·duc'tal *or* o'vi·du'cal
o·vif'er·ous
o'vi·form'
o'void *or* o·void'al
o'vo·plasm

o'vo·tes'tis
o'vo·vi·tel'lin
o'vu·lar
o'vu·late'
 ·lat'ed ·lat'ing
o'vu·la'tion
o'vu·la·to'ry
o'vule
o'vu·log'e·nous
ov'u·lum
 ·la
o'vum
 o'va
ox·a·cil'lin
ox'a·late'
ox·al'ic acid
ox'al·ism
ox·al·o·suc·cin'ic acid
ox'i·dant
ox'i·dase'
ox'i·da'sic
ox'i·da'tion
ox'i·da'tion–re·duc'tion
ox'i·da'tive
ox'ide
ox'i·diz'a·ble
ox'i·dize'
 ·dized' ·diz'ing
ox'i·diz'er
ox'ime
ox·im'e·ter
ox·o·ne'mi·a
ox·o·phen·ar'sine
ox'y·ac'id

ox'y·ce·phal'ic *or* ·ceph'a·lous
ox'y·ceph'a·ly
ox'y·e·coi'a
ox'y·gen
ox'y·gen·ase'
ox'y·gen·ate'
 ·at'ed ·at'ing
ox'y·gen·a'tion
ox'y·gen·a'tor
oxygen debt
ox'y·gen'ic *or* ox·yg'e·nous
ox'y·gen·ize'
 ·ized' ·iz'ing
ox'y·he'mo·glo'bin
ox'y·hy'dro·gen
ox·yn'tic
ox'y·op'ter
ox'y·phil *or* ·phile'
ox'y·phil'ic
ox'y·tet'ra·cy'cline
ox'y·to'ci·a
ox'y·to'cic
ox'y·to'cin
ox'y·u·ri'a·sis
 ·ses'
ox'y·u'rid
o·ze'na
o·ze'nous
o'zo·chro'ti·a
o'zone
o·zon'ic *or* o'zo·nous
o'zon·ide'
o'zon·i·za'tion

o′zon·ize′
  ·ized′ ·iz′ing
o′zon·iz′er
o′zo·nom′e·ter
o·zo′no·scope′
o′zo·sto′mi·a

P

pab′u·lar
pab′u·lum
pace
  paced pac′ing
pace′fol′low·er
pace′mak′er
pach′y·chei′li·a or
  ·chi′li·a
pach′y·der′ma or
  ·der′mi·a
pach′y·der′ma·tous
  or ·der′mous
pach′y·e′mi·a or
  ·y·he′mi·a
pach′y·me′ninx
  ·me·nin′ges
pa·chyn′sis
  ·ses
pa·chyn′tic
pach′y·o·nych′i·a
pach′y·tene′
Pa·ci′ni's
  corpuscles
pack′age
pack′er

pad
  pad′ded pad′ding
Padg′ett's
  dermatome
*For words
  beginning* pae– *see*
  PE–
Pag′et's disease
pa′go·pha′gi·a
pain′ful
pain′kill′er
paint
pair
  pairs
pal′a·tal
pal′ate
pa·lat′i·form′
pal′a·tine′
pal′a·to·glos′sus
  ·si
pal′a·tog′na·thous
pal′a·tog′ra·phy
  ·phies
pal′a·tor′rha·phy
  ·phies
pal·a′tum
  ·ta
pale
  pal′er pal′est
pa′le·en·ceph′a·lon
  or
  ·le·o·en·ceph′a·lon
pa′le·on·tol′o·gy
pa′le·o·stri·a′tal
pa′le·o·thal′a·mus
  ·mi′

pal′i·ki·ne′si·a or
  ·ci·ne′si·a
pal′i·nal
pal′in·drome
pal′in·dro′mi·a
pal′in·drom′ic
pal′in·gen′e·sis
  ·ses
pal′i·sade′
pal·la′di·um
pal′la·dous
pall′an·es·the′si·a
pal·les′cence
pal′li·ate′
  ·at′ed ·at′ing
pal′li·a′tion
pal′li·a′tive
pal′li·a′tor
pal′lid
pal′li·dal
pal′li·dum
pal′li·um
  ·li·ums or ·li·a
pal′lor
palm
pal′mar
pal·ma′ris
pal′mic
pal′mi·tate′
pal·mit′ic acid
pal′mi·tin
pal′mo·men′tal
  reflex
pal′mo·plan′tar
pal′mus
  ·mi
pal′pa·bil′i·ty

166

pal′pa·ble
pal′pate
·pat·ed ·pat·ing
pal·pa′tion
pal′pe·bral
pal′pe·brate′
·brat·ed ·brat·ing
pal′pi·tant
pal′pi·tate′
·tat·ed ·tat·ing
pal′pi·ta′tion
pal′sy
·sies
pal′sy
·sied ·sy·ing
pa·lu′dal
pal′u·dism
pal′y
pal′y·nol′o·gy
pam′per
pam·pin′i·form′
pam·ple′gi·a
pan′a·ce′a
pan′a·ce′an
pan′ag·glu′ti·nin
pan·at′ro·phy
·phies
pan·car·di′tis
pan·chro·mat′ic
pan·chro′mi·a
pan′cre·as
pan′cre·a′ta
pan′cre·a·tec′to·my
pan′cre·at′ic
pan′cre·at′i·co·du′o·de′nal
pan′cre·a·tin

pan′cre·a·ti′tis
pan′cre·a·to·du′o·de·nec′to·my
pan′cre·a·to·lyt′ic
 or ·cre·o·lyt′ic
pan′cre·a·tot′o·my
 or ·cre·at′o·my
pan′cre·ec′to·my
pan·dem′ic
pan·dic′u·la′tion
pan·en′do·scope′
Pa′neth's cells
pang
pan·gen′e·sis
pan·ge·net′ic
pan·glos′si·a
pan′hy·drom′e·ter
pan′hy·per·e′mi·a
pan′ic
·icked ·ick·ing
pan′ick·y
pan′ic–strick′en or
 –struck′
pan′i·dro′sis or
 ·hi·dro′sis
pa′nis
 (bread; SEE
 pannus, panus)
pan·mic′tic
pan·mix′i·a
pan·mix′is
pan′my·e·lo′sis
pan′neu·ri′tis
pan·nic′u·li′tis
pan·nic′u·lus
·li′

pan′nus
·ni
 (membrane; SEE
 panis, panus)
pa·nod′ic
pan·op′tic
pan·pho′bi·a or
 pan·o·pho′bi·a
pan·sex′u·al
pan′si·nus·i′tis or
 ·sin·u·i′tis
pan·sper′mi·a
pan·sys·tol′ic
pant
pant′ach·ro·mat′ic
pan·tal′gi·a
pan·ta·tro′phi·a or
 pan·tat′ro·phy
pan·thod′ic
pan′to·graph′
pan·to·then′ate
pan·to·then′ic acid
pa′nus
 (inflamed lymph
 node; SEE panis,
 pannus)
pap
 (food; SEE Pap
 test)
pa·pa′in
Pap′a·nic′o·la′ou's
 stain
pa·pav′er·ine′
pa·pa′ya or pa′paw
pa′per
pa·pes′cent

pa·pil′la
·lae
pap′il·lar′y
pap′il·late′ or ·lose′
pap′il·lec′to·my
·mies
pap′il·le·de′ma
·mas or ·ma·ta
pap′il·lo′ma
·ma·ta or ·mas
pap′il·lo′ma·tous
pa′po·va·vi′rus
pap′pose′ or ·pous
pap′pus
pap′pi
Pap test (or smear)
pap′u·lar
pap′ule
pap′u·lo·er′e·them′a·tous
pap′u·lo·pus′tu·lar
pap′u·lose′
pap′y·ra′ceous
par
par′a
·as or ·ae
par′a–a·mi′no·ben·zo′ic acid
par′a–an′es·the′si·a
par′a·bi·o′sis
·ses
par′a·bi·ot′ic
par′a·blast′
par′a·blas′tic
par′a·blep′si·a
par′a·ca′se·in
Par′a·cel′sus

par′a·cen·te′sis
·ses
par′a·cen·tet′ic
par·ac′me
par′a·cu′sis or par′a·cu′si·a or par′a·cou′sis
par′a·cyt′ic
(of cells; SEE parasitic)
par′a·di·chlo′ro·ben′zene
par′a·dox′
par′a·dox′i·cal
par′aes·the′si·a
par′af·fin
par′af·fine
·fined ·fin·ing
par′af·fi·no′ma
·mas or ·ma·ta
par′a·gan′gli·on
·gli·a
par′a·gen′i·tal
par′ag·no′men
·nom′i·na
par′a·graph′i·a
par′al′de·hyde
par′a·lex′i·a
par·al·lac′tic
par′al·lax′
par′al·lel′
par′al·lel·ism
par′al·lel·om′e·ter
pa·ral′o·gism
pa·ral′o·gis′tic
pa·ral′o·gize′
·gized′ ·giz′ing

pa·ral′y·sis
·ses′
paralysis ag′i·tans
par′a·lyt′ic
par′a·lyz′ant
par′a·ly·za′tion
par′a·lyze′
·lyzed′ ·lyz′ing
par′a·lyz′er
par′a·me′ci·um
·ci·a
par′a·med′ic
par′a·med′i·cal
pa·ram′e·ter
par′a·meth′a·di′one
par′a·met′ric
par′a·me·tri′tis
par′a·me·tri′um
·tri·a
par′·am·ne′sia
par′a·morph′
par′a·mor′phi·a
par′a·mor′phic or ·mor′phous
par′a·my·oc′lo·nus mul′ti·plex
par′a·noi′a
par′a·noi′ac
par′a·noid′
par′a·noi′dal
par′a·no′mi·a
par′a·nor′mal
par′a·phra′si·a
par′a·phra′sic
par·aph′y·sis
·ses
par′a·ple′gi·a

168

par′a·ple′gic
par′a·prax′i·a
par′a·prax′is
·es
par′a·proc·ti′tis
par′a·pro·fes′sion·al
par′a·psy·cho·log′i·cal
par′a·psy·chol′o·gist
par′a·psy·chol′o·gy
par′a·ros·an′i·line
par′a·quat′
par′a·site′
par′a·sit′ic or
·sit′i·cal
(of parasites; SEE paracytic)
par′a·sit′i·ci′dal
par′a·sit′i·cide′
par′a·sit′ism
par′a·sit′i·za′tion
par′a·sit·ize′
·ized′ ·iz′ing
par′a·si·to·gen′ic
par′a·si·to·log′i·cal
par′a·si·tol′o·gist
par′a·si·tol′o·gy
par′a·si·to′sis
·ses
par′a·sym′pa·thet′ic
par′a·sym′pa·tho′mi·met′ic
par′a·syn·ap′sis
·ses
par′a·tax′ic

par′a·tax′is or
·tax′i·es
par′a·ten′on
par′a·thi′on
par′a·thy′mi·a
par′a·thy′roid
par′a·troph′ic
pa·rat′ro·phy
·phies
par′a·ty′phoid
par·ax′i·al
par·ax′on
par′a·zo′an or
·zo′on
par′a·zone′
parch′ment crack′ling
par′ec·ta′si·a
par·ec′ta·sis
·ses′
par′e·gor′ic
pa·rei′ra (bra′va)
pa·ren′chy·ma
pa·ren′chy·mal or
par′en·chym·a′tous
par′ent·age
pa·ren′tal
par·en′ter·al
par′ent·hood′
par′ent·ing
pa·re′sis
·ses
par′es·the′si·a
par′es·thet′ic
pa·ret′ic

pa·ri′es′
pa·ri′e·tes′
pa·ri′e·tal
pa·ri′e·to·oc·cip′i·tal
pa·ri′e·to·squa·mo′sal
pa′ri pas′su
Par′is green
par′i·ty
par′kin·son·ism
Par′kin·son's disease
par′oc·cip′i·tal
par′o·don′tal
par′o·don·ti′tis
pa·role′
·roled′ ·rol′ing
pa·rol·ee′
par·ol′i·var′y
par·o·ni′ri·a or
·nei′ri·a
pa·rot′ic
par·ot′id
par′o·tin
par′o·tit′ic
par′o·ti′tis
par′ous
par′ox·ysm
par·ox·ys′mal
par′ri·ci′dal
par′ri·cide′
Par·rots′ disease
par′ry fracture
pars
par′tes
pars′–pla·ni′tis

par'tal
par·the·no·gen'e·sis
par·the·no·ge·net'ic
par'tial
par'ti·cle
par·tic'u·late
par·ti'tion
part'ner·ship'
part'–time'
par·tu'ri·en·cy
par·tu'ri·ent
par·tu'ri·fa'cient
par·tu·ri'tion
pa·ru'lis
  ·ru'li·des'
par'vule
pas'cal
pas'sage
pas'sion
pas'sive
pas'siv·ism
pas·siv'i·ty
paste
pas'ter
pas'teur·ism
pas'teur·i·za'tion
pas'teur·ize'
  ·ized' ·iz'ing
pas'teur·iz'er
Pas·teur' treatment
pas·tille' or pas'til
past'–point'ing
pas'ty
pa·ta'gi·um
  ·gi·a
patch test
pate

pat'e·fac'tion
pa·tel'la
  ·las or ·lae
pa·tel'lar
pa·tel'late
pa·tel'li·form'
pa'ten·cy
pat'ent
pa·ter'nal
pa·ter'ni·ty
pa·thet'ic or
  ·thet'i·cal
path'find'er
path'ic
path'o·gen or
  ·gene'
path'o·gen'e·sis
  ·ses'
path'o·ge·net'ic
path'o·gen'ic
path'o·ge·nic'i·ty
  ·ties
pa·thog'e·ny
  ·nies
pa·thog·no·mon'ic
path'o·log'i·cal or
  ·log'ic
pa·thol'o·gist
pa·thol'o·gy
  ·gies
pa·tho'sis
  ·ses
path'way'
pa'tience
pa'tient
pat'ri·ci'dal
pat'ri·cide'

pat'ri·lin'e·al
pat'ten
pat'tern
pat'u·lous
paunch
pause
  paused paus'ing
pave'ment·ing
pav'id
pa·vil'ion
Pav·lov'i·an
Pav'lov's method
pa'vor
peak
peak'ed
pea'nut' oil
pearl
pec'cant
pec'tase
pec'tate
pec'ten
  pec'ti·nes'
  (a comblike
  structure; SEE
  pectin)
pec'tic
pec'tin
  (a carbohydrate;
  SEE pecten)
pec'ti·nate' or
  ·nat'ed
pec'ti·na'tion
pec·tin'e·us
pec'tin·ous
pec'to·ral
pec'tus
  ·to·ra

170

ped'al
ped'er·ast'
ped'er·as'tic
ped'er·as'ty
 ·ties
pe'di·at'ric
pe'di·a·tri'cian *or*
 ·di·at'rist
pe'di·at'rics *or* pe'
 di·at'ry
ped'i·cel
ped'i·cel'late
ped'i·cle
pe·dic'u·lar
pe·dic'u·late
pe·dic'u·lo'sis
 ·ses
pe·dic'u·lous
 (lousy)
pe·dic'u·lus
 ·li'
 (a louse)
ped'i·cure'
ped'i·gree'
ped'i·lu'vi·um
pe'do·bar'o·ma·
 crom'e·ter
pe'do·don'tics
pe'do·don'tist
pe'do·log'ic *or*
 ·log'i·cal
pe·dol'o·gist
pe·dol'o·gy
pe·dom'e·ter
pe'do·mor'phism
pe'do·phile'
pe'do·phil'i·a

pe·dun'cle
pe·dun'cu·lar
pe·dun'cu·late *or*
 ·lat'ed
peel
pee'nash
peer review
peg
pe·jo·ra'tion
pe·jo'ra·tive
pe·lade'
pel'age
pel'a·gism
pel'ar·gon'ic acid
pel'id·no'ma *or* pe'
 li·o'ma
pel·la'gra
pel·la'grin
pel·la'grous
pel'let
pel·li·cle
pel·lic'u·lar *or* ·late
pel'li·to'ry
 ·ries
pel·lu'cid
pel·lu'cid'i·ty
pel·mat'ic
pel·mat'o·gram'
pe·lo·ther'a·py *or*
 pe·lop'a·thy
pel'vic
pel'vi·fix·a'tion
pel·vim'e·ter
pel'vi·os'co·py *or*
 pel·vos'co·py

pel'vi·per'i·to·ni'tis
 or
 ·vi·o·per'i·to·ni'tis
pel'vis
 ·vis·es *or* ·ves
pem'o·line'
pem'phi·goid'
pem'phi·gus
 ·gus·es *or* ·gi'
pen'cil
pen'del·luft'
pen'du·lar
pen'du·lous
pen'e·tra·bil'i·ty
 ·ties
pen'e·tra·ble
pen'e·trance
pen'e·trant
pen'e·trate'
 ·trat'ed ·trat'ing
pen'e·tra'tion
pen'e·trom'e·ter
pen'i·cil'la·mine'
pen'i·cil'late *or*
 ·cil'li·form'
pen'i·cil·la'tion
pen'i·cil'lin
pen'i·cil'li·nase'
pen'i·cil'lin–fast'
pen'i·cil'li·um
 ·li·ums *or* ·li·a
pen'i·cil'lus
 ·li
pe'nile
pe'nis
 ·nis·es *or* ·nes
pe·nis'chi·sis

pen'nate
pen'ni·form'
pen'ny·roy'al
pen'ny·weight'
pe'no·scro'tal
pen'sion neurosis
pen'ta·chlo'ro·phe'nol
pen'tad
pen'ta·dac'tyl
pen'ta·e·ryth'ri·tol'
pen·tal'o·gy
pen'tane
pen'ta·ploid'
pen'ta·quine'
pen'ta·tom'ic
pen'ta·va'lent
pen'ta·zo'cine
pen'to·bar'bi·tal' sodium
pen'to·san'
pen'tose
pen'to·side'
pent·ox'ide
pen'tyl
pen'tyl·ene·tet'ra·zol'
pep'los
pe'po
  ·pos
pep'per·mint'
pep'sin
pep'sin·ate'
  ·at·ed ·at·ing
pep·sin'o·gen
pep'tic

pep'ti·dase'
pep'tide
pep'tize
  ·tized ·tiz·ing
pep'tone
pep·ton'ic
pep'to·ni·za'tion
pep'to·nize'
  ·nized' ·niz·ing
pep'to·tox'in
pep'to·zyme'
per·ac'id
per an'num
  (annually)
per a'num
  (through the anus)
per·ar·tic'u·la'tion
per·bo'rate
per·bor'ic acid
per·cent' *or* per cent *or* per cent. *or* per cen'tum
per·cent'age
per·cen'tile
per'cept
per·cep'ti·bil'i·ty
  ·ties
per·cep'ti·ble
per·cep'tion
per·cep'tion·al
per·cep'tive
per·cep·tiv'i·ty
  ·ties
per·cep'tu·al
per·chlo'rate
per·chlo'ric acid
per·chlo'ride

per'co·late'
  ·lat'ed *or* ·lat'ing
per'co·la'tion
per'co·la'tor
per con·tig'u·um
per con·tin'u·um
per·cuss'
per·cus'sion
per·cus'sive
per·cus'sor
per·cu·ta'ne·ous
per di'em
per'fect
per·fec'tion·ism
per·fla'tion
per'fo·rans'
per'fo·rate'
  ·rat'ed ·rat'ing
per'fo·rate *or*
  ·rat'ed *or*
  ·fo·ra·ble *adj.*
per'fo·ra'tion
per'fo·ra'tive
per'fo·ra'tor
per·frig·er·a'tion
per'fu·sate'
per·fuse'
  ·fused' ·fus'ing
per·fu'sion
per·fu'sive
per'i·ac'i·nal *or*
  ·ac'i·nous
per'i·a'nal
per'i·ap'i·cal
per'i·ar·te'ri·al
per'i·ax'i·al
per'i·bron·chi'tis

per′i·car′di·al *or* ·di·ac′
per′i·car·di′tis
per′i·car′di·um ·di·a
per′i·chon′dri·al *or* ·dral
per′i·chon′dri·um ·dri·a
per′i·cra′ni·al
per′i·cra′ni·um ·ni·a
per′i·den·drit′ic
per′i·derm′
per′i·der′mal *or* ·der′mic
pe·rid′i·al
pe·rid′i·um ·i·a
per′i·en·ter′ic
per′i·en·ter·i′tis
per′i·ep·en′dy·mal
per′i·fis′tu·lar
per′i·kar′y·on ·y·a
per′i·ker·at′ic
pe·rim′e·ter
per′i·met′ric *or* ·met′ri·cal
pe·rim′e·try ·tries
per′i·my′si·um ·si·a
per′i·na′tal
per′i·ne′al (*of the perineum;* SEE peroneal)

per′i·ne′o·col′por·ec′to·my′o·mec′to·my ·mies
per′i·ne·om′e·ter
per′i·neph′ri·um
per′i·ne′um ·ne′a
per′i·neu′ri·al
per′i·neu′ri·um ·ri·a
pe′ri·od
pe·ri′o·date′
pe′ri·od′ic
pe′ri·o·dic′i·ty ·ties
per′i·o·don′tal
per′i·o·don′tic
per′i·o·don′tics *or* ·don′ti·a
per′i·o·don′tist
per′i·o·nych′i·um ·i·a
per′i·os′te·al
per′i·os′te·um ·te·a
per′i·os·tit′ic
per′i·os·ti′tis
per′i·o′tic
per′i·pa·tet′ic
pe·riph′er·ad′
pe·riph′er·al
pe·riph′er·y ·er·ies
per′i·phrase′
pe·riph′ra·sis ·ses′

per′i·phras′tic
pe·rip′lo·cin
per′i·por′tal *or* ·py′lic
per′i·re′nal
per′i·rhi′nal
per′i·scop′ic
per′ish
per′i·spon·dyl′ic
per′i·spon′dy·li′tis
pe·ris′so·dac′ty·lous
per′i·stal′sis ·ses
per′i·stal′tic
per′i·stome′
per′i·sto′mi·al
per′i·the′ci·al
per′i·the′ci·um ·ci·a
per′i·to·ne′al
per′i·to·ne′a·lize′ ·lized′ ·liz′ing
per′i·to·ne′um ·ne′a *or* ·ne′ums
per′i·to·ni′tis
per′i·to·ni·za′tion
per′i·to·nize′ ·nized′ ·niz′ing
per′i·ton′sil·lar
per′i·ton′sil·li′tis
pe·rit′ri·chous
per′i·xe·ni′tis
per·lèche′
per·lin′gual
per′ma·nent
per·man′ga·nate′
per′man·gan′ic acid

per′me·a·bil′i·ty
  ·ties
per′me·a·ble
per′me·ance
per′me·ant
per′me·ase′
per′me·ate′
  ·at′ed ·at′ing
per′me·a′tion
per′me·a′tive
per men′sem
per·mis′si·ble
per·mis′sion
per·mis′sive
per′mit n.
per·mit′
  ·mit′ted ·mit′ting
per′mu·ta′tion
per′mu·ta′tion·al
per·mute′
  ·mut′ed ·mut′ing
per·ni′cious
per′o·ne′al *(of the fibula;* SEE perineal)
per·o′ral
per os′
per·ox′i·dase′
per·ox′ide
  ·id·ed ·id·ing
per·ox′i·some′
per·ox′y·ac·e′tyl nitrate
per′pli·ca′tion
per pri′mam (in·ten′ti·o′nem)
per′salt′

per·sev′er·ate′
  ·at′ed ·at′ing
per·sev′er·a′tion
per·sev′er·a′tive
per·sist′ence
per·sist′ent
per·sist′er
per′son
per·so′na
  ·nas
per′son·al
per′son·al·is′tics
per′son·al′i·ty
  ·ties
per·son′i·fi·ca′tion
per′son–years′
per′spi·ra′tion
per·spir′a·to′ry
per·spire′
  ·spired′ ·spir′ing
per·suade′
  ·suad′ed ·suad′ing
per·sua′sion
per·sul′fate
per·sul′fide
Per′thes′ disease
per tu′bam
per·turb′
per′tur·ba′tion
per·tus′sal *or*
  ·tus′soid
per·tus′sis
per·vap′o·ra′tion
per·ver′sion
per′vert n.
per·vert′ v.
per·vert′ed

per′vi·ous
pes
  pe′des
pes′sa·ry
  ·ries
pes′si·mism
pes′si·mist
pes′si·mis′tic
pest′house′
pes′ti·ci′dal
pes′ti·cide′
pes·tif′er·ous
pes′ti·lence
pes′ti·lent
pes′ti·len′tial
pes′tis
pes′tle
  ·tled ·tling
pe·te′chi·a
  ·chi·ae
pe·te′chi·al
peth′i·dine′ hydrochloride
pet′i·o·late′
pet′i·ole′
pet·it′ mal′
pe′tri *(or* Pe′tri) dish
pet′ri·fac′tion *or*
  ·ri·fi·ca′tion
pet′ri·fac′tive
pet′ri·fy′
  ·fied′ ·fy′ing
pé′tris·sage′
pet′ro·chem′i·cal
pet′ro·la′tum
pe·tro′le·um

pet′ro·mas′toid
pe·tro′sa
 ·sae
pe·tro′sal
pet′ro·si′tis or
 ·rou·si′tis
pet′rous
PET scan′ner
pex′is
pe·yo′te or ·tl
Pey·ro·nie's′
 disease
Pfeif′fer's bacillus
pH
phac′oid
pha·co′ma
 ·mas or ·ma·ta
phac′o·met′a·co·re′
 sis
 ·ses
phac′o·scope′
phage
phag·e·de′na or
 ·dae′na
phag·e·den′ic
phag′o·cyte′
phag′o·cyt′ic
phag′o·cy·to′sis
 ·ses
phag′o·cy·tot′ic
phag′o·dy·na·mom′
 e·ter
phag′o·ma′ni·a
pha·ki′tis
pha·ko′ma
 ·mas or ·ma·ta
phal′a·cro′sis

phal′a·crot′ic
phal·ac′rous
phal′ange
pha·lan′ge·al or
 ·lan′gal
pha·lan′ges
 (sing. pha′lanx)
pha′lanx
 pha·lan′ges
phal′lic
phal′li·cism or
 phal′lism
phal′li·cist or phal′
 list
phal·li′tis
phal′lo·dyn′i·a
phal′lo·plas′ty
 ·ties
phal′lus
 ·li or ·lus·es
phan′er·o·gen′ic or
 ·ge·net′ic
phan′er·o′sis
 ·ses
phan′ic
phan′quone
phan·ta′si·a
phan′tasm
phan·tas′ma
 ·ma·ta
phan·tas′ma·go′ri·a
phan·tas′ma·go′ri·al
 or ·go′ric or ·ri·cal
phan·tas′ma·go′ry
 ·ries
phan·tas′mal or
 ·mic

phan′ta·sy
 ·sies
phan′tom
phan′to·mize′
 ·mized′ ·miz′ing
pharm′a·cal
phar′ma·ceu′ti·cal
 or ·ceu′tic
phar′ma·ceu′tics
phar′ma·cist
phar′ma·co·dy·
 nam′ic
phar′ma·co·dy·
 nam′ics
phar′ma·co·ge·net′
 ics
phar′ma·cog·no·sy
 ·sies
phar′ma·co·ki·
 net′ic
phar′ma·co·ki·net′
 ics
phar′ma·co·ki·net′
 ist
phar′ma·co·log′i·cal
 or ·log′ic
phar′ma·col′o·gist
phar′ma·col′o·gy
phar′ma·co·pe′ia or
 ·poe′ia
phar′ma·co·pe′ial
 or ·poe′ial
phar′ma·cy
 ·cies
pha·ryn′ge·al or
 ·gal
phar′yn·gis′mus

175

phar′yn·gi′tis
pha·ryn′go·a·myg′da·li′tis
phar′yn·gol′o·gy
phar·yn′go·scope′
phar′yn·gos′co·py
  ·pies
pha·ryn′go·tome′
phar′yn·got′o·my
  ·mies
phar′ynx
  phar′nyx·es or pha·ryn′ges
phase
  phased phas′ing
phase′–con′trast microscope
pha′sic
phat′nor·rha′gi·a
phe′na·caine′ hydrochloride
phe·nac′e·tin
phe·nan′threne
phe′nate
phen·cy′cli·dine′ hydrochloride
phe·net′i·dine′
phen′e·tole′
phen′go·pho′bi·a
phe′no·bar′bi·tal′
phe′no·cop′y
  ·ies
phe′nol
phe′no·late′
  ·lat′ed ·lat′ing
phe·no′lic
phe′no·log′i·cal

phe·nol′o·gist
phe·nol′o·gy
phe′nol·phthal′ein
phenol red
phe·nom′e·nol′o·gy
phe·nom′e·non′
  ·na
phe′no·thi′a·zine′
phe′no·type′
phe′no·typ′ic or ·typ′i·cal
phe·nox′ide
phe·nox′y
phe′no·zy′gous
phen′yl
phen′yl·al′a·nine′
phen′yl·am·ine′
phen′yl·bu′ta·zone′
phen′yl·ene′
phen′yl·ke′to·nu′ri·a
phen′yl·ke′to·nu′ric
phen′yl·pro·pa·nol′a·mine′ hydrochloride
phen′yl·py·ru′vic acid
phe′o·chro′mo·cy·to′ma
  ·mas or ·ma·ta
phe·re′sis
peth·ar′bi·tal′
phi
phi′al
phi′a·lid
phi·li′a·ter

phil′o·pro·gen′i·tive
phil′ter (a potion; SEE filter)
phil′trum
  ·tra
phi·mo′sis
  ·ses
phi·mot′ic
phleb·ar·te′ri·ec·ta′si·a
phle·bec′to·my
  ·mies
phleb·ec·to′pi·a or phle·bec′to·py
phle·bit′ic
phle·bi′tis
phle·boc′ly·sis
  ·ses′
phleb′o·gram′
phleb′oid
phleb′o·lith or ·lite
phleb′o·scle·ro′sis
  ·ses
phle·bot′o·mist
phle·bot′o·mize′
  ·mized′ ·miz′ing
phle·bot′o·my
phlegm
phleg·mat′ic or ·mat′i·cal
phleg′mon
phleg′mo·nous
phlegm′y
  ·i·er ·i·est
phlo·gis′tic
phlo′go·gen′ic or ·gog′e·nous

phlo′go·sin
phlo·rid′zi·nize′
  ·nized′ ·niz′ing
phlor′i·zin or phlo·rid′zin or phlo·rhi′zin
phlo′ro·glu′cin or ·glu′ci·nol or ·glu′col
phlox′ine
phlyc·te′na
  ·nae
phlyc·te′nar
phlyc·ten′u·lar
phlyc·ten′ule
phlyc·ten′u·lo′sis
pho′bi·a
pho′bic
pho·co·me′li·a
pho·co·me′lic
pho·nac′o·scope′
pho′na·cos′co·py
pho′nal
pho′nate
  ·nat·ed ·nat·ing
pho·na′tion
phone
pho′neme
pho·ne′mic
pho·net′ic
pho·net′ics
phon′ic
pho′nism
pho′no·car′di·o·gram′
pho′no·car′di·o·graph′

pho′no·car′di·og′ra·phy
  ·phies
pho′no·gram′
pho′no·log′i·cal or ·log′ic
pho·nol′o·gist
pho·nol′o·gy
pho·nom′e·ter
pho·nom′e·try
pho′non
pho′no·scope′
pho·rol′o·gy
pho′ro·tone′
phose
phos′gene
phos′pha·tase′
phos′phate
phos′phat′ic
phos′pha·tide′
phos′pha·tid′ic
phos′pha·ti·za′tion
phos′pha·tize′
  ·tized′ ·tiz′ing
phos′pha·tu′ri·a or phos·pha·phu′ri·a
phos′pha·tu′ric
phos′phene
phos′phide
phos′phine
phos′phite
phos′pho·cre′a·tine′
phos′pho·lip′id or ·lip′ide
phos·pho′ni·um
phos′pho·pro′tein

phos′pho·rate′
  ·rat′ed ·rat′ing
phos′pho·resce′
  ·resced′ ·resc′ing
phos′pho·res′cence
phos′pho·res′cent
phos′pho·ret′ed or ·ret′ted or ·phu·ret′ed or ·phu·ret′ted
phos·phor′ic
phos′pho·rism
phos′pho·rous adj.
phos′pho·rus n.
phos′pho·ryl·ase′
phos′pho·ryl·ate′
  ·at′ed ·at′ing
phos′pho·ryl·a′tion
pho·tal′gi·a
pho′tes·the′si·a
pho′tes·the′sis
phot
pho′tic
pho′tism
pho′to·ac·tin′ic
pho′to·au′to·troph′ic
pho′to·bi·ot′ic
pho′to·chem′i·cal
pho′to·chem′is·try
pho′to·chro′mo·gen
pho′to·co·ag′u·la′tion
pho′to·co·ag′u·la′tor
pho′to·der·ma·ti′tis
pho·tod′ro·my

pho′to·dy·nam′ic
pho′to·dy·nam′ics
pho′to·e·lec′tric
pho′to·e·lec·tric′i·ty
pho′to·e·lec′tron
pho′to·flu·o′ro·graph′ic
pho′to·flu·o·rog′ra·phy
pho′to·gene′
pho′to·gen′ic
pho′to·ki·ne′sis
·ses
pho′to·ki·net′ic
pho′to·lu′mi·nes′cence
pho′to·lu′mi·nes′cent
pho·tol′y·sis
·ses′
pho′to·lyt′ic
pho·tom′e·ter
pho′to·met′ric
pho·tom′e·try
·tries
pho′to·mi′cro·graph′
pho′to·mi′cro·graph′ic
pho′to·mi·crog′ra·phy
·phies
pho′to·mo′tor
pho′ton
pho·ton′o·sus
·si
pho′to·pe′ri·od

pho′to·pe′ri·od′ic
pho′to·pe′ri·od·ism
or ·pe′ri·o·dic′i·ty
pho·toph′i·lous or
  pho′to·phil′ic
pho′to·pho′bi·a
pho′to·pho′bic
pho′to·phore′
pho′to·pi·a
pho·to′pic
pho′to·re·cep′tion
pho′to·re·cep′tive
pho′to·re·cep′tor
pho′to·sen′si·tive
pho′to·sen′si·tiv′i·ty
pho′to·sen′si·ti·za′tion
pho′to·sen′si·tize′
  ·tized′ ·tiz′ing
pho′to·sen′si·tiz′er
pho′to·syn′the·sis
·ses′
pho′to·syn′the·size′
  ·sized′ ·siz′ing
pho′to·syn·thet′ic
pho′to·tac′tic
pho′to·tax′is
pho′to·ther′mal
pho′to·ther′mic
pho′to·ton′ic
pho′to·tot′o·nus
pho′to·trop′ic
pho·tot′ro·pism
phre·nec′to·my
phren′em·phrax′is
  ·es
phre·net′ic

phren′ic
phren′i·cec′to·my
  ·mies
phre·ni′tis
phren′o·car′di·a
phren′o·log′i·cal
phre·nol′o·gist
phre·nol′o·gy
phren′o·per′i·car·di′tis
phren′o·ple′gi·a
phren′o·sin
phren′sy
·sies
phren′sy
  ·sied ·sy·ing
phric′to·path′ic
phro·ne′ma
phryn′o·der′ma
pH–stat
phthal′ein
phthal′ic acid
phthal′in
phthal′yl·sul′fa·cet′a·mide′
phthi·ri′a·sis
·ses′
phthis′ic
phthis′i·cal
phthi′sis
·ses
phy·col′o·gy
phy′co·my′cete
phy′co·my·ce′tous
phy′la
  (sing. phy′lum)
phyl′lo·qui′none

phy′lo·gen′e·sis
·ses′
phy′lo·ge·net′ic or
·gen′ic
phy·log′e·ny
·nies
phy′lum
·la
phys′i·at′rics or
·at′ry
phys′i·at′rist
phys′ic
·icked ·ick·ing
phys′i·cal
phy·si′cian
phys′i·cist
phys′i·co·chem′i·cal
phys′i·co·gen′ic
phys′ics
phys′i·no′sis
phys′i·oc′ra·cy
phys′i·o·gen′e·sis
phys′i·og·nom′ic or
·nom′i·cal
phys′i·og′no·mist
phys′i·og′no·my
·mies
phys′i·o·log′i·cal or
·log′ic
phys′i·o·log′i·co·an′
a·tom′i·cal
phys′i·ol′o·gist
phys′i·ol′o·gy
phys′i·o·path′ic
phys′i·o·ther′a·pist
phys′i·o·ther′a·py
phy·sique′

phy′so·stig′ma
phy′so·stig′mine
phy′tase
phy′to·ag·glu′ti·nin
phy′to·be′zoar
phy′to·chem′is·try
phy′to·chin′in
phy′to·chrome′
phy′to·flag′el·late′
phy′to·gen′ic or
phy·tog′e·nous
phy′to·he′mag·glu′
ti·nin
phy′toid
phy′tol
phy′to·na·di′one
phy′to·path′o·log′ic
or ·log′i·cal
phy′to·pa·thol′o·gy
phy·toph′a·gous
phy·tos′ter·ol′
phy′to·tox′ic
phy′to·tox·ic′i·ty
·ties
phy′to·tox′in
pi
pi′a–ar·ach′ni·tis
pi′a–a·rach′noid or
pi′a·rach′noid
pi′al
pi′a ma′ter
pi′a·ma′tral
pi·an′
pi′ca
pick′ling
Pick's disease

pick·wick′i·an
syndrome
pi′co·cu′rie
pi′co·gram′
pic′o·line′
pi·cor′na·vi′rus
pic′rate
pic′ric acid
pic′ro·car′mine
pic′ro·for′mol
pic′rol
pic′ro·tox′in
pic′ro·tox′in·in
pic′to·graph′
pic′to·graph′ic
pie′bald′
piece
pi·e′dra
pierce
pierced pierc′ing
Pi·erre′ Ro·bin′
syndrome
pi′e·sim′e·ter or
·som′e·ter
pi·e′zo·chem′is·try
pi·e′zo·e·lec′tric or
·lec′tri·cal
pi·e′zo·e·lec·tric′i·ty
pi′e·zom′e·ter
pi·e′zo·met′ric
pi·e′zom′e·try
pi′geon breast
pi′geon–toed′
pig′ment
pig′men·tar′y
pig′men·ta′tion

pig′ment·ize′
  ·ized′ ·iz′ing
pig·men′tum ni′grum
pig′my
  ·mies
Pi·gnet's′ formula
pi·i′tis
pi′la
  ·lae
pi′lar *or* pil′ar·y
pi·las′ter
pi′le·ous
  (hairy)
piles
  (sing. pile*)*
pi′le·us
  (membrane)
pile′wort′
pi′li
  (sing. pi′lus*)*
pi′li·al
pi′li·ate
pi′li·a′tion
pil′i·form′
pi′li·mic′tion
pill
pil′lar
pill′box′
pil′let
pil′lion
pil′low
pill′–roll′ing
pi′lo·car′pine
pi′lo·cys′tic
pi′lo·e·rec′tion
pi′lo·mo′tor

pi′lo·ni′dal
pi′lose
pi′lo·se·ba′ceous
pi·los′i·ty
  ·ties
pil′u·lar
pil′ule
pi′lus
  ·li
pi·mel′ic acid
pim′e·li′tis
pim′e·lo′ma
  ·mas *or* ·ma·ta
pim′el·or′thop·ne′a
pi·men′ta
pi·men′to
  ·tos
pi·min′o·dine′ es′y·late
pim′ple
pim′ply
  ·pli·er ·pli·est
pin
  pinned pin′ning
pin′a·cy′a·nol′
pince·ment′
pince′–nez′
pin′cers
pinch
pinch′cock′
pine
pin′e·al
pin′e·al·ec′to·my
pin′e·a·lo·blas·to′ma
  ·mas *or* ·ma·ta
pin′e·a·lo·cyte′

pin′e·a·lo′ma *or*
  pin′e·a·lo·cy·to′ma
  ·mas *or* ·ma·ta
pi′nene
pin′e·o·blas·to′ma
  ·mas *or* ·ma·ta
pine tar
pin·guec′u·la *or*
  ·guic′u·la
pin′hole′ pupil
pin′i·form′
pink′eye′
pink′ie *or* pink′y
  pink′ies
pink′root′
pin′na
  ·nae *or* ·nas
pin′nal
pin′o·cy·to′sis
  ·ses
pin′o·some′
pin′o·ther′a·py
pin′prick′
Pins' sign
pint
pin′ta
pin′tid
pin′toid
pi′nus
pin′worm′
pi′o·ep′i·the′li·um
pi′o·ne′mi·a
pip′e·col′ic *(or* pip′e·co·lin′ic*)* acid
pi·per′a·zine′
pi·per′i·dine′
pip′er·ine′

pi·per′o·nal′
pi·pette′ *or* ·pet′
  ·pet′ted ·pet′ting
pi′po·bro′man
pi′po·sul′fan
pip·sis′se·wa
pi·qûre′
pir′i·form′
pir′o·men
pi′si·form′
pit
  pit′ted pit′ting
pitch
pitch′blende′
pit′fall′
pith
pith′e·coid′
pith′i·at′ic
pith·i′a·try
pith′ode
Pi·tres′′ sections
pi·tu′i·cyte
pi·tu′i·ta
  ·tae
pi·tu′i·tar·ism
pi·tu′i·tar′y
  ·tar′ies
pi·tu′i·tous
pit′y·ri·as′ic
pit′y·ri′a·sis
  ·ses′
pit′y·roid′
piv′ot
piv′ot·al
pix
pix′el

pla·ce′bo
  ·bos *or* ·boes
pla·cen′ta
  ·tas *or* ·tae
pla·cen′tal
pla·cen′ta·scan′
plac′en·ta′tion
pla·cen′tin
plac′en·ti′tis
plac′en·tog′ra·phy
  ·phies
pla·cen′toid
plac′en·to′ma
  ·mas *or* ·ma·ta
pla·cen′to·ther′a·py
plac′ode
plac′oid
plad′a·ro′ma
  ·mas *or* ·ma·ta
plad′a·ro′sis
  ·ses
pla·fond′
pla′gi·o·ce·phal′ic
pla′gi·o·ceph′a·ly *or*
  ·ceph′a·lism
plague
plain
  (*simple;* SEE plane)
plak′al·bu′min
pla′kins
plan
  planned plan′ning
pla′na
  (*sing.* pla′num)
plan′chet
Planck's constant

plane
  (*flat surface;* SEE plain)
plane
  planed plan′ing
pla′ni·gram′
pla·nig′ra·phy
  ·phies
pla·nim′e·ter
pla′ni·met′ric
pla·nim′e·try
plan′i·tho′rax
plank′ton
pla′no–con·cave′
pla′no–con·vex′
plan′ta
  ·tae
plan′tar
plan·tar·flex′ion
plan·tar′is
  ·es
plan·ta′tion
plan′ti·grade′
plan′u·la
  ·lae′
plan′u·loid′
pla′num
  ·na
plaque
plas′ma *or* plasm
plas′ma·blast′ *or*
  plas′ma·cy·to·blast′
plas′ma·crit′
plas′ma·cyte′
plas′ma·cy·to′ma
  ·mas *or* ·ma·ta

plas′ma·gel′
plas′ma·gene′
plas′ma·gen′ic
plas′mal
plas′ma·pher′e·sis
 ·ses′
plas′ma·phe·ret′ic
plas′ma·sol′
plas′ma·some′
plas·mat′ic *or* plas′mic
plas′ma·tor·rhex′is
 ·es
plas′mid
plas′min
plas·min′o·gen
plas′mo·cyte′
plas·mo′di·um
 ·di·a
plas′mo·gen
plas·mol′y·sis
 ·ses′
plas′mo·lyz′a·ble
plas′mo·lyze′
 ·lyzed′ ·lyz′ing
plas·mo·trop′ic
plas·mot′ro·pism
plas′te·in
plas′ter
plaster of Par′is
plas′tic
plas·tic′i·ty
plas′ti·cize′
 ·cized′ ·ciz′ing
plas′ti·ciz′er
plas′tid
plas′tron

plate
 plat′ed plat′ing
pla·teau′
 ·teaus′ *or* ·teaux′
plate′let
pla·tin′ic
plat′i·nous
plat′i·num
plat′y·ce′lous
plat′y·ce·phal′ic *or*
 ·ceph′a·lous
plat′yc·ne′mi·a
plat′yc·ne′mic
plat′y·hel′minth
plat′y·hel·min′thic
plat′y·o′pi·a
plat′y·op′ic
plat′y·pel′lic *or*
 ·pel′loid
plat′yr·rhine′
pla·tys′ma
 ·mas *or* ·ma·ta
pla·tys′mal
plec·trid′i·um
pled′get
pleg′a·pho′ni·a
ple′ia·des′
plei·o·trop′ic
plei·ot′ro·py *or*
 ·ro·pism
ple′o·chro′ic
ple·och′ro·ism
ple′o·mas′ti·a *or*
 ·ma′zi·a
ple′o·mor′phic *or*
 ·phous
ple′o·mor′phism

ple′o·nasm
ple′o·nas′tic
ple′on·os·te·o′sis
ple·op′to·phor′
ple·ro·cer′coid
ple·ro′sis
ples·sim′e·ter
ples′si·met′ric
ples′sor
pleth′o·ra
ple·thor′ic
ple·thys′mo·gram′
ple·thys′mo·graph′
ple·thys′mo·graph′ic
pleth′ys·mog′ra·phy
pleu′ra
 ·rae
pleu′ral
pleu·rec′to·my
 ·mies
pleu′ri·sy
pleu·rit′ic
pleu′ro·cele′
pleu·roc′ly·sis
pleu·ro·dyn′i·a
pleu·ro·pa·ri′e·to·pex′y
pleu·ro·pneu·mo′ni·a
pleu·ro·pul′mo·nar′y
pleu·rot′o·my
 ·mies
plex′al
plex′i·form′

plex·im′e·ter or
  plex·om′e·ter
plex′o·gen′ic
plex′or
plex′us
  ·us·es or ·us
pli′a·bil′i·ty
pli′a·ble
pli′ca
  ·cae
pli′cate or ·cat·ed
pli′cate′
  ·cat′ed ·cat′ing
pli·ca′tion
plic′a·ture
pli·cot′o·my
plinth or plint
ploi′dy
plom·bage′
plo′sive
plo·to·ly′sin
plo·to·tox′in
pluck reflex
plug
  plugged plug′ging
plug′ger
plum·ba′go
  ·gos
plum′bic
plum′bism
plum′bo·ther′a·py
plum′bum
plu′mose
plu·mos′i·ty
plump
plump′er
plump′ish

plu′ri·cep′tor
plu′ri·dys·crin′i·a
plu′ri·glan′du·lar
plu′ri·par′i·ty
plu′ri·po·ten′tial or
  plu·rip′o·tent
plu′ri·po·ten′ti·al′
  i·ty
  ·ties
plu·to·ma′ni·a
plu·to′ni·um
PMS
pne′o·car′di·ac
pne′o·dy·nam′ics
pne·om′e·try
pne′o·scope′
pneu′ma
pneu′mal
pneu′mar·thro′sis
pneu·mat′ic
pneu·mat′ics
pneu′ma·ti·za′tion
pneu′ma·tize′
  ·tized′ ·tiz′ing
pneu′ma·to·cele′
pneu′ma·tol′o·gy
pneu′ma·tom′e·ter
pneu·mat′o·scope′
pneu′ma·to′sis
  ·ses
pneu′ma·tu′ri·a or
  ·ma·ti·nu′ri·a
pneu′ma·type′
pneu·mec′to·my
  ·mies
pneu′mo·ba·cil′lus
  ·cil′li

pneu′mo·ceph′a·lus
pneu′mo·coc′cal or
  ·coc′cic
pneu′mo·coc′cus
  ·coc′ci
pneu′mo·co·ni·o′sis
  ·ses
pneu′mo·en·ceph′a·
  lo·gram′
pneu′mo·en·ceph′a·
  log′ra·phy
pneu′mo·gas′tric
pneu′mo·gram′
pneu′mo·graph′
pneu·mog′ra·phy
pneu′mo·mas·sage′
pneu′mo·nec′to·my
  ·mies
pneu′mo·ni·a
pneu·mon′ic
pneu·mo·ni′tis
pneu′mon·o·cyte′
pneu′mo·nop′a·thy
pneu′mo·no·pex′y
pneu′mo·no′sis
  ·ses
pneu′mo·per′i·car′
  di·um
pneu′mo·per′i·to·
  ne′um
pneu′mo·per′i·to·ni′
  tis
pneu′mo·tach′o·
  gram′

pneu·mo·tach'o·graph' *or* ·tach'y·graph' *or* ·ta·chom'e·ter
pneu·mo·tho'rax
·rax·es *or* ·ra·ces'
pneu·mo·u'ri·a
pneu·mo·ven'tri·cle
pneu·mo·ven·tric'u·log'ra·phy
pneu'sis
pni'go·pho'bi·a
pock'et
pock'mark'
pock'marked'
pock'y
·i·er ·i·est
po·dag'ra
po·dag'ral *or* ·ric
po·dal'ic
pod·ar·thri'tis
pod·e·de'ma
po'di·at'ric
po·di'a·trist
po·di'a·try
po·dis'mus
pod·o·bro'mi·dro'sis
pod'o·cyte'
pod'o·dy'na·mom'e·ter
pod'o·dyn'i·a
pod'o·gram'
po·dol'o·gy
pod'o·phyl'lin
pod'o·phyl'lum
·li

pod'o·spasm' *or* ·spas'mus
po'go·ni'a·sis
po·go'ni·on
poi'ki·lo·cyte'
poi'ki·lo·cy·to'sis *or* poi·kil'o·cy·the'mi·a
poi'ki·lo·der'ma
·mas *or* ·ma·ta
poi·kil'o·therm'
poi·kil'o·ther'mal *or* ·ther'mic
poi·kil'o·ther'mism
point
point'er
poin'til·lage'
poise
Poi·seuille's' law
poi'son
poi'son·ous
po'lar
po·lar·im'e·ter
po·lar'i·met'ric
po·lar·im'e·try
po·lar'i·scope'
po·lar'i·scop'ic
po·lar'i·ty
·ties
po·lar·iz'a·ble
po·lar·i·za'tion
po·lar·ize'
·ized' ·iz'ing
po·lar·iz'er
po·lar'o·gram'
po·lar'o·graph'ic
po·lar·og'ra·phy

pole
pol'i·clin'ic
pol'i·cy
·cies
po·li·o'
po·li·o·clas'tic
po·li·o·dys'tro·phy *or* ·dys·tro'phi·a
po·li·o·en·ceph·a·li'tis
po·li·o·my·e·li'tis
po·li·o·plasm'
po·li·o'sis
po·li·o·vi'rus
pol'ish
pol'i·sog'ra·phy
pol'it·zer·i·za'tion
pol'la·ki·dip'si·a
pol·la·ki·u'ri·a *or* ·ki·su'ri·a
pol'len
pol'len·o'sis *or* pol'li·no'sis
·ses
pol'lex
pol'li·ces'
pol'li·cal
pol·lut'ant
pol·lute'
·lut'ed ·lut'ing
pol·lut'er
pol·lu'tion
po·lo'ni·um
pol·ox'a·mer
po'lus
·li
pol'y·ad·e·ni'tis

pol′y·ad′e·nous
pol′y·al·ge′si·a
pol′y·am′ide
pol′y·an′dric
pol′y·an′drous
pol′y·an′dry
pol′y·bas′ic
pol′y·blast′
pol′y·chrest′
pol′y·chro·mat′ic
 or pol′y·chrome′
pol′y·chro·mat′o·
 phil or ·phile′
pol′y·chro·mat·o·
 phil′i·a
pol′y·clin′ic
pol′y·crot′ic
pol·yc′ro·tism
pol′y·cy′clic
pol′y·cy·the′mi·a
pol′y·dac′tyl
pol′y·dac′tyl·ism or
 ·dac′ty·ly
pol′y·dac′ty·lous
pol′y·dip′si·a
pol′y·dis·per′soid
pol′y·e·lec′tro·lyte′
pol′y·em′bry·o·ny
 ·nies
pol′y·ene′
pol′y·eth′yl·ene′
po·lyg′a·mist
po·lyg′a·mous
po·lyg′a·my
pol′y·genes′
pol′y·gen′ic

pol′y·graph′
pol′y·graph′ic
pol′y·graph′ist or
 po·lyg′ra·pher
po·lyg′y·nous
po·lyg′y·ny
pol′y·he′dral
pol′y·his′tor
pol′y·hy′brid
pol′y·hy·dram′ni·os
pol′y·hy′dric or
 ·hy·drox′y
pol′y I′:C′
pol′y·in·fec′tion
pol′y·men′or·rhe′a
 or ·y·me′ni·a
pol′y·mer
po·lym′er·ase′
pol′y·mer′ic
po·lym′er·ism
po·lym′er·i·za′tion
po·lym′er·ize′
 ·ized′ ·iz′ing
po·lym′i·tus
pol′y·morph′
pol′y·mor′phism
pol′y·mor′pho·cel′
 lu·lar
pol′y·mor′pho·cyte′
pol′y·mor′pho·nu′
 cle·ar
pol′y·mor′phous or
 ·phic
pol′y·my·oc′lo·nus
pol′y·myx′in
pol′y·ne′sic

pol′y·neu′ral
pol′y·neu·rit′ic
pol′y·neu·ri′tis
pol′y·neu·rop′a·thy
 ·thies
pol′y·nu′cle·ar
pol′y·o′ma virus
pol′yp
pol′y·pep′tide
po·lyph′a·gous
pol′y·phar′ma·cy
 ·cies
pol′y·phase′
pol′y·pha′sic
pol′y·phy·let′ic
pol′yp·ide′
pol′y·ploid′
pol′y·ploi′dy
po·lyp′o·rous
pol′y·po′sis
 ·ses
pol′yp·ous
pol′y·pro′pyl·ene′
pol′y·ri′bo·some′
pol′yr·rhe′a
pol′y·sac′cha·ride′
pol′y·sar′cous
pol′y·se·ro·si′tis
pol′y·some′
pol′y·so′mic
pol′y·sor′bate
pol′y·sper′mi·a or
 ·sper′mism
pol′y·sper′my
 ·mies
pol′y·ster·ax′ic

pol′y·sty′rene
pol′y·sul′fide
pol′y·tro′phi·a or
  po·lyt′ro·phy
pol′y·troph′ic
pol′y·trop′ic
pol′y·typ′ic
pol′y·un·sat′u·
  rat′ed
pol′y·u′ri·a
pol′y·u′ric
pol′y·va′lence
pol′y·va′lent
pol′y·vin′yl
pol′y·vin′yl·pyr·rol′
  i·done′
po·made′ or
  ·ma′tum
po′man·der
pome′gran′ate
pom′phus
  ·phi
po′mum
pon′der·a·ble
pon′der·al
pon′do·stat′u·ral
po′no·graph′
pons
  pon′tes
pon′tic
pon·tic′u·lus
  ·li
pon′tile or ·tine
pon′to·bul′bar
pon′to·cer′e·bel′lar
pool
pop′eyed′

pop′les
pop·lit′e·al
pop·lit′e·us
pop′pied
pop′py
  ·pies
pop·u·la′tion
por′ad·e·ni′tis
por′al
por′ce·lain
por′ce·la′ne·ous or
  ·cel·la′ne·ous
por′cine
pore
  (tiny opening; SEE
  pour)
por′en·ceph′a·li′tis
po′ri
  (sing. po′rus)
po′ri·a
  (sing. po′ri·on)
po′ri·o·ma′ni·a
po′ri·on
  ·ri·a
por′no·graph′ic
por·nog′ra·phy
po′ro·cele′
po′ro·ceph′a·li′a·sis
  or ·ceph′a·lo′sis
  ·ses′
po·ro′ma
  ·ma·ta
po′ro·plas′tic
po·ro′sis
  ·ses
po·ros′i·ty
  ·ties

po·rot′ic
po′rous
por′phin or ·phine
por′pho·bi·lin′o·gen
por′phyr′a·tin
por′phyr′i·a
por′phy·rin
por′phy·ri·nu′ri·a
por′phy·ri·za′tion
por·ri′go
por·rop′si·a
por′ta
  ·tae
por′ta·ca′val
por′tal
por′ti·o
  por′ti·o′nes
por′to·gram′
por·tog′ra·phy
  ·phies
port′–wine′ mark
po′rus
  ·ri
po·si′tion
po·si′tion·al
po·si′tion·er
pos′i·tive
pos′i·tron′
po′so·log′ic
po·sol′o·gy
pos·sessed′
pos·ses′sion
post′a·bor′tal
post′ax′i·al
post·ca′va
  ·vae
post·ca′val

post'ci·bal
pos·te'ri·or
pos'ter·o·an·te'ri·or
post'fe'brile
post'gan'gli·on'ic
pos·thi'tis
post'hu·mous
post'hy'oid
post'hyp·not'ic
post'ma·ture'
post'ma·tu'ri·ty
post'–mor'tem
post'na'sal
post'na'tal
post'op'er·a·tive
post·or'bit·al
post'par'tum
post'po'li·o' syndrome
post'pran'di·al
post'pu·bes'cent
post'–term' infant
pos'tu·late
pos'tu·late'
 ·lat'ed ·lat'ing
pos'tur·al
pos'ture
po'ta·bil'i·ty
 ·ties
po'ta·ble
Po·tain's' sign
pot'ash'
pot·as·se'mi·a
po·tas'sic
po·tas'si·o·mer·cu'ric

po·tas'si·um
po·ta'tion
pot'bel'ly
 ·lies
po'tence
po'ten·cy
 ·cies
po'tent
po·ten'tial
po·ten'ti·ate'
 ·at'ed ·at'ing
po·ten'ti·a'tion
po·ten'ti·a'tor
po·ten'ti·om'e·ter
poth'e·car'y
 ·car'ies
po'tion
po·to·ma'ni·a
Pott's disease
pouch
pou·drage'
poul'tice
 ·ticed ·tic·ing
pound
pound'al
pour
 (to flow; SEE pore)
pov'er·ty
po'vi·done'
po'vi·done'–i'o·dine'
pow'der
pow'der·y
pow'er
pox
pox'vi'rus
P'–pul'mo·na'le

prac'ti·ca·bil'i·ty
 ·ties
prac'ti·ca·ble
prac'ti·cal
prac'ti·cal'i·ty
 ·ties
prac'tice
 ·ticed ·tic·ing
prac·ti'tion·er
For words beginning prae– *see* PRE–
prag'mat·ag·no'si·a
prag·mat'ic *or* ·mat'i·cal
prag·mat'ics
prag'ma·tism
prag'ma·tist
pran'di·al
pra'se·o·dym'i·um
prax'i·ol'o·gy
prax'is
pra'zo·sin hydrochloride
pre'a·dult'
pre·ag'o·nal
pre·ax'i·al
pre·can'cer·ous
pre·cau'tion
pre·ca'va
 ·vae
pre·ca'val
prec'e·dence *or* ·den·cy
prec'e·dent
pre'cept
pre·chord'al

pre·cip′i·tant
pre·cip′i·tate′
　·tat′ed ·tat′ing
pre·cip′i·ta′tion
pre·cip′i·tin
pre·cip′i·tin′o·gen
pre·cip′i·tum
pre·clin′i·cal
pre′co·cene
pre·co′cious
pre·coc′i·ty
pre′cog·ni′tion
pre·cog′ni·tive
pre·con′scious
pre·crit′i·cal
pre·cu′ne·us
pre·cur′sor
pre·cur′so·ry
pre′de·cease′
　·ceased′ ·ceas′ing
pred′e·ces′sor
pre′de·cid′u·al
pre′de·ter′mi·nate
pre′de·ter′mine
　·mined ·min·ing
pre·dict′a·ble
pre·dic′tor
pre′di·gest′
pre′di·ges′tion
pre′dis·pose′
　·posed′ ·pos′ing
pre′dis·po·si′tion
pred·nis′o·lone′
pred′ni·sone′
pree′mie
pre·en′zyme′
pre′ex·ci·ta′tion

pre′ex·ist′ or –ex·
　ist′ or ·ëx·ist′
pre′ex·ist′ence or
　–ex·ist′ence
pre′ex·ist′ent or
　–ex·ist′ent
pre′for·ma′tion
pre·fron′tal
pre′gan·gli·on′ic
preg′nan·cy
　·cies
preg′nane·di′ol
preg′nant
pre·hal′lux
pre·heat′
pre·hen′sile
pre′hen·sil′i·ty
　·ties
pre·hen′sion
Prei′ser's disease
pre·lim′i·nar′y
　·nar′ies
pre′lum
pre′ma·lig′nant
pre·mar′i·tal
pre·ma·ture′
pre′ma·tu′ri·ty
pre·max·il′la
　·lae
pre·max′il·lar′y
pre′med′
pre·med′i·cal
pre·med′i·cant
pre·med′i·cate′
　·cat′ed ·cat′ing
pre′med·i·ca′tion
pre·men′stru·al

pre·men′stru·um
　·stru·a
pre·mo′lar
pre·mo·ni′tion
pre·mon′i·to′ry
pre·mu·ni′tion
pre·mu′ni·tive
pre·mu′nize′
　·nized′ ·niz′ing
pre′nar·co′sis
　·ses
pre·na′tal
pre·op′er·a′tion·al
pre·op′er·a·tive
pre·op′er·a·to′ry
prep
　prepped prep′ping
pre·pack′age
　·aged ·ag·ing
pre·pal′a·tal
prep′a·ra′tion
pre·par′a·tive
pre·par′a·tor
pre·par′a·to′ry
pre·pare′
　·pared′ ·par′ing
pre·pa′tent
pre·pon′der·ance or
　·an·cy
pre·pon′der·ant
pre·po′ten·cy
　·cies
pre·po′tent
pre′puce
pre·pu′tial
pre·pu′ti·um
　·ti·a

pre·req'ui·site
pres'by·at'rics
pres·by·cu'sis *or*
·cou'sis
·ses
pres'by·ope'
pres'by·o'pi·a
pres'by·op'ic
pres'by·ti·at'rics
pre'school'
pre·school'er
pre·scribe'
·scribed' ·scrib'ing
pre·scrip'tion
pre·scrip'tive
pre'se·nil'i·ty
·ties
pres'ent *adj. and n.*
pre·sent' *v.*
pre·sen·ta'tion
pre·sent'a·tive
pre·serv'a·ble
pres'er·va'tion
pre·ser'va·tive
pre·serve'
·served' ·serv'ing
pre·serv'er
pre·set'
·set' ·set'ting
pre·so'mite
press
pres·som'e·ter
pres'sor
pres'so·re·cep'tive
or ·sen'si·tive
pres'so·re·cep'tor
pres'sure

pressure point
pres'sur·i·za'tion
pres'sur·ize'
·ized' ·iz'ing
pres'sur·iz'er
pre·ster'num
·nums *or* ·na
pre·sump'tive
pre·sup'pu·ra'tive
pre·sys'to·le
pre'sys·tol'ic
pre'teen'
pre'term'
pre'test' *n.*
pre·test' *v.*
pre'thy'roid *or*
·thy·roi'de·al *or*
·thy·roi·de'an
prev'a·lence
prev'a·lent
pre·vent'
pre·vent'a·ble *or*
·i·ble
pre·vent'er
pre·ven'tion
pre·ven'tive *or*
·vent'a·tive
pre·ven·tric'u·lus
·li
pre'vi·ous
(prior)
pre'vi·us
(obstructing)
pre'zone'
pre·zo'nu·lar
pri·ap'ic
pri'a·pism

pri'a·pi'tis
pri·a'pus
·pi *or* ·pus·es
prick'ing
prick'le
·led ·ling
prick'ly
·li·er ·li·est
pri'ma·cy
·cies
pri'mal
pri'ma·quine'
pri'ma·ry
pri'mate
prime
primed prim'ing
prim'er
pri'mi·grav'i·da
·das *or* ·dae'
pri·mip'a·ra
·a·ras *or* ·a·rae'
pri'mi·par'i·ty
·ties
pri·mip'a·rous
pri'mite
pri·mi'ti·ae
prim'i·tive
pri·mor'di·al
pri·mor'di·um
·di·a
prim'u·la
pri'mum non no'
ce·re
prin'ceps
·ci·pes'
prin'ci·pal
(chief)

prin′ci·ple
  (ingredient)
Prinz′met′al's
  angina
prism
pris′ma
  ·ma·ta
pris·mat′ic
pris′moid
pris′mop·tom′e·ter
  or pris′op·tom′e·ter
pris′mo·sphere′
Prit′i·kin diet
pri′va·cy
  ·cies
pri′vate
pri·va′tion
priv′i·lege
  ·leged ·leg·ing
prob′a·bil′i·ty
  ·ties
prob′a·ble
pro′ac·cel′er·in
pro·ac·ti·no·my′cin
pro·ac′ti·va′tor
pro′al
pro′an·ti·throm′bin
pro′band
pro′bang
pro·ba′tion
pro·ba′tion·ar′y *or* ·tion·al
probe
  probed prob′ing
pro′bit
prob′lem

prob′lem·at′ic
pro·bos′cis
  ·cis·es *or* ·ci·des′
pro′caine
pro·car′y·ote′
pro·car′y·ot′ic
pro·ca·tarc′tic
pro·ca·tarx′is
pro·ce′dur·al
pro·ce′dure
pro·ce′li·a
pro·ce′lous
pro·cen′tri·ole′
pro·ce·phal′ic
proc′ess
pro·ces′sus
pro·chei′lon
pro·chlor·per′a·zine′
pro·chon′dral
pro·ci·den′ti·a
pro·cre·ate′
  ·at·ed ·at·ing
pro·cre·a′tion
pro·cre·a′tive
proc·tag′ra
proc·tec′to·my
  ·mies
proc′teu·ryn′ter
proc·ti′tis
proc′to·cele′
proc·toc′ly·sis
  ·ses′
proc′to·coc·cy·pex′y
proc′to·dae′al *or* ·de′al

proc′to·dae′um ·dae′a
proc′to·de′um ·de′a
proc′to·log′ic *or* ·log′i·cal
proc·tol′o·gist
proc·tol′o·gy
proc′to·scope′
proc′to·scop′ic
proc·tos′co·py
  ·pies
proc′to·sig′moi·dos′co·py
  ·pies
proc′to·ste·no′sis
  ·ses
pro·cum′bent
pro′cur·va′tion
pro·dro′mal *or* ·drom′ic
pro′drome
pro′drug′
prod′uct
pro·duc′tion
pro·duc′tive
pro·e′mi·al
pro·en′zyme
pro·es′trus *or* ·trum
pro·fes′sion
pro·fes′sion·al
pro′fi·bri·nol′y·sin
pro′file
  ·filed ·fil·ing
pro′fi·lom′e·ter
pro·fla′vine

pro·flu′vi·um
·vi·a or ·vi·ums
pro·fun′da
·dae
pro·fun′da·plas′ty
or ·do·plas′ty
pro·fun′dus
pro·fuse′
pro·fu′sion
pro·gen′i·tive
pro·gen′i·tor
prog′e·ny
·nies
pro·ger′i·a
pro·ges·ta′tion·al
pro·ges′ter·oid′
pro·ges′ter·one′
pro·ges′tin
pro·ges′to·gen
pro·glot′tid
pro·glot′tis
·ti·des′
prog′na·thism
prog′na·thous or
prog·nath′ic
prog·nose′
·nosed′ ·nos′ing
prog·no′sis
·no′ses
prog·nos′tic
prog·nos′ti·cate′
·cat′ed ·cat′ing
prog·nos′ti·ca′tion
prog·nos′ti·ca′tive
prog′nos′ti·cian
pro′gon·o′ma
·mas or ·ma·ta

pro′gram
·grammed or
·gramed
·gram·ming or
·gram·ing
prog′ress n.
pro·gress′ v.
pro·gres′sion
pro·gres′sive
pro·jec′tile
pro·jec′tion
pro·jec′tive
pro·ji′cient
pro·kar′y·ote′
pro·kar′y·ot′ic
pro·lac′tin
pro·la·mine′ or
·min
pro′lan
pro′lapse or pro·
lap′sus
pro·lapse′
·lapsed′ ·laps′ing
pro·lep′sis
·ses
pro·lep′tic
pro·lif′er·ate′
·at′ed ·at′ing
pro·lif′er·a′tion
pro·lif′er·ous
pro·lif′ic
pro·lif′i·ca·cy
pro′line
pro·mas′ti·gote′
prom′a·zine′
hydrochloride
pro·me′thi·um

pro′mine
prom′i·nence
prom′i·nent
prom′i·nen′ti·a
·ti·ae
prom′on·to·ry
·ries
pro·mot′er
pro·my′e·lo·cyte′
pro·na′si·on
pro′nate
·nat·ed ·nat·ing
pro·na′tion
pro·na′tor
prone
pro·neph′ric
pro·neph′ros
·roi
pro′no·grade′
pro·nom′e·ter
pro·nounced′
pro·nu′cle·ar
pro·nu′cle·us
·cle·i′
pro·o′tic
prop′a·ga·ble
prop′a·gate′
·gat·ed ·gat·ing
prop′a·ga′tion
prop′a·ga′tive
prop′a·ga′tor
pro·pal′i·nal
pro′pane
pro·pa·nol′
pro′pene
pro·per′din
pro′phase′

pro′phy·lac′tic
pro′phy·lax′is
  ·lax′es
pro′pi·o·nate′
pro′pi·on′ic acid
pro·por′tion
pro·pos′i·tus
  ·ti
pro·pox′y·phene′
pro·pran′o·lol′
pro·pri′e·tar′y
  ·tar′ies
pro′pri·o·cep′tive
pro′pri·o·cep′tor
pro′pri·o·spi′nal
pro′pro′tein
prop·to′sis
  ·ses
pro·pul′sion
pro·pul′sive or
  ·pul′so·ry
pro′pyl
pro′pyl·ene′
pro·pyl′ic
pro′pyl·thi′o·u′ra·cil
pro re na′ta
pror′rha·phy
pro·sect′
pro·sec′tion
pro·sec′tor
pros′en·ce·phal′ic
pros′en·ceph′a·lon
  ·la
pros′o·dem′ic
pros′o·pag·no′si·a

pros′o·pal′gi·a
pros′o·pal′gic
pros′o·po·a·nos′chi·sis
  ·ses′
pros′o·po·ple′gi·a
pros′o·po·ple′gic
pros′o·pos′chi·sis
  ·ses′
pros′o·po·tho′ra·cop′a·gus
  ·gi
pros′ta·cy′clin or
  ·cline
pros′ta·glan′din
pros′tate or pros·tat′ic
  (gland; SEE
  prostrate)
pros′ta·tec′to·my
  ·mies
pros′ta·tism
pros′ta·ti′tis
pros′ta·to·cys·ti′tis
pros′ta·to·cys·tot′o·my
  ·mies
pros′ta·tot′o·my
  ·mies
pros′ta·to·ve·sic′u·li′tis
pros·the′sis
  ·ses′
pros·thet′ic
pros·thet′ics
pros′thi·on′
pros′tho·don′tic

pros′tho·don′tics or
  ·don′ti·a
pros′tho·don′tist
pros′tho·ker′a·to·plas′ty
  ·ties
pros′ti·tu′tion
pros′trate
  ·trat·ed ·trat·ing
  (prone; SEE
  prostate)
pros·tra′tion
pro·tac·tin′i·um
pro′ta·gon
pro′tal
pro′ta·mine′
pro′tan·o′pi·a
pro′tan·op′ic
pro′te·an (protein
  derivative; SEE
  protein)
pro′te·ase′
pro·tect′
pro·tec′tion
pro·tec′tive
pro·tec′tor
pro′te·id or ·ide′
pro′tein
  (nitrogenous
  substance; SEE
  protean)
pro′tein·ase′
pro′tein·ate′
pro′tein·og′en·ous
pro′tein·oid′
pro′tein·ol′o·gy

pro′tein·o′sis
  ·ses
pro·tein′o·ther′a·py
pro′tein·u′ri·a
pro·ten′si·ty
pro′te·o·clas′tic
pro′te·ol′y·sis
  ·ses′
pro′te·o·lyt′ic
pro′te·o·pex′ic or
  ·pec′tic
pro′te·o·pex′y
pro′te·ose′
pro′test n.
pro′teus
pro·throm′bin
pro′tist
pro·tis′tan
pro·tis·tol′o·gist
pro·tis·tol′o·gy
pro′ti·um
pro′to·col′
pro′to·di′a·stol′ic
pro′ton
pro′to·path′ic
pro′to·plasm
pro′to·plas′mic
pro′to·plast′
pro′to·plas′tic
pro′to·por′phy·rin
pro′to·spasm
pro′to·troph′ic
pro′to·typ′al or
  pro′to·typ′ic or
  ·typ′i·cal
pro′to·type′

pro·tox′ide
pro′to·zo′an
pro′to·zo′ic
pro′to·zo′on
  ·zo′a
pro·tract′
pro·tract′i·ble
pro·trac′tion
pro·trac′tor
pro·trude′
  ·trud′ed ·trud′ing
pro·trud′ent
pro·tru′sion
pro·tru′sive
pro·tryp′sin
pro·tu′ber·ance
pro·tu′ber·ant
pro·tu′ber·ate′
  ·at′ed ·at′ing
proud flesh
pro′vi·rus
pro·vi′sion·al
pro·vi′ta·min
prov′o·ca′tion
pro·voc′a·tive
pro·voke′
  ·voked′ ·vok′ing
prox′i·mad′
prox′i·mal
prox′i·mate
prox′i·mo·a·tax′i·a
prox′i·mo·lin′gual
pro′zone
pro·zy′mo·gen
pru′ri·ence or
  ·en·cy
pru′ri·ent

pru·rig′i·nous
pru·ri′go
pru·rit′ic
pru·ri′tus
Prus′sak's space
Prus′sian blue
prus′si·ate′
prus′sic acid
psal·te′ri·um
  ·ri·a
psam′mo·car′ci·
  no′ma
  ·mas or ·ma·ta
psam·mo′ma
  ·mas or ·ma·ta
psam′mous
psau·os′co·py
psel′a·phe′si·a or
  pse·laph′e·sis
psel′lism
pseu′dar·thro′sis
pseu′des·the′si·a
pseu′do·an′gi·na
pseu′do·cele′ or
  ·coele′
pseu′do·cy·e′sis
  ·ses
pseu′do·gene′
pseu′do·her·maph′
  ro·dite′
pseu′do·her·maph′
  ro·dit′ic
pseu′do·her·maph′
  ro·dit′ism or
  ·ro·dism
pseu′do·mo′nad
pseu′do·mo′nas

pseu'do·morph'
pseu·do·mor'phous
 or ·mor'phic
pseu'do·preg'
 nan·cy
  ·cies
pseu'do·preg'nant
psi
psi'lo·cin
psi'lo·cy'bin
psi·lo'sis
psit'ta·co'sis
pso'as
pso·i'tis
pso'ra
psor'a·len
pso'ri·as'i·form'
pso·ri'a·sis'
 ·ses' (skin disease;
 SEE siriasis)
pso'ri·at'ic
pso'roph·thal'mi·a
psych
 psyched psych'ing
psy'cha·go'gy
psy·chal'gi·a or
 psy·chal·ga'li·a
psy·chal'gic
psych'as·the'ni·a
psych'as·then'ic
psy'che
psy·che·de'li·a
psy·che·del'ic
psy·chi·at'ric
psy·chi'a·trist
psy·chi'a·try

psy'chic or psy'chi·
 cal
psy'chism
psy'cho
psy'cho·a·cous'tic
 or ·cous'ti·cal
psy'cho·a·cous'tics
psy'cho·ac'tive
psy'cho·a·nal'y·sis
 ·ses'
psy'cho·an'a·lyst
psy'cho·an'a·lyt'ic
 or ·lyt'i·cal
psy'cho·an'a·lyze'
 ·lyzed' ·lyz'ing
psy'cho·bi·ol'o·gy
psy'cho·chem'i·cal
psy'cho·cor'ti·cal
psy'cho·di·ag·nos'
 tics
psy'cho·dra'ma
psy'cho·dra·mat'ic
psy'cho·dy·nam'ic
psy'cho·dy·nam'ics
psy'cho·ed'u·ca'
 tion·al
psy'cho·gen'der
psy'cho·gen'e·sis
 ·ses'
psy'cho·ge·net'ic
psy'cho·gen'ic
psy'cho·gram'
psy'cho·graph'
psy'cho·ki·ne'sis
 ·ses
psy'cho·ki·net'ic
psy'cho·lag'ny

psy'cho·lep'sy
psy'cho·lep'tic
psy'cho·lin·guis'tics
psy'cho·log'i·cal or
 ·log'ic
psy·chol'o·gist
psy·chol'o·gize'
 ·gized' ·giz'ing
psy·chol'o·gy
 gies
psy'cho·met'ric or
 ·met'ri·cal
psy·cho·me·tri'cian
psy'cho·met'rics
psy·chom'e·trist
psy·chom'e·try
 ·tries
psy'cho·mi·met'ic
psy'cho·mo'tor
psy'cho·neu·ro'sis
 ·ro'ses
psy'cho·neu·rot'ic
psy'cho·path'
psy'cho·path'ic
psy'cho·path·o·log'
 i·cal
psy'cho·pa·thol'o·
 gist
psy'cho·pa·thol'
 o·gy
psy·chop'a·thy
psy'cho·phar·ma·
 ceu'ti·cals
psy'cho·phar·ma·
 co·log'i·cal
psy'cho·phar·ma·
 col'o·gist

psy′cho·phar·ma·col′o·gy
psy′cho·phys′i·cist
psy′cho·phys′ics
psy′cho·phys′i·o·log′i·cal
psy′cho·phys·i·ol′o·gy
psy′cho·sen′so·ry
psy′cho·sex′u·al
psy′cho·sex·u·al′i·ty
psy′cho·sine′
psy·cho′sis
·ses
(mental disorder;
SEE *sycosis)*
psy′cho·so′cial
psy′cho·so·mat′ic
psy′cho·sur′ger·y
psy′cho·tech′nics
psy′cho·ther′a·peu′tic
psy′cho·ther′a·peu′tics
psy′cho·ther′a·pist
psy′cho·ther′a·py
psy·chot′ic
psy·chot′o·gen
psy·chot′o·mi·met′ic
psy′cho·ton′ic
psy′cho·tox′ic
psy′cho·trine′
psy′cho·trop′ic
psy·chro·al′gi·a
psy·chrom′e·ter
psy·chrom′e·try

psy′chro·ther′a·py
·pies
psyl′li·um
ptar′mic
ptar′mus
pte′ri·on
pte·ro′ic acid
pter′o·yl·mon′o·glu·tam′ic acid
pte·ryg′i·al
pte·ryg′i·um
·i·ums *or* ·i·a
pter′y·goid′
pter′y·go·max·il·la′re
(fissure point)
pter′y·go·max·il·lar′y
(of the pterygoid process)
pter′y·go·pal′a·tine′
pti·lo′sis
·ses
ptis′an
pto′maine
pto′sis
·ses
pto′tic
pty′a·lin
pty′a·lism
pty′a·lith′
pty′a·lize′
·lized′ ·liz′ing
pty′a·lo·lith′
pty·oc′ri·nous
ptys′ma·gogue′
pu·bar′che

pu′ber·tal
pu′ber·ty
pu·ber′u·lent
pu′bes
pu·bes′cence
pu·bes′cent
pu′be·trot′o·my
·mies
pu′bic
pu·bi·ot′o·my
·mies
pu′bis
pu′bes
pu′bo·fem′o·ral
pu′bo·pros·tat′ic
pu′bo·ves′i·cal
pu·den′dal
pu·den′dum
·den′da
pudg′y
·i·er ·i·est
pu′dic
pu′er·ile
pu′er·il·ism
pu·er·il′i·ty
·ties
pu·er′per·al
pu·er′per·al·ism
pu·er′per·ant
pu·er·pe′ri·um
puff
puff′ball′
puff′y
·i·er ·i·est
pu′li·cide′ *or* pu·lic′i·cide′

pul′ley
  ·leys
pul′lu·late′
  ·lat·ed ·lat·ing
pul′lu·la′tion
pul′mo·a·or′tic
pul′mo·lith′
pul′mom′e·try
  ·tries
pul′mo·nar′y or
  pul·mon′ic
pul′mo·nec′to·my
  ·mies
pul′mon·ol′o·gy
pul′mo·tor
pulp
pul′pal
pul′pi·fac′tion
pul′pi·fy′
  ·fied′ ·fy′ing
pul·pi′tis
pul′po·dis′tal
pul′po·don′ti·a
pulp′ous
pulp′y
  ·i·er ·i·est
pul′que
pul′sant
pul′sate
  ·sat·ed ·sat·ing
pul′sa·tile
pul·sa′tion
pul′sa·tive
pul′sa·tor
pul′sa·to′ry
pulse
  pulsed puls′ing

pul·sim′e·ter
pul′sion
pul·som′e·ter
pul′sus
pul·ta′ceous
pul′ver·iz′a·ble or
  pul′ver·a·ble
pul′ver·i·za′tion
pul′ver·ize′
  ·ized′ ·iz′ing
pul′ver·iz′er
pul·ver′u·lence
pul·ver′u·lent
pul·vi′nar
pul′vi·nate′
pum′ice
pu·mi′ceous
pump
pu′na
punch′–drunk′
punched′–out′
punc′tate or ·tat·ed
punc·ta′tion
punc′ti·form′
punc′to·graph′
punc′tum
  ·ta
punc′ture
  ·tured ·tur·ing
pun′gen·cy
pun′gent
pu′ny
  ·ni·er ·ni·est
pu′pil
pu′pil·lar′y
pu·pil·lom′e·ter

pu′pil·los′co·py
  ·pies
pu·pil′lo·sta·tom′e·ter
pur′blind′
pure
pure line
pur·ga′tion
pur′ga·tive
purge
  purged purg′ing
pu′ri·fi·ca′tion
pu·rif′i·ca·to·ry
pu′ri·fi′er
pu′ri·form′
pu′ri·fy′
  ·fied′ ·fy′ing
pu′rine
pu′ri·ty
Pur·kin′je′s cells
pu′ro·mu′cous
pu′ro·my′cin
pur·pu′ra
pur·pu′ric
pur′pu·rif′er·ous or
  pur′pu·rip′a·rous
  or pur′pu·rig′e·
  nous
pur′pu·rin
purr
pur′sy
  ·si·er ·si·est
pu′ru·lence or
  ·len·cy
pu′ru·lent
pus

push′–up′ *or* push′ up′
pus′sy
  ·si·er ·si·est
pus′tu·lant
pus′tu·lar *or* ·tu·lous
pus′tu·late′
  ·lat′ed ·lat′ing
pus′tu·la′tion
pus′tule
pus′tu·lo·crus·ta′ceous
pus′tu·lo′sis
  ·ses
pu·ta′men
  ·tam′i·na
pu′tre·fac′tion
pu′tre·fac′tive
pu′tre·fy′
  ·fied′ ·fy′ing
pu·tres′cence
pu·tres′cent
pu·tres′ci·ble
pu·tres′cine
pu′trid
pu·trid′i·ty
put′ty
P wave
py′ar·thro′sis
  ·ses
pyc·nom′e·ter
py′e·lec′ta·sis *or* ·e·lec·ta′si·a
py′e·lit′ic
py′e·li′tis
py′e·lo·cal′i·ce′al *or* ·cal′y·ce′al
py′e·lo·gram′
py′e·log′ra·phy
py′e·lo′ne·phri′tis
py′e·los′to·my
  ·mies
py′e·lot′o·my
  ·mies
py·em′e·sis
py·e′mi·a
py·e′mic
py′en·ceph′a·lus
py′gal
py·gal′gi·a
pyg·mae′an *or* ·me′an
pyg′moid
Pyg′my *or* pyg′my
  ·mies
pyg′my·ism
py·gop′a·gus
py′ic
py′in
pyk′nic
pyk′no·dys·os·to′sis
pyk·no·mor′phous
pyk·no′sis
  ses
pyk·not′ic
py′la
py′lar
py′lon
py′lo·rec′to·my
py·lor′ic
py·lo′rus
  ·ri
py′o·ceph′a·lus
py′o·der′ma
py′o·der′mic
py′o·gen′e·sis
py′o·gen′ic
py′oid
py′o·phy′so·me′tra
py′or·rhe′a *or* ·rhoe′a
pyorrhea al·ve′o·la′ris
py′or·rhe′al *or* ·rhoe′al
py·o′sis
py′o·stat′ic
pyr′a·mid
py·ram′i·dal
pyr·a·mid′ic *or* ·mid′i·cal
pyr′a·mis
  py·ram′i·des′
py·re′thrin
py·re′thrum
py·ret′ic
py·ret′i·co′sis
py·ret′o·gen
py′re·to·ther′a·py
  ·pies
Py′rex
  (trademark)
py·rex′i·a
py·rex′i·al *or* ·rex′ic
pyr′i·dine′
pyr′i·dox′al
pyr′i·dox′a·mine′
pyr′i·dox′ine

pyr′i·form′
py·rim′i·dine′
pyr′i·thi′a·mine
py′ro·cat′e·chol′ *or*
·cat′e·chin
py′ro·gal′lic acid
py′ro·gal′lol
py′ro·gen
py′ro·gen′ic *or* py·
rog′e·nous
py′ro·lig′ne·ous
py·rol′y·sis
·ses′
py′ro·lyt′ic
py′ro·ma′ni·a
py′ro·ma′ni·ac′
py′ro·ma·ni′a·cal
py′rone
py′ro·nine′
py′ro·nin′o·phil′i·a
py′ro·pho′bi·a
py′ro·phos′phate
py′ro·phos·phor′ic
acid
py′ro·punc′ture
py′ro·scope′
py·ro′sis
py·rot′ic
py′ro·tox′in
py·rox′y·lin *or* ·line
pyr′role
py′ru·vate
py·ru′vic acid
py′tho·gen′e·sis
·ses′
py′tho·gen′ic

py·u′ri·a

Q

qat
Q fever
Q–Tips, Q–tips
(trademarks)
Quaa′lude
(trademark)
quack
quack′er·y
quack′sal′ver
quad′ran′gle
quad·ran′gu·lar
quad′rant
quad·ran′tal
quad′rate
quad′ri·ceps′
quad′ri·cip′i·tal
quad′ri·gem′i·nal
quad′ri·lat′er·al
quad′ri·ple′gi·a
quad′ri·ple′gic
quad′ri·sect′
quad′ri·va′lent
quad′ru·ped′
quad·ru′pe·dal
quad·ru′ple
quad·ru′plet
quale
qual′i·fi·ca′tion
qual′i·fy′
·fied′ ·fy′ing
qual′i·ta′tive

qual′i·ty
·ties
qualm
quan′ta
(sing. quan′tum*)*
quan·tim′e·ter
quan′ti·tate′
·tat′ed ·tat′ing
quan′ti·ta′tive
quan′ti·ty
·ties
quan′tum
·ta
quar′an·tin′a·ble
quar′an·tine′
·tined′ ·tin′ing
quart
quar′tan
quar′ter
quar′tile
quar′ti·sect′
quartz
quas·sa′tion
quas′si·a
qua′ter·na′ry
qua′ver·y
quea′sy
·si·er ·si·est
que·bra′cho
·chos
Queck′en·stedt's′
sign
quench
quer·cet′ic
quer′ce·tin
quer′u·lent *or*
quer′u·lous

ques′tion
ques′tion·a·ble
quick′en
quick′lime′
quick′sil′ver
qui·es′cence
qui·es′cent
qui′et
quill
quil·lai′a *or* quil·laj′a *or* quil·lai′
quin′a·crine hydrochloride
Quinck′e's puncture
quin′ic acid
quin′i·dine′
qui′nine
quin′oid
qui·noi′dine
quin′o·line′
qui·none′
qui·non′i·mine′
quin′o·noid′
quin′qua·ge·nar′i·an
quin′que·va′lence *or* ·va′len·cy
quin′que·va′lent
quin′sy
quin′tan
quin·tu′ple
quin·tu′plet
quit′tor
quiv′er
quo′ta
quo·tid′i·an

quo′tient
Q wave

R

rab′bet
·bet·ted ·bet·ting
rab′bit fever
rab′bit·pox′
ra′bi·ate
rab′i·ci′dal
rab′id
ra·bid′i·ty
ra′bies
race
ra·ce′mate′
ra·ceme′
ra·ce′mic
rac′e·mism
rac′e·mi·za′tion
rac′e·mose′
ra′chi·o·camp′sis
ra′chi·op′a·gus
ra′chi·o·tome′
ra′chi·re·sis′tance
ra′chis
ra′chis·es *or* ra′chi·des′
ra·chis′chi·sis
·ses
ra·chit′ic
ra·chi′tis
rach′i·tome′
ra′cial
ra·clage′

rad
ra′dar·ky·mog′ra·phy
·phies
ra·dec′to·my
·mies
ra′di·al
ra′di·an
ra′di·ant
ra′di·ate′
at′ed ·at′ing
ra′di·a′tion
ra′di·a′tion·al
ra′di·a′tive
ra′di·a′tor
rad′i·cal
(group of atoms; SEE radicle)
rad′i·ces′
(sing. rad′ix)
rad′i·cle
(rootlet; SEE radical)
ra·dic′u·lar
ra·dic′u·lec′to·my
·mies
ra·dic′u·lo·my′e·lop′a·thy
ra′di·i′
(sing. ra′di·us)
ra′di·o·ac′tive
ra′di·o·ac·tiv′i·ty
·ties
ra′di·o·au′to·graph′
ra′di·o·au′to·graph′ic

ra·di·o·au·tog′ra·phy
  ·phies
ra′di·obe′
ra′di·o·bi′o·log′i·cal
ra′di·o·bi·ol′o·gist
ra′di·o·bi·ol′o·gy
ra′di·o·car′bon
ra′di·o·chem′i·cal
ra′di·o·chem′is·try
ra′di·ode′
ra′di·o·el′e·ment
ra′di·o·fre′quen·cy
ra′di·o·gen′ic
ra′di·o·gram′
ra′di·o·graph′
ra′di·og′ra·pher
ra′di·o·graph′ic
ra′di·og′ra·phy
ra′di·o·im′mu·no·as′say
ra′di·o·i′o·dine′
ra′di·o·i′so·tope′
ra′di·o·i′so·top′ic
ra′di·o·log′ic or ·log′i·cal
ra′di·ol′o·gist
ra′di·ol′o·gy
ra′di·o·lu′cen·cy
  ·cies
ra′di·o·lu′cent
ra·di′o·lus
ra′di·ol′y·sis
  ·ses′
ra′di·o·lyt′ic
ra′di·om′e·ter
ra′di·o·met′ric
ra′di·om′e·try
  ·tries
ra′di·o·mi·met′ic
ra′di·on′
ra′di·o·ne·cro′sis
ra′di·o·nu′clide
ra′di·o·pac′i·ty
  ·ties
ra′di·o·paque′
ra′di·o·phar′ma·ceu′ti·cal
ra′di·o·pro·tec′tive
ra′di·o·scop′ic
ra′di·os′co·py
  ·pies
ra′di·o·sen′si·tive
ra′di·o·sen′si·tiv′i·ty
  ·ties
ra′di·o·stron′ti·um
ra′di·o·ther′a·py
  ·pies
ra′di·o·ther′my
ra′di·o·tho′ri·um
ra′di·o·trac′er
ra′di·o·trop′ic
ra′di·um
ra′di·us
  ·di·i′ or ·us·es
rad′ix
  rad′i·ces′ or rad′ix·es
ra′don
raf′fi·nose′
rage
rag′o·cyte′
rag′weed′
rail′li·e·ti·ni′a·sis

raise
  raised rais′ing
râle or rale
ral′ly
  ·lied ·ly·ing
ra′mal
ra′mi
  (sing. ra′mus)
ram′i·fi·ca′tion
ram′i·form′
ram′i·fy′
  ·fied′ ·fy′ing
ram′i·sec′tion or ·sec′to·my
ra′mol·lisse·ment′
ra′mose or ra′mous
ramp
ram′part
ram′u·lose′
ram′u·lus
  ·li
ra′mus
  ·mi
ran′cid
ran·cid′i·fy′
  ·fied′ ·fy′ing
ran·cid′i·ty
ran′dom
ran′dom·i·za′tion
ran′dom·ize′
  ·ized′ ·iz′ing
range
  ranged rang′ing
ra′nine
rank

ran'kle
  ·kled ·kling
ran'u·la
ran'u·lar
ra·nun'cu·lus
  ·lus·es *or* ·li'
rape
  raped rap'ing
rape'seed'
ra'phe
rap·port'
rap'ture of the deep
rap'tus
  ·ti
rar'e·fac'tion
rar'e·fac'tive
rar'e·fy'
  ·fied' ·fy'ing
ra·scet'a
rash
ra'sion
ras'pa·to'ry
rasp'y
  rasp'i·er ·i·est
rat'bite' fever
rate
ra'tio
  ·tios
ra'tion
ra'tion·al
ra'tion·ale'
ra'tion·al'i·ty
  ·ties
ra'tion·al·i·za'tion
ra'tion·al·ize'
  ·ized' ·iz'ing

rat mite
rats'bane'
rat'tle
  ·tled ·tling
rat'tle·snake'
rau'cous
rau·wol'fi·a
rave
  raved rav'ing
ray
ray'age
Ray'leigh test
Ray·naud's' disease
re'ab·sorb'
re'ab·sorp'tion
re·act'
re·act'ance
re·act'ant
re·ac'tion
re·ac'tion·al
re·ac'ti·vate'
  ·vat'ed ·vat'ing
re·ac'ti·va'tion
re·ac'tive
re·ac·tiv'i·ty
  ·ties
re·ac'tor
read
  read read'ing
re'ad·just'
re'ad·mis'sion
re'ad·mit'tance
read'out'
read'through'
re·a'gent
re·a·gin'
re·a·gin'ic

re·al'gar
re'a·lign'ment
re·al'i·ty
  ·ties
ream'er
re'am·pu·ta'tion
re·an'i·mate'
  ·mat'ed ·mat'ing
re·an'i·ma'tion
re'ar·range'
  ·ranged' ·rang'ing
rea'son
rea'son·a·ble
re'as·sur'ance
re'as·sure'
  ·sured' ·sur'ing
re'at·tach'ment
Re'au·mur' scale
re'base
re·bel'lious
re'bound' *n.*
re·bound' *v.*
re·cal'ci·fi·ca'tion
re·cal'ci·trant
re·call'
re'ca·pit'u·late'
  ·lat'ed ·lat'ing
re'ca·pit'u·la'tion
re·cede'
  ·ced'ed ·ced'ing
re·ceive'
  ·ceived' ·ceiv'ing
re·ceiv'er
re'cep·tac'u·lum
  ·la
re·cep'tive
re·cep'tor

201

re′cess
re·ces′sion
re·ces′sive
re·ces′sus
re·cid′i·va′tion
re·cid′i·vism
re·cid′i·vist
re·cid′i·vis′tic *or* re·cid′i·vous
rec′i·pe
re·cip′i·ence *or* ·i·en·cy
re·cip′i·ent
re·cip′i·o·mo′tor
re·cip′ro·cal
re·cip′ro·ca′tion
rec′i·proc′i·ty
 ·ties
rec′li·na′tion
re·cline′
 ·clined′ ·clin′ing
rec′og·ni′tion
rec′og·niz′a·ble
rec′og·nize′
 ·nized′ ·niz′ing
re·coil′
rec′ol·lec′tion
 (memory)
re′–col·lec′tion
 (collection of fluid)
re·com′bi·nant
re·com′bi·na′tion
re′com·po·si′tion
re′com·pres′sion
re′con
re′con·di′tion

re′con·sti·tute′
 ·tut′ed ·tut′ing
re′con·sti·tu′tion
re′con·struct′
re′con·struc′tion
re′con·struc′tive
re·con′tour
rec′ord *n., adj.*
re·cord′ *v.*
re·cord′er
re·cov′er
re·cov′er·y
 ·er·ies
rec′re·a′tion
rec′re·ment
rec′re·men′tal
rec′re·men·ti′tious
re′cru·desce′
 ·desced′ ·desc′ing
re′cru·des′cence
re′cru·des′cent
re·cruit′ment
rec′tal
rec′ti·fi′a·ble
rec′ti·fi·ca′tion
rec′ti·fi′er
rec′ti·fy′
 ·fied′ ·fy′ing
rec′ti·lin′e·ar *or* ·lin′e·al
rec′to·ab·dom′i·nal
rec′to·cele′
rec′to·sig′moid
rec·tos′to·my
rec′to·u·re′thral
rec′to·u′ter·ine
rec′to·vag′i·nal

rec′to·ves′i·cal
rec′tum
 ·tums *or* ·ta
rec′tus
 ·ti
re·cum′ben·cy
re·cum′bent
re·cu′per·ate′
 ·at′ed ·at′ing
re·cu′per·a′tion
re·cu′per·a′tive
re·cur′
 ·curred′ ·cur′ring
re·cur′rence
re·cur′rent
re·cur′vate
re′cur·va′tion
re·curve′
 ·curved′ ·curv′ing
re·cy′cle
 ·cled ·cling
red blood cell
Red Cross
re′di·a
 ·di·ae
red·in′te·grate′
 ·grat′ed ·grat′ing
red·in′te·gra′tion
red lead
red′out′
re′dox
re′–dress′
re·dresse·ment′
re·duce′
 ·duced′ ·duc′ing
re·duc′er

re·duc′i·bil′i·ty
·ties
re·duc′i·ble
re·duc′tase
re·duc′tion
re·dun′dance
re·dun′dan·cy
·cies
re·dun′dant
re·du′pli·cate′
·cat′ed ·cat′ing
re·du′pli·ca′tion
re·du′pli·ca′tive
re·du′vi·id
re·ed′u·cate′ or
re–ed′u·cate′ or
re·ëd′u·cate′
·cat′ed ·cat′ing
re·ed′u·ca′tion or
re–ed′u·ca′tion
re·ed′u·ca′tive or
re–ed′u·ca′tive
reef
re′en·force′ or
re′–en·force′ or
re′ën·force′
·forced′ ·forc′ing
re·en′try or re–en′try or re·ën′try
·tries
re·ep′i·the′li·al·i·za′tion
re·ep′i·the′li·al·ize′ or re–ep′i·the′li·al·ize′ or re·ëp′i·the′li·al·ize′
·ized′ ·iz′ing

re′ex·am′i·na′tion or re′–ex·am′i·na′tion or re′ëx·am′i·na′tion
re′ex·am′ine or re′–ex·am′ine or re′ëx·am′ine
re·fect′
re·fec′tion
re·fer
·ferred′ ·fer′ring
ref′er·ence
re·fer′ral
re′fill n.
re·fill′ v.
re·fill′a·ble
re·fine′
·fined′ ·fin′ing
re·flect′
re·flec′tion
re·flec′tion·al
re·flec′tor
re′flex
re·flex′o·gen′ic or re′flex·og′e·nous
re·flex′o·graph
re·flex′o·log′ic
re·flex·ol′o·gist
re·flex·ol′o·gy
re′flux
re·fract′
re·frac′tile
re·frac′tion
re·frac′tion·ist
re·frac′tive
re·frac·tiv′i·ty
·ties

re·frac·tom′e·ter
re·frac′tor
re·frac′to·ry
re·frac′ture
·tured ·tur·ing
re·fran′gi·bil′i·ty
·ties
re·fran′gi·ble
re·fresh′
re·frig′er·ant
re·frig′er·ate′
·at′ed ·at′ing
re·frig′er·a′tion
re·frig′er·a′tive or ·er·a·to′ry
re·frin′gent
re·fu′sion
re·gen′er·a·cy
re·gen′er·ate′
·at′ed ·at′ing
re·gen′er·ate adj.
re·gen′er·a′tion
re·gen′er·a′tive
re·gime′ or ré·gime′
reg′i·men
re′gi·o
re′gi·o′nes
re′gion
re′gion·al
reg′is·ter
reg′is·tered
reg′is·trant
reg′is·trar
reg′is·tra′tion
reg′is·try
·tries

reg′nan·cy
re′gress n.
re·gress′ v.
re·gres′sion
re·gres′sive
reg′u·la·ble
reg′u·lar
reg′u·lar′i·ty
 ·ties
reg′u·late′
 ·lat′ed ·lat′ing
reg′u·la′tion
reg′u·la′tive or
 ·la·to′ry
reg′u·la′tor
re·gur′gi·tant
re·gur′gi·tate′
 ·tat′ed ·tat′ing
re·gur′gi·ta′tion
re′ha·bil′i·tate′
 ·tat′ed ·tat′ing
re′ha·bil′i·ta′tion
re′ha·bil′i·ta′tive
re′ha·la′tion
re·hy′drate′
 ·drat′ed ·drat′ing
re′hy·dra′tion
Rei′chert's
 cartilage
re′im·plan·ta′tion
re′in·fec′tion
re′in·force′
 ·forced′ ·forc′ing
re′in·force′ment
re′in·forc′er
re′in·fu′sion
re′in·oc·u·la′tion

re′in·te·gra′tion
re′in·ver′sion
re·ject′
re·jec′tion
re·jec′tive
re·ju′ve·nate′
 ·nat′ed ·nat′ing
re·ju′ve·na′tion
re·ju′ve·na′tor
re·ju′ve·nes′cence
re·ju′ve·nes′cent
re·lapse′
 ·lapsed′ ·laps′ing
re·laps′er
re·lat′a·ble
re·late′
 ·lat′ed ·lat′ing
re·la′tion
re·la′tion·al
re·la′tion·ship′
rel′a·tive
re·lax′
re·lax′ant
re′lax·a′tion
re·lax′er
re·lax′in
re·learn′
re·lease′
 ·leased′ ·leas′ing
re·li′a·bil′i·ty
re·li′a·ble
re·li′ance
re·li′ant
re·lief′
re·lieve′
 ·lieved′ ·liev′ing

re·line′
 ·lined′ ·lin′ing
REM
 REMs
 (eye movement)
rem
 (radiation dosage)
re·mains′
re·me′di·al
re·me′di·a′tion
re·me′di·a′tion·al
rem′e·dy
 ·dies
 ·died ·dy·ing
re·min′er·al·i·za′tion
rem′i·nis′cence
re·mis′sion
re·mis′sive
re·mit′
 ·mit′ted ·mit′ting
re·mit′ta·ble
re·mit′tence
re·mit′tent
rem′nant
re·mote′
 ·mot′er ·mot′est
re·mov′a·ble
re·mov′al
re·move′
 ·moved′ ·mov′ing
ren
 re′nes
re′nal
re·nic′u·lus
 ·li
ren′i·fleur′

ren′i·form′
re′nin
ren′i·por′tal
ren′i·punc′ture
ren′net
ren′nin
re′no·gram′
re·nog′ra·phy
re′no·pri′val
re′o·vi′rus
rep
re·pair′
re·pair′a·ble
re·pand′
rep′a·ra·ble
re·par′a·tive
re·peat′a·ble
re·peat′er
re·pel′lent or ·lant
re·pel′ler
re′per·cus′sion
re′per·cus′sive
rep′e·ti′tion
re·pet′i·tive
re·place′
 ·placed′ ·plac′ing
re·place′ment
re·plant′
re′plan·ta′tion
re·plen′ish·er
re·plete′
re·ple′tion
rep′li·ca
rep′li·ca·ble
rep′li·cate′
 ·cat′ed ·cat′ing

rep′li·cate n.
rep′li·ca′tion
rep′li·con′
re·po′lar·i·za′tion
re·port′a·ble
re·pose′
 ·posed′ ·pos′ing
re·po·si′tion
re·pos′i·tor
re·pos′i·to′ry
 ·ries
re·press′
re·press′er or
 ·pres′sor
re·press′i·ble
re·pres′sion
re′pro·duce′
 ·duced′ ·duc′ing
re′pro·duc′i·ble
re′pro·duc′tion
re′pro·duc′tive
rep′tile
re·pul′sion
re·quire′
 ·quired′ ·quir′ing
re·search n.
re·search′ v.
re·search′er or
 ·search′ist
re·sect′
re·sec′tion
re·sec′to·scope′
re·ser′pine
re·serve′
 ·served′ ·serv′ing
res′er·voir′

re·set′
 ·set′ ·set′ting
re·set′ n.
re·shape′
 ·shaped′ ·shap′ing
res′i·den·cy
 ·cies
res′i·dent
re·sid′u·al
res′i·due′
re·sid′u·um
 ·u·a
re·sil′ience or
 ·ien·cy
re·sil′ient
res′in
res′in·ate′
 ·at′ed ·at′ing
res′in·oid′
res′in·ous or res′
 in·y
res′ ip′sa lo′qui·tor
re·sist′ance
re·sist′ant
re·sis′tor
res′o·lu′tion
re·solve′
 ·solved′ ·solv′ing
re·sol′vent
res′o·nance
res′o·nant
res′o·nate′
 ·nat′ed ·nat′ing
res′o·na′tor
re·sorb′
re·sorb′ent
res·or′cin·ol′

re·sorp′tion
re·sorp′tive
re·spir′a·bil′i·ty
  ·ties
re·spir′a·ble
res′pi·ra′tion
res′pi·ra′tion·al
res′pi·ra′tor
res′pi·ra·to′ry
re·spire′
  ·spired′ ·spir′ing
res′pi·rom′e·ter
re·sponse′
re·spon′si·bil′i·ty
  ·ties
re·spon′si·ble
re·spon′sive
rest cure
re′ste·no′sis
rest′ful
res′ti·form′
res′ti·tu′tion
res′ti·tu′tive
res′tive
res′to·ra′tion
re·stor′a·tive
re·store′
  ·stored′ ·stor′ing
re·strain′
re·strain′a·ble
re·straint′
re·sult′
re·sult′ant
re·su′pi·nate′
re·su′pi·na′tion
re′su·pine′
res′ur·rec′tion·ism
res′ur·rec′tion·ist
re·sus′ci·tate′
  ·tat′ed ·tat′ing
re·sus′ci·ta′tion
re·sus′ci·ta′tive
re·sus′ci·ta′tor
re·tain′er
re·tard′ant
re·tard′ate
re′tar·da′tion
re·tard′a·tive or
  ·a·to′ry
re·tard′ed
re·tard′er
retch
re′te
  ·ti·a
re·ten′tion
re·ten′tive
re·tic′u·lar
re·tic′u·late′
  ·lat′ed ·lat′ing
re·tic′u·late or
  ·lat′ed adj.
re·tic′u·la′tion
re·tic′u·li′tis
re·tic′u·lo·cyte′
re·tic′u·lo·cyt′ic
re·tic′u·lo·en′do·
  the′li·al
re·tic′u·lo·his′ti·o·
  cy·to′ma
re·tic′u·lo·his′ti·o·
  cy·to′sis
re·tic′u·loid′
re·tic′u·lo′sis
re·tic′u·lum
  ·la
re′ti·form′
ret′i·na
  ·nas or ·nae′
ret′i·nac′u·lar
ret′i·nac′u·lum
  ·u·la
ret′i·nal
ret′ine
ret′in·ene′
ret′i·ni′tis
retinitis pig′men·
  to′sa
ret′i·no·cho·roi·di′·
  tis
ret′i·noid′
ret′i·nol′
ret′i·nop′a·thy
  ·thies
ret′i·no·scope′
ret′i·no·scop′ic
ret′i·nos′co·py
  ·pies
ret′i·sper′sion
re′to·per′i·the′li·um
re·tort′
re·tract′
re·tract′a·bil′i·ty
  ·ties
re·tract′a·ble
re·trac′tile
re′trac·til′i·ty
re·trac′tion
re·trac′tor
re·treat′ism
re·trench′ment

re·triev′al
ret′ro·ac′tion
ret′ro·au·ric′u·lar
ret′ro·bul′bar
ret′ro·ce′cal
ret′ro·cede′
 ·ced′ed ·ced′ing
ret′ro·cer′vi·cal
ret′ro·ces′sion
ret′ro·coch′le·ar
ret′ro·co′lic
ret′ro·col′lic
ret′ro·col′lis
ret′ro·cur′sive
ret′ro·de·vi·a′tion
ret′ro·dis·place′ment
ret′ro·flex′ or
 ·flexed′
ret′ro·flex′ion or
 ·flec′tion
ret′ro·grade′
 ·grad′ed ·grad′ing
ret′ro·gress′
ret′ro·gres′sion
ret′ro·len′tal
ret′ro·lin′gual
ret′ro·mas′toid
ret′ro·pla′si·a
ret′ro·spec′tion
ret′ro·spec′tive
ret′ro·ver′sion
re·trude′
re·tru′sion
re·un′ient
re·ver′ber·a′tion

re·ver′sal
re·verse′
 ·versed′ ·vers′ing
re·vers′i·bil′i·ty
re·vers′i·ble
re·ver′sion
re·ver′sion·ar′y or
 ·sion·al
re·vert′
re·ver′tant
re·view′
re·vi′tal·i·za′tion
re·vi′tal·ize′
 ·ized′ ·iz′ing
re·vive′
 ·vived′ ·viv′ing
re·viv′i·fi·ca′tion
re·viv′i·fy′
 ·fied′ ·fy′ing
rev′i·vis′cent
rev′o·lute′
re·vul′sant
re·vul′sion
re·vul′sive
Reye's syndrome
rhab′doid
rhab′do·my·o′ma
 ·mas or ·ma·ta
rhab′do·vi′rus
rhag′a·des
rha·gad′i·form′
rham′nose
rheg′ma
rhe′ni·um
rhe′o·base′
rhe′o·bas′ic
rhe′o·log′i·cal

rhe·ol′o·gist
rhe·ol′o·gy
rhe·om′e·ter
rhe′o·met′ric
rhe′o·stat′
rhe′o·stat′ic
rhe′o·tac′tic
rhe′o·tax′is
rhe′o·trop′ic
rhe·ot′ro·pism
rhe′sus
rheum
rheu·mat′ic
rheu′ma·tism
rheu′ma·toid′
rheu′ma·tol′o·gist
rheu′ma·tol′o·gy
rheum′y
 ·i·er ·i·est
Rh factor
rhig′o·lene′
rhi′nal
rhi′nen·ce·phal′ic
rhi′nen·ceph′a·lon′
 ·la
rhi·ni′tis
rhi′no·lar·yn·gol′o·gist
rhi′no·lar·yn·gol′o·gy
rhi·nol′o·gist
rhi·nol′o·gy
rhi′no·phar′yn·gi′tis
rhi′no·plas′tic
rhi′no·plas′ty
 ·ties

rhi′nor·rhe′a
rhi′no·scope′
rhi·nos′co·py
  ·pies
rhi′no·vi′rus
rhi′zoid
rhi·zoi′dal
rhi′zo·mel′ic
rhi′zo·pod′
rhi·zot′o·my
  ·mies
rho
rho′da·mine′
rho′dic
rho′di·um
rho′do·phy·lac′tic
rho′do·phy·lax′is
rho·dop′sin
rhom′ben·ceph′a·lon
rhom′bic
rhom′bo·cele′
rhom′boid
rhom·boi′dal
rhom·boi′de·us
  ·de·i′
rhon′chal *or* ·chi·al
rhon′chus
  ·chi
rho′ta·cism
rhy·pa·ri′a
rhythm
rhyth′mic *or* ·mi·cal
rhyth·mic′i·ty
  ·ties

rhyt′i·dec′to·my
  ·mies
rhyt′i·do·plas′ty
  ·ties
rhyt′i·do′sis
  ·ses
rib′bon
rib cage
ri′bo·fla′vin *or* ·vine
ri′bo·nu′cle·ase′
ri′bo·nu·cle′ic acid
ri′bose
ri′bo·so′mal
ri′bo·some′
ri′cin
ric′in·o·le′ic acid
ric′in·o′le·in
rick′ets
rick·ett′si·a
  ·si·ae′ *or* ·si·as
rick·ett′si·al
rick·ett′si·o′sis
rick′et·y
ric′tal
ric′tus
rid
  rid *or* rid′ded rid′ding
rid′er's bone
ridge
  ridged ridg′ing
ri·fam′pin *or* ·fam′pi·cin
rif′a·my′cin
right′–hand′ed
right heart

rig′id
ri·gid′i·fi·ca′tion
ri·gid′i·fy′
  ·fied′ ·fy′ing
ri·gid′i·ty
  ·ties
rig′or
rig′or mor′tis
rim
  rimmed rim′ming
ri′ma
  ·mae
ri·mose′ *or* ·mous
ri·mos′i·ty
Ring′er's solution
ring finger
ring′–knife′
ring′worm′
rinse
  rinsed rins′ing
ri′pa
ri·par′i·an
ripe
rip′en
risk
ri·so′ri·us
  ·ri·i′
ri′sus
rit′u·al
rit′u·al·is′tic
ri′val·ry
  ·ries
riz′i·form′
rob′o·rant
ro·bust′
Ro·chelle′ salt

Rock'y Moun'tain
  spot'ted fever
rod
ro·dent'i·cide'
ro·do·nal'gi·a
roent'gen
roent'gen·ize'
  ·ized' ·iz'ing
roent'gen·o·gram'
roent'gen·o·
  graph'ic
roent'gen·og'ra·phy
roent'gen·o·log'ic
roent'gen·ol'o·gist
roent'gen·ol'o·gy
roent'gen·o·ther'
  a·py
Ro·lan'do's angle
roll'er
ro·man'o·pex'y
ro·man'o·scope'
Rom'berg's sign
ron·geur'
Rönt'gen
roof
room'ing–in'
root canal
root'let
Ror'schach test
ro·sa'ce·a
ros·an'i·line
ro'sa·ry
  ·ries
rose fever
ro·se'o·la
ro·se'o·lous

ro·sette'
ros'in
ros·tel'lar
ros'tel·late'
ros·tel'lum
  ·tel'la
ros'tral
ros'trate
ros'trum
  ·trums or ·tra
rot
  rot'ted rot'ting
ro'ta·me'ter
ro'ta·ry
ro'tat·a·ble
ro'tate
  ·tat·ed ·tat·ing
ro·ta'tion
ro·ta'tion·al
ro'ta·tor
  ro·ta·to'res
ro'ta·to'ry
ro'ta·vi'rus
rö'teln or roe'theln
ro'te·none'
rot'ten
ro·tund'
rouge
rough'age
rough'en
rou·leau'
  ·leaux' or ·leaus'
round'worm'
rub
  rubbed rub'bing
rub'ber
rub'ber–dam

rub'ber·y
ru'be·an'ic acid
ru·be'do
ru·be·fa'cient
ru·be·fac'tion
ru·bel'la
ru·be'o·la
ru·bes'cence
ru·bes'cent
ru·bid'i·um
ru·big'i·nous or
  ·nose'
ru'bor
ru'bric
ru'bro·spi'nal
ruc'tus
ru'di·ment
ru'di·men'ta·ry or
  ·men'tal
ru'di·men'tum
  ·ta
rue
ru'fous
ru'ga
  ·gae
ru'gate
ru'gi·tus
ru'gose or ·gous
ru·gos'i·ty
  ·ties
rule
ru'mi·nate'
  ·nat'ed ·nat'ing
ru'mi·na'tion
rump
Rumpf's symptom

run
  ran
  run
  run′ning
run′a·round′
ru′pi·a
ru′pi·al
rup′ture
  ·tured ·tur·ing
rush
Rus′sell's viper
rust
ru·ta·my′cin
ru·then′ic
ru·the′ni·ous
ru·the′ni·um
ruth′er·ford
ru·ti·do′sis
ru′tin

S

sab·a·dil′la
Sa′bin's vaccine
sab′u·lous
sa·bur′ra
sa·bur′ral
sac
sac·cade′
sac·cad′ic
sac′cate
sac·cha·rase′
sac′cha·rate′
sac·char′ic
sac′cha·ride′

sac·char′i·fi·ca′tion
sac·char′i·fy′
  ·fied′ ·fy′ing
sac·cha·rim′e·ter
sac′cha·rin
  (sugar substitute)
sac′cha·rine
  (sugarlike)
sac·cha·rin′i·ty
  ·ties
sac·cha·ro·ga·lac′tor·rhe·a
sac·cha·ro·lyt′ic
sac·cha·rom′e·ter
sac·cha·ro·my′ces
  ·my·ce′tes
sac′cha·rose′
sac′cu·lar
sac′cu·late′ *or*
  ·lat′ed
sac′cu·la′tion
sac′cule
sac′cu·lus
  ·li′
sac′cus
  ·ci
sa′crad
sa′cral
sa′cral·i·za′tion
sac′ral·ize′
  ·ized′ ·iz′ing
sa′cro·coc·cyg′e·al
sa′cro·coc′cyx
  ·coc·cy′ges *or*
  ·cyx·es
sa′cro·il′i·ac′
sa′cro·il′i·i′tis

sa′cro·lum′bar
sa′cro·sci·at′ic
sa′cro·spi′nal
sa·crot′o·my
sa′cro·u′ter·ine
sa′cro·ver′te·bral
sa′crum
  ·cra *or* ·crums
sac′to·sal′pinx
sad′dle
saddle block
  (anesthesia)
sad′ism
sad′ist
sa·dis′tic
sad′o·mas′o·chism
sad′o·mas′o·chist
sad′o·mas′o·chis′tic
safe′guard′
saf′flow′er
saf′fron
saf′ra·nine′ *or* ·nin
saf′role
sag′it·tal
sa′go
  ·gos
Saint An′tho·ny's
fire
Saint Vi′tus' *(or*
Vi′tus's*)* dance
sal
  sal′es
sa·laam′ convulsion
sa·la′cious
sal ammoniac
sal′i·cin
sa·lic′y·late′

sal'i·cyl'ic acid
sal'i·cyl'ism
sal'i·cyl·ize'
  ·ized' ·iz'ing
sa'lient
sal'i·fi'a·ble
sal'i·fy'
  ·fied' ·fy'ing
sa·lim'e·ter
sa'line
sa·lin'i·ty
sal'i·nom'e·ter
sa·li'va
sal'i·vant
sal'i·var'y
sal'i·vate'
  ·vat'ed ·vat'ing
sal'i·va'tion
sal'i·va'tor
sal'i·va·to'ry
Salk vaccine
sal'low
sal'mo·nel'la
  ·nel'lae or ·nel'la
  or ·nel'las
sal'mo·nel·lo'sis
  ·ses
sal'pin·gec'to·my
  ·mies
sal'pin·gem·
  phrax'is
sal·pin'gi·an
sal·pin'gi·on
sal·pin·gi'tis
sal·pin'go·cele'
sal·pin'gog'ra·phy
sal·pin'go·li·thi'a·sis

sal'pin·gol'y·sis
sal·pin'go-o'o·pho·
  rec'to·my
  ·mics
sal·pin'go-o'o·pho·
  ri'tis
sal·pin'go-o·oph'o·
  ro·cele'
sal·pin'go·pex'y
sal·pin'go·pha·ryn'
  ge·al
sal·pin'go·plas'ty
sal·pin·gos'to·my
  ·mies
sal·pin·got'o·my
  ·mies
sal'pinx
  sal·pin'ges
salt
sal·ta'tion
sal'ta·to'ri·al
sal'ta·to'ry
salt'-free' diet
salt'pe'ter
salt'wa'ter
sa·lu'bri·ous
sa·lu'bri·ty
sal'u·re'sis
sal'u·tar'y
sal'vage
  ·vaged ·vag·ing
salve
  salved salv'ing
sal'vi·a
sal vo·la'ti·le'
sa·ma'ri·um

sam'ple
  ·pled ·pling
san'a·tive
san'a·to'ri·um
  ·ri·ums or ·ri·a
san'a·to'ry
san'dal·wood' oil
san'da·rac'
sand'–blind'
sand fly
sane
San·fi·lip'po's
  disease
san·guic'o·lous
san·guif'er·ous
san'gui·mo'tor or
  ·mo'to·ry
san'gui·na'ri·a
san'gui·nar'y
san'guine
san·guin'e·ous
san·guin'o·lent
san·gui·su'ga
san'i·cle
sa'ni·es'
sa'ni·o·pu'ru·lent
sa'ni·o·se'rous
sa'ni·ous
san'i·tar'i·an
san'i·tar'i·um
  ·i·ums or ·i·a
san'i·tar'y
san'i·ta'tion
san'i·ti·za'tion
san'i·tize'
  ·tized' ·tiz'ing
san'i·tiz'er

211

san′i·ty
san·ton′i·ca
san′to·nin
sap
  sapped sap′ping
sa·phe′na
sa·phe′nous
sap′id
sa·pid′i·ty
sa′po
sap′o·na′ceous
sa·pon′i·fi′a·ble
sa·pon′i·fi·ca′tion
sa·pon′i·fi′er
sa·pon′i·fy′
  ·fied′ ·fy′ing
sap′o·nin
sa′por
sa·po·rif′ic *or* sa′-
  por·ous
sap′phism
sa·pre′mi·a *or*
  ·prae′mi·a
sa·pre′mic
sa·pro′bic
sap′ro·don′ti·a
sap′ro·gen′ic *or* sa·
  prog′e·nous
sa·proph′a·gous
sap′ro·phyte′
sap′ro·phyt′ic
sap′ro·zo′ic
sar′a·pus
sar·ci′na
  ·nas *or* ·nae
sar′cine
sar′co·blast′

sar′co·car′ci·no′ma
  ·mas *or* ·ma·ta
sar′coid
sar′coid·o′sis
  ·ses
sar·col′o·gy
sar·col′y·sis
sar′co·lyte′
sar′co·lyt′ic
sar·co′ma
  ·mas *or* ·ma·ta
sar·co′ma·toid′
sar·co′ma·to′sis
  ·ses
sar·co′ma·tous
sar′co·mere′
sar′co·plasm
sar′co·plast′
sar·cot′ic
sar′cous
sar·don′ic
sa′rin
sar′sa·pa·ril′la
sar·to′ri·us
sas′sa·fras′
sas′sy bark *or* sas′-
  sy·wood′
sat′el·lite′
sat′el·li·to′sis
sa·tia·bil′i·ty
sa′tia·ble
sa′ti·ate′
  ·at′ed ·at′ing
sa′ti·ate *adj.*
sa·ti·a′tion
sa·ti′e·ty
sat′u·ra·bil′i·ty

sat′u·ra·ble
sat′u·rant
sat′u·rate *adj.*
sat′u·rate′
  ·rat′ed ·rat′ing
sat′u·ra′tion
sat′u·ra′tor
sat′ur·nine′
sat′ur·nism
sat′yr
sat′y·ri′a·sis
sat′y·ro·ma′ni·a
sau′cer
sau′cer·i·za′tion
sau′cer·ize′
  ·ized′ ·iz′ing
sau′na
sau·ri′a·sis
sau′ro·der′ma
sav′in *or* ·ine
saw
sax·if′ra·gant
sax′i·tox′in
scab
  scabbed scab′bing
scab′by
  ·bi·er ·bi·est
sca′bi·cide′
sca′bies
sca·bi·et′ic
sca′bi·ous
sca·bri′ti·es
scab′rous
sca′la
  ·lae
sca·lar′i·form′
scald

scale
    scaled scal'ing
sca·lene'
sca'le·nec'to·my
sca'le·not'o·my
    ·mies
sca·le'nus
    ·ni
scal'er
scal'i·ness
scall
sca'lo·gram'
scalp
scal'pel
scal'pri·form'
scal'prum
scal'y
    ·i·er ·i·est
scam'mo·ny
    ·nies
scan
    scanned scan'ning
scan'di·um
scan'na·ble
scan'ner
scan·so'ri·us
scant
scant'y
    ·i·er ·i·est
scaph'o·ce·phal'ic
    or ·o·ceph'a·lous
scaph'oid
scaph'oi·di'tis
scap'u·la
    ·lae' or ·las
scap'u·lar

scap'u·lar'y
    ·ies
scap'u·lec'to·my
scap'u·lo·cla·vic'u·lar
sca'pus
    ·pi
scar
    scarred scar'ring
scar·a·bi'a·sis
scarf'skin'
scar'i·fi·ca'tion
scar'i·fi·ca'tor
scar'i·fi'er
scar'i·fy'
    ·fied' ·fy'ing
scar·la·ti'na
scar·la'ti·nal
scar·lat'i·nel'la
scar'let fever
Scar'pa's fluid
Scars'dale diet
scat·a·cra'tia
sca·te'mi·a
scat·o·log'i·cal
sca·tol'o·gy
sca·to'ma
sca·toph'a·gy
sca·tos'co·py
    ·pies
scat'ter
scat'u·la
scav'en·ger
sce·lal'gi·a
scent'ed
Scha'fer's method
Schäf'fer's reflex

Schaf'fer's test
schar'lach R
sched'ule
    ·uled ·ul·ing
sche'ma
    ·ma·ta
sche·mat'ic
sche·mat'o·gram'
sche·mat'o·graph'
scheme
Schick test
Schil'ling blood
    count
schin·dy·le'sis
schis·to·ce'li·a
schis'to·some'
schis'to·so·mi'a·sis
    ·ses'
schiz·am'ni·on
schiz·ax'on
schiz·en·ceph'a·ly
schiz'o
    schiz'os
schiz'o·gen'e·sis
    ·ses'
schi·zog'o·ny
schiz'oid
schiz'o·my·cete'
schiz'o·my·ce'tous
schiz'o·my·co'sis
schiz'ont
schiz'o·phrene'
schiz'o·phre'ni·a
schiz'o·phren'ic
schiz'o·thy'mi·a
schiz'o·thy'mic
school

Schult′ze's cells
Schultz reaction
Schwann's cell
sci·age′
sci·at′ic
sci·at′i·ca
sci′ence
sci·en·tif′ic
sci′en·tist
sci′er·o′pi·a
scil′la
scin′ti·gram′
scin·tig′ra·phy
scin′til·late′
  ·lat′ed ·lat′ing
scin′til·la′tion
scin′til·la′tor
scin′til·lom′e·ter
scin′ti·pho′to·graph′
scin′ti·pho·tog′ra·phy
scin′ti·scan′
scin′ti·scan′ner
sci′on
scir′rhoid
scir′rhous
  (of scirrhus)
scir′rhus
  ·rhi or ·rhus·es
  (cancerous tumor)
scis′sile
scis′sion
scis′sors
scis·su′ra
  ·rae
scis′sure

scle′ra
  ·ras or ·rae
scle′ral
scle·rec·ta′si·a
scle·rec·to·ir′i·do·di·al′y·sis
  ·ses
scle′re·de′ma
scle·ri′a·sis
scle·rit′is
scler·o·der′ma
scler·o·der′ma·tous
scle·rog′e·nous
scler′oid
scle·ro′ma
  ·ma·ta
scle·rom′e·ter
scle′ro·pro′tein
scle·rose′
  ·rosed′ ·ros′ing
scle·ro′sis
  ·ses
scle′ro·thrix′
scle·ro′tial
scle·rot′ic
scle·ro′ti·um
  ·ti·a
scle·rot′o·my
  ·mies
scle′ro·trich′i·a
scle′rous
scob′i·nate
sco′lex
  sco·le′ces or sco′li·ces′
sco′li·o·ky·pho′sis
sco′li·om′e·ter

sco·li·o′sis
  ·ses
sco′li·o·som′e·try
sco′li·ot′ic
sco·lop′sis
scom′broid
scoop
sco·pa′ri·us
scope
sco·pol′a·mine
sco·pom′e·ter
sco′po·mor′phin·ism
scor′a·cra′ti·a
scor·bu′tic or ·bu′ti·cal
scor·bu′ti·gen′ic
scor′di·ne′ma
score
  scored scor′ing
scor′ings
scor′pi·on
sco·to′ma
  ·ma·ta or ·mas
sco·tom′a·tous
sco·tom′e·ter
sco′to·mi·za′tion
sco′to·pho′bin
sco·to′pi·a
sco·to′pic
sco·tos′co·py
scout film
scrape
  scraped scrap′ing
scrap′er
scratch test
screen′ing

screen memory
screw'worm'
scribe
   scribed scrib'ing
scro·bic'u·late
scro·bic'u·lus
   ·li
scrof'u·la
scrof'u·lo·der'ma
scrof'u·lous
scro'tal
scro·tec'to·my
scro·ti'tis
scro'to·cele'
scro'tum
   ·ta *or* ·tums
scrub
   scrubbed scrub'bing
scru'ple
Scul·te'tus' bandage
scum
   scummed scum'ming
scum'my
   ·mi·er ·mi·est
scurf
scur'vy
scu'ta
   (*sing*. scu'tum)
scu'tate
scute
scu'ti·form'
scu'tum
   ·ta
scyb'a·lous

scyb'a·lum
   ·la
scy'phoid *or* scy'phi·form'
seal
   (*tight closure;* SEE seel)
seal'ant
search'er
sea'sick'
seat'worm'
se·ba'ceous
se·bac'ic acid
se·bif'er·ous *or* ·bip'a·rous
seb'or·rhe'a *or* ·rhoe'a
seb'or·rhe'ic *or* ·rhoe'ic
se'bum
se·cer'nent
se·clu'sion
se'co·bar'bi·tal'
sec'ond
sec'ond·ar'y
sec'ond–de·gree' burn
se·cret'a·gogue' *or* ·cret'o·gogue'
se·crete'
   ·cret'ed ·cret'ing
se·cre'tin
se·cre'tion
sec're·to·mo'tor *or* ·mo'to·ry
se·cre'to·ry
sec·tar'i·an

sec'tile
sec'ti·o
   sec'ti·o'nes
sec'tion
sec'tor
sec·to'ri·al
se·cun'di·grav'i·da
   ·das *or* ·dae
sec'un·dines'
se'cun·dip'a·ra
   ·ras *or* ·rae'
se·cure'
se·cu'ri·ty
   ·ties
se·date'
   ·dat'ed ·dat'ing
se·da'tion
sed'a·tive
sed'en·tar'y
sed'i·ment
sed'i·men'ta·ry
sed'i·men·ta'tion
sed'i·men·ta'tor
see
   saw seen see'ing
seed
seel
   (*to blind;* SEE seal)
seg'ment
seg·men'tal
seg'men·tar'y
seg'men·ta'tion
seg'ment·er
seg·men'tum
   ·ta
seg're·gate'
   ·gat'ed ·gat'ing

215

seg′re·ga′tion
seg′re·ga′tor
Seid′litz powders
seis·es·the′si·a
seis′mes·the′si·a
seis′mo·car′di·og′ra·phy
seis′mo·ther′a·py
sei′zure
se·junc′tion
se·lec′tion
se·lec′tive
se·lec·tiv′i·ty
se·lec′tor
se·le′nic
se·le′ni·um
se·le′no·dont′
self
  selves
self′–a·buse′
self′–cen′tered
self′–con′cept
self′–con′fi·dence
self′–con′scious
self′–con·trol′
self′–di·ges′tion
self′–dis·cov′er·y
self′–hyp·no′sis
self′–im′age
self′–in·duced′
self′–in·duc′tance
self′–in·flict′ed
self′–knowl′edge
self′–lim′it·ed
self′–med′i·cate′
  ·cat′ed ·cat′ing

self′–stim′u·la′tion
self′–wise′
sel′la
  ·lae *(depression;*
    SEE cella)
sel′lar
se·man′tics
se′mei·og′ra·phy
sem′el·in′ci·dent
se′men
  sem′i·na
sem′i·cir′cu·lar
sem′i·co′ma
sem′i·con′scious
sem′i·lu′nar
sem′i·nal
sem′i·na′tion
sem′i·nif′er·ous
se′mi·ol′o·gy
se′mi·ot′ic *or*
  ·ot′i·cal
se′mi·ot′ics
sem′i·per′me·a·ble
sem′i·pri′vate
sem′i·sul′cus
sem′i·syn·thet′ic
sem′i·sys′tem·at′ic
  (or sem′i·triv′i·al)
  name
sem′i·ten′di·nous
se·ne′ci·o
  ·ci·os
sen′e·ga
se·nes′cence
se·nes′cent
se′nile

se·nil′i·ty
  ·ties
se′ni·um
sen′na
se·nog′ra·phy
se·no′pi·a
sen′sate
sen·sa′tion
sen·sa′tion·al
sense
  sensed sens′ing
sense da′tum
  sense da′ta
sen′si·bil′i·ty
  ·ties
sen′si·ble
sen·sif′er·ous
sen·sig′e·nous
sen′si·tin′o·gen
sen′si·tive
sen′si·tiv′i·ty
  ·ties
sen′si·ti·za′tion
sen′si·tize′
  ·tized′ ·tiz′ing
sen′si·tiz′er
sen′si·tom′e·ter
sen′si·to·met′ric
sen′si·tom′e·try
sen′so·mo·bil′i·ty
sen′sor
sen′so·ri·mo′tor
sen′so·ri·neu′ral
sen·so′ri·um
  ·ri·ums *or* ·ri·a
sen′so·ri·vas′cu·lar
  or ·vas′o·mo′tor

sen′so·ry *or* sen·so′ri·al
sen′su·al
sen′su·al·ism
sen′su·al′i·ty
sen′tience *or* ·tien·cy
sen′tient
sen′ti·ment
sen′ti·nel node
sep′a·rate
  ·rat′ed ·rat′ing
sep′a·rate *n., adj.*
sep′a·ra′tion
sep′a·ra′tor
sep′sis
  ·ses
sep′ta
  (sing. sep′tum)
sep′tal
sep′tate
sep·tec′to·my
sep′tic
sep′ti·ce′mi·a
sep′ti·ce′mic
sep′ti·cine
sep′ti·co·py·e′mi·a
sep′ti·co·py·e′mic
sep′to·mar′gi·nal
sep·tom′e·ter
sep′to·tome′
sep·tot′o·my
sep′tu·a·ge·nar′i·an
sep′tu·lum
  ·la
sep′tum
  ·tums *or* ·ta

sep·tu′plet
se′quel
se·que′la
  ·lae
se′quence
  ·quenced ·quenc·ing
se·quen′tial
se·ques′ter
se·ques′trant
se′ques·tra′tion
se′ques·trec′to·my *or* ·trot′o·my
  ·mies
se·ques′trum
  ·trums *or* ·tra
se′ra
  (sing. se′rum)
se′ral·bu′min
se′ra·phe·re′sis
  ·ses
se·rem′pi·on
ser′en·dip′i·tous
ser′en·dip′i·ty
se′ri·al
ser′i·cin
se′ries
ser′i·flux′
ser′ine
se′ri·os′co·py
ser′i·scis′sion
se′ro·fast′
se′ro·log′ic *or* ·log′ical
se·rol′o·gist
se·rol′o·gy
se′ro·neg′a·tive
se′ro·pos′i·tive

se′ro·prog·no′sis
se′ro·pu′ru·lent
se·ro′sa
  ·sas *or* ·sae
se·ro′sal
se′ro·san·guin′e·ous
se·ros′i·ty
  ·ties
se′ro·to′nin
se′ro·type′
se′rous
se′ro·var′
ser·pig′i·nous
ser·pi′go
ser′rate *adj.*
ser·rate′
  ·rat′ed ·rat′ing
ser·ra′tion *or* ser′ra·ture
serre·fine′
serre·noeud′
ser′ru·late *or* ·lat′ed
ser′ru·la′tion
se′rum
  ·rums *or* ·ra
se′rum·al
ser′vo·mech′a·nism
ser′vo·mo′tor
ses′a·me′
ses′a·moid′
ses′qui·ho′ra
ses′qui·ter′pene
ses′sile
set
  set set′ting

se′ta
·tae
se·ta′ceous
set′back′
se·tif′er·ous *or* se·tig′er·ous
se′ton
se′tose
set′up
sev′er
se′vo·flu′ran
sew′age
sex·a·ge·nar′i·an
sex′–con·di′tioned
sex·dig′i·tate′
sex·duc′tion
sex hygiene
sex′i·va′lent
sex′–lim′it·ed
sex′–link′age
sex′–linked′
sex·ol′o·gist
sex·ol′o·gy
sex′tan
sex·tu′plet
sex′u·al
sex′u·al′i·ty
sex′u·al·ize′
·ized′ ·iz′ing
Sé·za·ry′ cell
shad′ow
shad′ow–cast′ing
shad′ow·graph′
shad′ow·y
shaft
sha·green′ patch

shake
shook shak′en shak′ing
shakes
shak′ing palsy
shak′y
·i·er ·i·est
sha′man
·mans
sha′man·ism
shank
shape
shaped shap′ing
shark′skin′
shear
shears
sheath
sheaths
sheathe
sheathed sheath′ing
shed
shed shed′ding
sheet
shelf
shel·lac′ *or* ·lack′
·lacked′ ·lack′ing
shell shock
shell′shocked′
shi·at′su (massage)
shield
shift
shi·gel′la
·gel′lae *or* ·gel′las
shi′gel·lo′sis
shi·kim′ic acid
shin′bone′
shin′gles

shin′plas′ter
shin′splints′
shiv′er
shiv′er·y
shock therapy
shoe
shoot
shot shoot′ing
short′en
short′head′
short′head′ed
short′–lived′
short′sight′ed
short′–wind′ed
shot′–feel′
shot′gun′ experiment
shoul′der
shoulder blade
show
show′er
shreds
shriv′el
·eled *or* ·elled
·el·ing *or* ·el·ling
shud′der
shunt
si′ag·an·tri′tis *or* ·a·gon′an·tri′tis
si·al′a·den
si·al′a·gog′ic
si·al′a·gogue′
si·al′ic
si′a·lin
(salivary tetrapeptide)

si′a·line
  (of saliva)
si′a·lism or si′a·lis′mus
si′a·li·tis
si′a·log′ra·phy
  ·phies
si′a·loid′
si′a·lo·li·thi′a·sis
  ·ses′
si′a·lo·li·thot′o·my
si′a·los′che·sis
si′a·lot′ic
Si′a·mese′ twins
sib
sib′i·lant
sib′i·la′tion
sib′i·lus
sib′ling
sib′ship′
sic′ca·tive
sic·cha′si·a
sic′co·la′bile
sic′cus
sick bay
sick′bed′
sick call
sick′en
sick′ish
sick′le–cell′ anemia
sick·le′mi·a
sick′ly
  ·li·er ·li·est
sick′ly
  ·lied ·ly·ing
sick′ness
sick′room′

side effect
sid′er·a′tion
sid′er·ism
sid′er·o·cyte′
sid′cr·o·pe′ni·a
sid′er·oph′i·lous
sid′er·o′sis
sid′er·o·some′
sid′er·ot′ic
SIDS
sie′mens
Sie′mens′ syndrome
sieve
  sieved siev′ing
sigh
sight′ed
sig′ma
sig′ma·tism
sig′moid or sig·moid′al
sig·moi′do·proc·tos′to·my
sig·moid′o·scope′
sig·moid′o·scop′ic
sig′moid·os′co·py
  ·pies
sig′moid·ot′o·my
sign
sig′na
sig′nal
sig′na·ture
sig·nif′i·cant
sig′ni·fy′
  ·fied′ ·fy′ing
sign language

sig′num
  ·na
Si·las′tic
  (trademark)
si′lent
si′lex
sil′i·ca
sil′i·cate
si·li′ceous or ·cious
si·lic′ic
si·li′ci·um
sil′i·co·flu′o·ride′
sil′i·con
  (element)
sil′i·cone′
  (plastic compound)
sil′i·co·sid′er·o′sis
sil′i·co′sis
  ·ses
sil′i·qua
sil′i·quose′ or ·quous
sil′ver nitrate
sim′es·the′si·a
sim′i·an crease
sim′i·lar
si·mil′i·a si·mil′i·bus cu·ran′tur
sim′i·ous
sim′ple
  ·pler ·plest
si′mul
sim′u·lant
sim′u·late′
  ·lat′ed ·lat′ing
sim′u·la′tion
sim′u·la′tive

sim′u·la·tor
si′nal
sin′a·pism
sin·cip′i·tal
sin′ci·put′
  sin·cip′i·ta *or* sin′ci·puts
sin′ew
sin′ew·y
sin′gle–cell′ protein
sin·gul·ta′tion
sin·gul′tous
sin′is·ter
sin′is·trad′
sin′is·tral
sin′is·tral′i·ty
sin′is·tro·car′di·a
sin′is·troc·u′lar
sin′is·tro·dex′tral
sin′is·trorse′
sin′is·tro·tor′sion
sin′is·trous
si′no·a′tri·al
si′no·gram′
si·nog′ra·phy
sin′ter
sin′u·os′i·ty
  ·ties
sin′u·ous *or* ·u·ate
si′nus
si′nus·al
si′nus·i′tis
si′nu·soid′
si′nu·soi′dal
si′nu·soi′dal·i·za′tion

si′nus·ot′o·my
si′phon
si′phon·age
si′phon·al *or* si·phon′ic
si·ren′i·form′
si′ren·o·me′li·a
si′ren·om′e·lus
  ·li
si·ri′a·sis
  ·ses′ *(sunstroke;* SEE *psoriasis)*
sir′up
sis′mo·ther′a·py
sis′ter
site
sit′i·eir′gi·a
si·tol′o·gy
  (dietetics; SEE *cytology)*
si′to·pho′bi·a
si·tos·ter′ol′
si′to·tox′ism
si·tot′ro·pism
sit′u·ate′
  ·at′ed ·at′ing
sit′u·a′tion
sit′u·a′tion·al
si′tus
sitz bath
skat′ole
skein
skel′e·tal
skel′e·ti·za′tion
skel′e·tog′e·nous
skel′e·ton
skene′o·scope′

Skene's glands
ske·ni′tis
ske′o·cy·to′sis
skew
ski′a·graph′ *or* ·gram′
ski·ag′ra·phy
ski′a·scope′
ski·as′co·py
ski′a·sco·tom′e·try
skim
  skimmed skim′ming
skin
skin′fold′ thick′ness
Skin′ner box
skin′ny
  ·ni·er ·ni·est
skin test
skull′cap′
sky blue
slab′–off′
slake
  slaked slak′ing
slant
slave
slav′er
sleep
  slept sleep′ing
sleep′walk′er
slide
  slid slid′ing
slim′y
  ·i·er ·i·est
sling

slip
  slipped slip′ping
slit
  slit slit′ting
slit′lamp′
slope
  sloped slop′ing
slough
slow
slows
sludge
sludg′ing
sluice′way′
slur
  slurred slur′ring
slur′ry
  ·ries
slyke
small′pox′
smear
smeg′ma
smeg·mat′ic
smeg′mo·lith′
smell
  smelled *or* smelt
  smell′ing
smi′lax
smog
smoke
smok′er
smooth
smoth′er
smudg′ing
snake′bite′
snap
  snapped snap′ping

snare
  snared snar′ing
sneeze
  sneezed sneez′ing
Snel′len's test
snore
  snored snor′ing
snor′er
snow′–blind′
snuff
snuf′fles
soap′bark′
soap′suds′
sob
  sobbed sob′bing
so′ber
so′cial
so′cial·i·za′tion
so′cial·ize′
  ·ized′ ·iz′ing
so′ci·o·a·cu′sis
so′ci·o·bi′o·log′i·cal
so′ci·o·bi·ol′o·gist
so′ci·o·bi·ol′o·gy
so′ci·o·e′co·nom′ic
so′ci·o·log′i·cal *or*
  ·log′ic
so′ci·ol′o·gist
so′ci·ol′o·gy
so′ci·o·met′ric
so′ci·om′e·try
so′ci·o·path′
so′ci·o·path′ic
so′ci·op′a·thy
so′ci·o·psy′cho·log′-
  i·cal
sock′et

so′da
so′dic
so′di·o·ci′trate
so′di·um
so′do·ko′sis
so′do·ku′
sod′om·ite′ *or* sod′-
  om·ist
sod′om·ize′
  ·ized′ ·iz′ing
sod′om·y
sof′ten
soix′ante′–neuf′
sol
so·la·na′ceous
so′la·nine′ *or* ·nin
so′la·noid′
  (potatolike; SEE
  solenoid*)*
so·la′num
so′lar
so·lar′i·um
  ·i·a
so′lar·i·za′tion
so′lar·ize′
  ·ized′ ·iz′ing
solar plexus
sol′ate
  ·at·ed ·at·ing
sol·a′tion
sol′der
sole
so·le′no·glyph′
so′le·noid′
  (electrical coil; SEE
  solanoid*)*
so′le·noi′dal

221

sole′plate′
so′le·us
·le·i
sol′id
sol·id′i·fi·ca′tion
sol·id′i·fy′
·fied′ ·fy′ing
sol′i·dism
sol′i·dus
·i·di′
sol′ip·sism
sol′ip·sist
sol′ip·sis′tic
sol′i·tar′y
sol·u·bil′i·ty
·ties
sol·u·bi·li·za′tion
sol·u·bi·lize′
·lized′ ·liz′ing
sol′u·ble
sol′ute
so·lu′tion
solv·a·bil′i·ty
·ties
solv′a·ble
sol′vate
·vat·ed ·vat·ing
sol·va′tion
sol′ven·cy
sol′vent
sol·vol′y·sis
so′ma
so′ma·ta
so′ma·tal′gi·a
so′mat·as·the′ni·a
so′mat·es·the′si·a
so·mat′ic

so·mat′i·cism
so·mat′i·cize′
·cized′ ·ciz′ing
so′mat·i·za′tion
so′ma·tize′
·tized′ ·tiz′ing
so′ma·to·cep′tor
so′ma·to·log′ic or
·log′i·cal
so′ma·tol′o·gist
so′ma·tol′o·gy
so′ma·to·me′din
so′ma·to·path′ic
so′ma·to·plasm′
so′ma·to·plas′tic
so′ma·to·pleu′ral
so′ma·to·pleure′
so′ma·to·psy′chic
so′ma·to·psy·cho′sis
so′ma·tos′co·py
so′ma·to·stat′in
so′ma·to·trop′ic
so′ma·to·tro′pin or
·tro′phin
so′ma·to·type′
so′mite
so·mit′ic or so′mi·tal
som·nam′bu·lant
som·nam′bu·late′
·lat′ed ·lat′ing
som·nam′bu·la′tion
som·nam′bu·la′tor
som·nam′bu·lism
som·nam′bu·list
som·nam′bu·lis′tic

som′ni·fa′cient
som·nif′er·ous
som·nil′o·quist
som·nil′o·quy
som′no·lence or
·len·cy
som′no·lent
som·no·len′ti·a
sone
son′ic
son′i·cate′
·cat′ed ·cat′ing
son′i·ca′tion
son′i·ca′tor
so·ni′tus
son′o·gram′
son′o·graph′
so·nog′ra·phy
so·no·lu′cent
so·nor′i·ty
·ties
so·no′rous
so·phis′ti·cate′
·cat′ed ·cat′ing
so·phis′ti·ca′tion
so′por
sop′o·rif′ic or
·rif′er·ous
sorb
sor′be·fa′cient
sor′bent
sor′bic acid
sor′bi·tol′
sor′bose
sor′des
sore
sor′er sor′est

so·ro'che
so·ro'ri·a'tion
sorp'tion
souf'fle
sound wave
spa
space
spac'er
spa'cial
spa·gyr'ic
spag'y·rist
spall·a'tion
span
   spanned span'ning
span'dex
Span'ish fly
spank
spar'er
spar'ga·no'sis
  ·ses
spar·ga'num
  ·na
sparge
  sparged sparg'ing
spark
spar'te·ine'
spasm
spas·mod'ic or
  ·mod'i·cal
spas'mo·gen
spas'mo·gen'ic
spas·mol'o·gy
spas'mo·lyg'mous
spas'mus
spas'tic
spas·tic'i·ty
  ·ties

spa'tial
spa'ti·al'i·ty
spa'ti·um
spat'u·la
spat'u·lar
spat'u·late adj.
spat'u·late'
  ·lat'ed ·lat'ing
spat'u·la'tion
spear'mint'
spe'cial·ism
spe'cial·ist
spe'cial·is'tic
spe'cial·i·za'tion
spe'cial·ize'
  ·ized' ·iz'ing
spe'cial·ty
  ·ties
spe'cies
spe·cif'ic
spec'i·fic'i·ty
  ·ties
spe·cil'lum
  ·la
spec'i·men
spec'ta·cles
spec'tra
  (sing. spec'trum)
spec'tral
spec'tro·chem'i·cal
spec'tro·chem'is·try
spec'tro·gram'
spec'tro·graph'
spec'tro·graph'ic
spec·trom'e·ter
spec'tro·met'ric
spec·trom'e·try

spec'tro·pho·tom'e·ter
spec'tro·pho'to·met'ric
spec'tro·pho·tom'e·try
spec'tro·scope'
spec'tro·scop'ic
spec·tros'co·pist
spec·tros'co·py
spec'trum
  ·tra or ·trums
spec'u·lar
spec'u·lum
  ·la or ·lums
speech
speed
spell
sperm
sper'ma·ce'ti
sper'ma·ry
  ·ries
sper·mat'ic
sper'ma·tid
sper'ma·ti'tis
sper·mat'o·cys·tec'to·my
sper·mat'o·cyte'
sper·mat'o·gen'e·sis
sper·mat'o·ge·net'ic
sper·mat'o·go'ni·al
sper·mat'o·go'ni·um
  ·ni·a
sper'ma·toid'
sper·mat'or·rhe'a

sper·mat′o·zo′al *or*
·zo′an *or* ·zo′ic
sper·mat′o·zo′on
·zo′a
sper′ma·tu′ri·a
sperm′i·ci′dal
sperm′i·cide′
sper′mi·dine′
sper′mi·duct′
sperm′ine
sper′mi·o·gen′e·sis
·ses′
sper·mo·go′ni·um
·ni·a
sper′mo·lith′
sper′mo·phle′bec·ta′si·a
sper′mo·plasm′
sper′mous
sphac′e·late′
·lat′ed ·lat′ing
sphac′e·la′tion
sphac′e·lism
sphac′e·lous
(gangrenous)
sphac′e·lus
(gangrenous mass)
spha′gi·as′mus
sphe′ni·on
·ni·a
sphe′no·eth′moid
or sphen·eth′moid
sphe′noid *or* sphe·noi′dal
sphe′noi·dot′o·my
sphe′no·max′il·lar′y
sphe·not′ic

sphe′no·tre′si·a
sphe′no·tribe′
spher′al
sphere
spher′i·cal *or*
spher′ic
sphe′ro·cyte′
sphe′ro·cyt′ic
sphe′ro·cy·to′sis
sphe′roid *or* sphe·roid′al
sphe·rom′e·ter
spher′o·plast′
spher′u·lar
spher′ule
sphinc′ter
sphinc′ter·al
sphinc·ter·al′gi·a
sphinc′ter·is′mus
sphinc′ter·os′co·py
sphinc′ter·o·tome′
sphin′go·ga·lac′to·side′
sphin′go·lip′id
sphin′go·lip′i·do′sis
·ses
sphin′go·my′e·lin
sphin′go·sine′
sphyg′mic
sphyg′mo·bo·lom′e·ter
sphyg′mo·gram′
sphyg′mo·graph′
sphyg′mo·graph′ic
sphyg·mog′ra·phy
sphyg′moid

sphyg′mo·ma·nom′e·ter
sphyg·mom′e·ter
sphyg′mo·pal·pa′tion
sphyg′mo·scope′
sphyg′mus
sphy·rot′o·my
spi′ca
·cae
spic′u·late′ *or* ·lar
spic′ule
spic′u·lum
·la
spi′der
spi′der–burst′
spi·ge′li·an line
spike
spiked spik′ing
spill
spilled *or* spilt
spill′ing
spill′age
spill′way′
spi·lo′ma
·mas *or* ·ma·ta
spi′lo·pla′ni·a
spi′lo·plax′i·a
spi′na
·nae
spina bif′i·da
spi′nal
spi′nant
spi′nate
spin′dle
spine
spi·nif′u·gal

224

spi·nip′e·tal
spinn′bar·keit′
spi′no·gle′noid
spi′nose
spi′no·tec′tal
spi′nous
spin·thar′i·scope′
spin′ther·ism
spin′ther·om′e·ter
spin′y–head′ed worm
spi′ra·cle
spi·rac′u·lar
spi·rad′e·no′ma
 ·mas or ·ma·ta
spi′ral
spi′ra·my′cin
spi′reme
spi·ril′li·ci′dal
spi′ril·lol′y·sis
spi′ril·lo′sis
spi·ril′lum
 ·la
spir′it
spir′it·u·ous or spir′i·tous
spir′i·tus fru·men′ti
spi′ro·chet′al
spi′ro·chete′ or ·chaete′
spi′ro·che·tol′y·sis
spi′ro·che·to′sis
 ·ses
spi′ro·che·tot′ic
spi′ro·gram′
spi′ro·graph′
spi′ro·graph′ic

spi′roid
spi′ro–in′dex
spi·rom′e·ter
spi·ro·met′ric
spi·rom′e·try
spi′ro·phore′
spi′ro·scope′
spi·ros′co·py
spi′ru·roid′
spis′sat·ed
spis′si·tude
spit
 spit spit′ting
spit′al *(hospital)*
spit′ter
spit′tle *(saliva)*
splanch′nap·o·phys′e·al
splanch′na·poph′y·sis
splanch′nic
splanch′no·cele′ *(visceral hernia)*
splanch′no·coele′ *(pleuroperitoneal cavity)*
splanch′no·lith′
splanch·nol′o·gy
splanch′no·pleu′ral
splanch′no·pleure′
splanch′nop·to′sis or ·to′si·a
splanch·not′o·my
S–plas′ty
splay
splay′foot′
 ·feet′

splay′foot′ed
spleen
sple·nec′to·mize′
 ·mized′ ·miz′ing
sple·nec′to·my
 ·mies
sple′nec·to′pi·a or sple·nec′to·py
sple·nel·co′sis
sple·net′ic or ·net′i·cal
sple·ni′al
splen′ic
sple·ni′tis
sple′ni·um
sple′ni·us
 ·ni·i′
splen′i·za′tion or ·i·fi·ca′tion
sple′no·gen′ic or sple·nog′e·nous
sple′noid
sple·nol′o·gy
sple′no·meg′a·ly
sple′no·por·tog′ra·phy
sple·no′sis
sple·not′o·my
splen′u·lus
 ·li′
splice
 spliced splic′ing
splint
splin′ter
split
 split split′ting
spo·dog′e·nous

spod′o·gram′
spo·doph′o·rous
spoke′shave′
spon·da′ic
spon′dee
spon′dy·lal′gi·a
spon′dy·lit′ic
spon′dy·li′tis
spon′dy·lo′lis·the′sis
spon′dy·lo′sis
spon′dy·lo·syn·de′sis
sponge
   sponged spong′ing
spon′gi·form′
spon′gin
spon′gi·o·blast′
spon′gi·oid′
spon′gi·o·plasm′
spon′gy
   ·gi·er ·gi·est
spon·ta′ne·ous
spool
spoon
spoon′–feed′
   –fed′–feed′ing
spoon′ful′
   ·fuls′
spo·rad′ic
spo·ran′gi·al
spo·ran′gi·um
   ·gi·a
spore
   spored spor′ing
spo′ri·ci′dal
spo·rif′er·ous

spo′ro·cyst′
spo′ro·gen′e·sis
   ·ses′
spo′ro·gen′ic
spo·rog′e·nous
spo·rog′o·ny
spo′ront
spo′ro·phore′
spo·roph′o·rous
spo′ro·tri·cho′sis
   ·ses
spo′ro·zo′an
spo′ro·zo′ic *or* ·zo′al
spo′ro·zo′ite
spo′ro·zo′on
   ·zo′a
sport
sports medicine
spor′u·lar
spor′u·late′
   ·lat′ed ·lat′ing
spor′u·la′tion
spor′ule
spot
   spot′ted spot′ting
sprain
spray
spread
   spread spread′ing
spread′er
spring
sprue
spue
   spued spu′ing
spur
spu′ri·ous

spu′tum
   ·ta
squa′ma
   ·mae
squa′mate
squa·ma′tion
squa·ma·ti·za′tion
squa′mo·cel′lu·lar
squa′mo·pa·ri′e·tal
squa·mo′sal
squa′mo·tem′po·ral
squa′mous *or* ·mose
squar′rose *or* ·rous
squat
   squat′ted *or* squat squat′ting
squeeze
   squeezed squeez′ing
squeeze bottle
squill
squint′–eyed′
stab
   stabbed stab′bing
sta′bile
   (heat–resistant; SEE stable*)*
sta·bil′i·ty
   ·ties
sta′bi·li·za′tion
sta′bi·lize′
   ·lized′ ·liz′ing
sta′bi·liz′er
sta′ble
   ·bler ·blest
   (unchanging; SEE stabile*)*

stac·ca'to
  ·tos
stac·tom'e·ter
sta'di·um
  ·di·a
staff
  staffs
  (personnel)
staff
  staves
  (instrument)
staff of Aes·cu·la'pi·us
stage
  staged stag'ing
stag'ger
stag'gers
stag'nant
stag'nate
  ·nat·ed ·nat·ing
stag·na'tion
stain'a·ble
stair'case'
sta·lac'tite
stal'ag·mom'e·ter
sta·lag'mon
stalk
stam'i·na
stam'i·nal
stam'mer
stanch
stand
  stood stand'ing
stand'ard
stand'ard·i·za'tion
stand'ard·ize'
  ·ized' ·iz'ing

stand'still'
Stan'ford–Bi·net' test
stan'nic
stan'nous
sta·pe'di·al
sta·pe'di·o·te·not'o·my
  ·mies
sta·pe'di·us
  ·di·i
sta'pes
  sta'pes or sta·pe'des
staph
staph'y·le
staph'y·lec'to·my
staph'y·le·de'ma
staph'y·line
staph'y·li'nus
sta·phy'li·on'
staph'y·lo·coc'cal
  or ·coc'cic
staph'y·lo·coc'cus
  ·coc'ci
staph'y·lol'y·sin
staph'y·lo'ma
staph'y·lom'a·tous
staph'y·lo·plas'tic
staph'y·lo·plas'ty
  ·ties
staph'y·lor'rha·phy
  ·phies
staph'y·lo·tox'in
sta'ple
  ·pled ·pling
star

starch blocker
starch'y
  ·i·er ·i·est
stare
  stared star'ing
Star'ling's law
start'er
star'tle
  ·tled ·tling
star·va'tion
starve
  starved starv'ing
sta'sis
  ·ses
stat
state
stat'ic
stat'im
sta'tion
sta'tion·ar'y
sta·tis'tic
sta·tis'ti·cal
sta·tis'tics
stat'o·cyst'
stat'o·cys'tic
stat'o·lith'
stat'o·lith'ic
sta·tom'e·ter
stat'ure
sta'tus
  ·tus·es
sta·tu'vo·lence
stat'volt'
staunch
stau'ri·on
stax'is

227

steal
  (blood diversion; SEE steel)
steam bath
steam′y
  ·i·er ·i·est
ste·ap′sin
ste′a·rate′
ste·ar′ic
ste′a·rin *or* ·rine
ste′a·rop′tene
ste′a·tite′
ste′a·tit′ic
ste′a·ti′tis
ste′a·tog′e·nous
ste′a·tol′y·sis
  ·ses′
ste′a·to·pyg′i·a
ste′a·to·pyg′ic *or* ·py′gous
ste′a·tor·rhe′a
steel
  (metal; SEE steal)
ste′ge
steg·no′sis
  ·ses
steg·not′ic
stel′la
  ·lae
stel′lar
stel′late *or* ·lat·ed
stel′lu·la
stem
sten′i·on
  ·i·a
sten′o·com·pres′sor
sten′o·co·ri′a·sis

sten′o·pe′ic *or* ·pa′ic
ste·nose′
  ·nosed′ ·nos′ing
ste·no′sis
  ·ses
sten′o·therm′
sten′o·ther′mal *or* ·ther′mous *or* ·ther′mic
ste·not′ic
Sten′sen's ducts
stent
ste·pha′ni·al
ste·pha′ni·on
step′page
ste·ra′di·an
ster′co·bi′lin
ster′co·ra′ceous
ster·cu′li·a
ster′cus
  ·co·ra
ster′e·o·ar·throl′y·sis
ster′e·o·chem′is·try
ster′e·o·gram′
ster′e·o·graph′
ster′e·o·i′so·mer
ster′e·o·i′so·mer′ic
ster′e·o′i·som′er·ism
ster′e·o·met′ric *or* ·met′ri·cal
ster′e·om′e·try
ster′e·op′sis
ster′e·o·scope′
ster′e·o·scop′ic

ster′e·os′co·py
ster′e·o·tax′i·a
ster′e·o·tax′ic *or* ·tac′tic
ster′e·o·tax′is
  ·tax′es
ster′e·o·typ′y
ster′ic
ster′i·lant
ster′ile
ste·ril′i·ty
  ·ties
ster′i·li·za′tion
ster′i·lize′
  ·lized′ ·liz′ing
ster′i·liz′er
ster′na
  (sing. ster′num)
ster′nal
ster′no·cos′tal
ster′noid
ster·nos′chi·sis
  ·ses′
ster′num
  ·nums *or* ·na
ster′nu·ta′tion
ster′nu·ta′tive
ster′nu·ta′tor
ster·nu′ta·to′ry
  ·ries
ster′oid
ste·roi′dal
ste·roi′do·gen·e′sis
  ·ses′
ster′ol
ster′tor
ster′to·rous

steth'o·graph'
steth'o·kyr'to·graph' or ·cyr'to·graph'
steth'o·my'o·si'tis or ·my·i'tis
steth'o·scope'
steth'o·scop'ic or ·scop'i·cal
ste·thos'co·py
sthe'ni·a
sthen'ic
sthen·om'e·ter
sthen·om'e·try
stib'i·al·ism
stib'i·a'tion
stib'ine
stib'i·um
stich'o·chrome'
stict·ac'ne
stiff'en
sti'fle
 ·fled ·fling
stig'ma
 ·mas or stig·ma'ta
stig'mal
stig·mas'ter·ol'
stig·mat'ic or ·mat'i·cal
stig'ma·tism
stig'ma·ti·za'tion
stig'ma·tize'
 ·tized' ·tiz'ing
stig'ma·tom'e·ter
stil'bene
stil·bes'trol
still'birth'
still'born'
stil'li·cid'i·um
stim'u·lant
stim'u·late'
 ·lat'ed ·lat'ing
stim'u·lat'er or ·la'tor
stim'u·la'tion
stim'u·la'tive
stim'u·lus
 ·u·li'
sting
 stung
 sting'ing
sting'er
sting'ray' or sting'a·ree'
stip'ple
 ·pled ·pling
stir'rup
stitch
sto·chas'tic
stock culture
stock'i·nette' or ·net'
stock'ing
stoi'chi·ol'o·gy
stoi'chi·o·met'ric
stoi'chi·om'e·try
stoke
Stokes'–Ad'ams syndrome
sto'ma
 ·ma·ta or ·mas
sto·mac'a·ce
stom'ach
stom'ach·ache'
stom'a·chal
stom·a·chal'gi·a
sto·mach'ic or ·mach'i·cal
sto'ma·chos'co·py
stomach pump
sto'ma·tal
sto'ma·tal'gi·a
sto·mat'ic
sto'ma·ti'tis
sto'ma·to·dyn'i·a
sto'ma·to·gas'tric
sto'ma·to·log'i·cal
sto'ma·tol'o·gy
sto'ma·to·my·co'sis
sto'ma·to·ne·cro'sis
sto'ma·tor·rha'gi·a
sto'ma·to'sis
sto'ma·tous
sto'mo·dae'um
 ·dae'a
sto'mo·de'al
sto'mo·de'um
 ·de'a
stone'–blind'
stone'–deaf'
stool
stoop
stop
 stopped stop'ping
stop'page
stor'age
sto'rax
store
 stored stor'ing
sto'ri·form'
storm

stoss′ther·a·py
STP
stra·bis′mal *or* ·mic
stra·bis·mom′e·ter
 or stra·bom′e·ter
stra·bis′mus
straight
 (not curved; SEE
 strait*)*
straight′en
straight′jack′et
strain
strain′er
strait
 (passageway; SEE
 straight*)*
strait′jack′et
stra·mo′ni·um
strand
stran′gle
 ·gled ·gling
stran′gu·late′
 ·lat′ed ·lat′ing
stran′gu·la′tion
stran′gu·ry
strap
 strapped strap′ping
stra′ta
 (sing. stra′tum*)*
stra′tal
strat′i·fi·ca′tion
strat′i·form′
strat′i·fy′
 ·fied′ ·fy′ing
strat′i·graph′ic
stra·tig′ra·phy

stra′tum
 ·ta *or* ·tums
straw′ber′ry mark
streak
stream
stream′line′ flow
stream of
 consciousness
strength
strength′en
stren′u·ous
strep
streph′o·sym·bo′li·a
strep′i·tus
 ·ti′
strep′ti·ce′mi·a
strep′to·an′gi·na
strep′to·ba·cil′lus
strep′to·coc′cal *or*
 ·coc′cic
strep′to·coc′cus
 ·coc′ci
strep′to·ki′nase
strep′to·my′ces
strep′to·my′cin
strep′to·thri′cin
stress
stress fracture
stretch
stretch′er
stri′a
 ·ae
stri′ate
 ·at·ed ·at·ing
stri·a′tion
stri·a′tum
 ·ta

strick′en
stric′ture
stric′tured
stric′tur·o·tome′
stric′tur·ot′o·my
stri′dent
stri′dor
strid′u·lous *or* ·lant
string
stri′o·ccl′lu·lar
strip
 stripped strip′ping
stripe
strip′per
stro·bi′la
 ·lae
stro·bi′lar
stro·bi·la′tion
stro′bile
stro′bi·loid′
stro·bi′lus
 ·li
stro′bo·scope′
stro′bo·scop′ic *or*
 ·scop′i·cal
stroke
 stroked strok′ing
stro′ma
 ·ma·ta
stro′mal *or* stro·
 mat′ic
stro′muhr
stron′gyle
stron′gy·loi·di′a·sis
 or ·loi·do′sis
stron′gy·lo′sis
 ·ses

stron'ti·a
stron'tic
stron'ti·um
stro·phan'thin
struc'tur·al
struc'ture
stru'ma
　·mae
stru·mec'to·my
stru'mi·form'
stru·mi·pri'vous
stru·mi'tis
stru'mose'
stru'mous
Strüm'pell's sign
strych'nine
strych'nin·ism
strych'nism
strych'nize'
　·nized' ·niz'ing
Stry'ker frame
Stu'art factor
Stu'dent's t test
stud'y
　·ies
stump
stun
　stunned stun'ning
stung
stunt
stupe
stu'pe·fa'cient or
　·fac'tive
stu'pe·fac'tion
stu'pe·fy'
　·fied' ·fy'ing
stu'por

stu'por·ous
stut'ter
stut'ter·er
sty or stye
　sties
style
sty'let
sty'li·form'
sty·lis'cus
sty'lo·hy'oid
sty'loid
sty'lo·man·dib'u·lar
sty'lo·mas'toid
sty'lo·max'il·lar'y
sty'lus
　·lus·es or ·li
stype
styp'sis
styp'tic
styp·tic'i·ty
sty'rene
sty'rone
sub·ab·dom'i·nal
sub·ac'e·tate'
sub·ac'id
sub'a·cid'i·ty
sub'a·cute'
sub'as·trin'gent
sub'a·tom'ic
sub·au'ral
sub'au·ric'u·lar
sub·chlo'ride
sub·class'
sub·cla'vi·an
sub·clin'i·cal
sub·con'scious

sub·cor'ti·cal
sub·cos'tal
sub·cra'ni·al
sub'cul'ture
sub'cu·ta'ne·ous
sub·cu'tis
sub·duce'
　·duced' ·duc'ing
sub·duct'
sub·duc'tion
sub·du'ral
sub'ep·i·der'mal
su·ber'ic acid
su'ber·in
sub'fam'i·ly
　·lies
sub'ge'nus
　·gen'er·a or
　·ge'nus·es
su·bic'u·lum
sub'in·vo·lu'tion
sub·ja'cen·cy
sub·ja'cent
sub'ject n.
sub·ject' v.
sub·jec'tive
sub'jec·tiv'i·ty
sub·le'thal
sub'li·mate'
　·mat'ed ·mat'ing
sub'li·ma'tion
sub·lime'
　·limed' ·lim'ing
sub·lim'i·nal
sub·lin'gual
sub·mar'gin·al

sub·max·il′la
  ·lae or ·las
sub·max′il·lar′y
sub·merge′
  ·merged′ ·merg′ing
sub·mer′gi·ble
sub·merse′
  ·mersed′ ·mers′ing
sub·mer′sion
sub′mi·cro·scop′ic
sub·nor′mal
sub′nor·mal′i·ty
  ·ties
sub′oc·cip′i·tal
sub·or′bit·al
sub·or′der
sub·ox′ide
sub′per·i·car′di·al
sub·phy′lum
  ·la
sub′pla·cen′ta
sub·pul′mo·nar′y
sub·scap′u·lar
sub·scrip′tion
sub·side′
  ·sid′ed ·sid′ing
sub·sid′ence
sub·son′ic
sub·spe′cial·ty
sub·spe′cies
sub·spi′nous
sub′stage
sub′stance
sub·stand′ard
sub·stan′ti·a
  ·ti·ae
sub·stit′u·ent

sub·sti·tute′
  ·tut′ed ·tut′ing
sub′sti·tu′tion
sub′sti·tu′tion·al or
  ·tion·ar′y
sub′sti·tu′tive
sub′strate
sub′stra′tum
  ·ta
sub·struc′tur·al
sub·struc′ture or
  ·tion
sub·tem′po·ral
sub′tile
sub′til·in
sub′tle
  ·tler ·tlest
sub·to′tal
  ·taled or ·talled
  ·tal·ing or ·tal·ling
sub·tract′
sub·trac′tion
sub′um·bil′i·cal
sub·un′gual
sub′vo·lu′tion
suc′ce·da′ne·ous
suc′ce·da′ne·um
  ·ne·a
suc′ci·nate′
suc·cin′ic acid
suc′ci·nyl·cho′line
suc·cor·rhe′a
suc·cumb′
suc′cus
  ·ci
  (fluid)

suc·cuss′
  (to shake)
suc·cus′sion
suck
suck′le
  ·led ·ling
su′crase
su′crose
su′cro·se′mi·a
su′cro·su′ri·a
suc′tion
suc·to′ri·al
Su·dan′
su·da′tion
su·da·to′ri·um
  ·ri·a
su·da′to·ry
  ·ries
sud′den infant
  death syndrome
su·dor′al
su·do·re′sis
su·dor·if′er·ous
su·dor·if′ic
su′do·rip′a·rous
su′et
suf′fer
suf′fo·cant
suf′fo·cate′
  ·cat′ed ·cat′ing
suf′fo·ca′tion
suf·fuse′
  ·fused′ ·fus′ing
suf·fu′sion
sug′ar
sug′ar·coat′
sug·gest′

sug·gest′i·bil′i·ty
sug·gest′i·ble
sug·ges′tion
sug·ges′tive
sug′gil·la′tion
su′i·ci′dal
su′i·cide′
 ·cid′ed ·cid′ing
su′i·ci·dol′o·gy
suint
sul′cal
sul′cate or ·cat·ed
sul′cus
 ·ci
sul′fa
sul′fa·di′a·zene′
sul′fa·mer′a·zine′
sul′fa·nil′a·mide′
sul′fa·nil′ic acid
sul′fa·pyr′i·dine′
sulf′ars·phen′a·mine′
sul′fate
sul′fa·thi′a·zole′
sul·fa′tion
sulf·he′mo·glo′bin
sul′fide
sul′fi·nyl
sul′fite
sul·fon′a·mide′
sul′fo·nate′
 ·nat·ed ·nat′ing
sul′fone
sul·fon′ic
sul·fo′ni·um
sul′fon·meth′ane
sul′fo·nyl

sul′fo·sal′i·cyl′ic acid
sulf·ox′ide
sul′fur
sul′fu·rate′
 ·rat′ed ·rat′ing
sul·fu·ra′tion
sul′fu·ret n.
sul′fu·ret′
 ·ret′ed or ·ret′ted
 ·ret′ing or ·ret′ting
sul·fu′ric
sul′fu·ri·za′tion
sul′fu·rize′
 ·rized′ ·riz′ing
sul′fu·rous
sul′fur·yl
sul′lage
For words beginning sulph– *see* SULF–
su′mac or ·mach
sum·ma′tion
sum′mit
sun′burn′
sunk′en
sun′lamp′
sun′screen′
sun′stroke′
sun′tan′
su′per·al′i·men·ta′tion
su′per·cil′i·ar′y
su′per·cil′li·um
 ·li·a
su′per·e′go
 ·gos

su′per·fam′i·ly
 ·lies
su′per·fe′cun·da′tion
su′per·fe·ta′tion
su′per·fi′cial
su′per·fi·ci·al′is
su′per·fi′ci·es′
su′per·im′preg·na′tion
su′per·in·duce′
 ·duced′ ·duc′ing
su′per·in·duc′tion
su′per·in·fec′tion
su·pe′ri·or
su·pe′ri·or′i·ty
su′per·ja′cent
su′per·na′tant
su′per·nate′
su′per·nor′mal
su′per·nu′mer·ar′y
su′per·nu·tri′tion
su′per·ox′ide
su′per·par′a·site′
su′per·par·a·sit′ism
su′per·sat′u·rate′
 ·rat′ed ·rat′ing
su′per·sat′u·ra′tion
su′per·scrip′tion
su′per·se·cre′tion
su′per·sed′ent
su′per·sen′si·tive
su′per·sen′si·tiv′i·ty
su′per·soft′
su′per·son′ic
su′per·struc′ture
su′per·ten′sion

su′per·ven′ient
su′per·ve·nos′i·ty
su′per·ven′tion or ·ven′ience
su′per·vir′u·lent
su′per·vi′sor
su′pi·nate′
　·nat·ed ·nat·ing
su′pi·na′tion
su′pi·na′tor
su·pinc′
sup′ple·ment
sup′ple·men′tal
sup′ple·men′ta·ry
sup′ple·men·ta′tion
sup·ply′
　·plied′ ·ply′ing
sup·port′
sup·port′er
sup·pos′i·to′ry
sup·press′
sup·pres′sant
sup·pres′sion
sup·pres′sive
sup′pu·rant
sup′pu·rate′
　·rat·ed ·rat·ing
sup′pu·ra′tion
sup′pu·ra′tive
su′pra·cla·vic′u·lar
su′pra·lim′i·nal
su′pra·max·il′lar·y
su′pra·mo·lec′u·lar
su′pra·or′bit·al
su′pra·pu′bic
su′pra·re′nal
su′pra·scap′u·lar

su′pra·vag′i·nal
su′ra
su′ral
sur′al·i·men·ta′tion
sur′di·ty
sur′face
　·faced ·fac·ing
sur′face–ac′tive
sur·fac′tant
sur′geon
Surgeon Gen′er·al
　Surgeons General
　or Surgeon
　Generals
sur′ger·y
　·ger·ies
sur′gi·cal
sur′ro·gate′
　·gat·ed ·gat·ing
sur′sum·duc′tion
sur′sum·ver′sion
sur·veil′lance
sur·vey′ing
sur·vey′or
sus·cep′ti·bil′i·ty
　·ties
sus·cep′ti·ble
sus·cep′tive
sus′cep·tiv′i·ty
sus′ci·tate′
　·tat·ed ·tat·ing
sus′ci·ta′tion
sus·pend′ed
sus·pen′sion
sus·pen′soid
sus·pen′so·ry
　·ries

sus′pi·ra′tion
sus·pire′
　·pired′ ·pir′ing
sus·pi′ri·ous
sus′te·nance
sus·ten·tac′u·lar
sus·ten·tac′u·lum
　·la
su·sur′rant
su·sur′rate
　·rat·ed ·rat·ing
su′sur·ra′tion
su·sur′rus
su·tu′ra
su′tur·al
su·tu·ra′tion
su′ture
　·tured ·tur·ing
Sved′berg (unit)
swab
　swabbed swab′bing
swab′ber
swad′dle
　·dled ·dling
swage
　swaged swag′ing
swal′low
swal′low·a·ble
Swan′–Ganz′
　catheter
swarm′ing
swathe
　swathed swath′ing
sway′back′
sway′backed′

sweat
    sweat *or* sweat′ed
    sweat′ing
sweat′y
    ·i·er ·i·est
Swed′ish massage
sweet′en·er
swell
    swelled
    swelled *or* swol′len
    swell′ing
swell′head′
swim′mer's itch
swine flu
switch
swob
    swobbed swob′bing
swol′len
swoon
sy·co′ma
sy·co′si·form′
sy·co′sis
    ·ses
    (hair disease; SEE
    psychosis*)*
syl′la·bus
    ·bus·es *or* ·bi′
syl·lep′sis
    ·ses
syl·lep′tic
syl′van
syl·vat′ic
Syl′vi·an angle
sym′bi·ont′
sym′bi·on′tic
sym′bi·o′sis
sym′bi·ot′ic
sym·bleph′a·ron′

sym′bol
    ·boled *or* ·bolled
    ·bol·ing *or* ·bol·ling
sym·bol′ic *or*
    ·bol′i·cal
sym′bol·ism
sym′bol·i·za′tion
sym′bol·ize′
    ·ized′ ·iz′ing
sym·met′ri·cal *or*
    ·met′ric
sym′me·try
    ·tries
sym′pa·thec′to·my
    ·mies
sym′pa·thet′ic
sym′pa·thet′i·co·to′ni·a
sym·path′i·cop′a·thy
    ·thies
sym′pa·thize′
    ·thized′ ·thiz′ing
sym′pa·thiz′er
sym′pa·tho·lyt′ic
sym′pa·tho·mi·met′ic
sym′pa·thy
    ·thies
sym·pex′i·on
    ·i·a
sym·phys′i·al *or*
    ·e·al
sym·phys′i·ec′to·my
sym·phys′i·ot′o·my
sym′phy·sis
    ·phy·ses′

sym·po′di·a
sym′port
symp′tom
symp′to·mat′ic
symp′tom·a·tize′
    ·tized′ ·tiz′ing
symp′tom·a·tol′o·gy
symp′to·ma·to·lyt′ic *or*
    ·to·mo·lyt′ic
symp′tom·ize′
    ·ized′ ·iz′ing
sym·to′sis
syn′aes·the′si·a
syn′aes·thet′ic
syn·apse′
    syn·ap′ses
syn·ap′sis
    ·ses
syn·ap′tic
syn·ap′to·some′
syn′ar·thro′di·al
syn′ar·thro′sis
    ·ses
sync *or* synch
syn′chon·dro′sis
syn′chon·drot′o·my
syn′chro·nal
syn·chron′ic
syn′chro·nism
syn′chro·nis′tic
syn′chro·nous
syn′chro·ny
syn′chro·tron′
syn′chy·sis
syn·cli′nal

syn'clo·nus
  ·ni'
syn'co·pal
syn'co·pe
syn·cre'ti·o
syn·cre·tize'
  ·tized' ·tiz'ing
syn·cy'ti·al
syn·cy'ti·um
  ·ti·a
syn·dac'tyl or ·tyle
syn·dac'tyl·ism
syn'de·sis
  ·ses
syn'des·mec'to·my
  ·mies
syn'des·mo'sis
  ·ses
syn'des·mot'ic
syn'drome
syn·drom'ic
syn·ech'i·a
  ·i·ae
syn'e·col'o·gy
syn·er'e·sis
  ·ses'
syn·er·get'ic
syn·er'gic
syn'er·gism
syn'er·gist
syn'er·gis'tic
syn'er·gy
  ·gies
syn'es·the'si·a
syn'es·thet'ic
syn·gam'ic or syn'ga·mous

syn'ga·my
syn·gen'e·sis
  ·ses'
syn'ge·net'ic
syn'i·ze'sis
syn·kar'y·on
syn'ki·ne'sis
  ·ses
syn'ki·net'ic
syn'o·nym
syn·op'si·a
  (fusion of the eyes)
syn·op'sis
  ·ses
  (summary)
syn·op'to·phore'
syn·o'vi·a
syn·o'vi·al
syn·o·vi'tis
syn·o'vi·um
syn·tac'tic or
  ·tac'ti·cal
syn·tac'tics
syn'ta·sis
syn·tax'is
syn·tex'is
syn'the·sis
  ·ses'
syn'the·size'
  ·sized' ·siz'ing
syn·thet'ic or
  ·thet'i·cal
syn·ton'ic
syph'i·lid
syph'i·lis
syph'i·lit'ic
syph'i·loid'

syph'i·lol'o·gist
syph'i·lol'o·gy
syph'i·lo'ma
syph'i·lo·pho'bi·a
sy·ringe'
  ·ringed' ·ring'ing
sy·rin'go·coele'
sy·rin'goid
sy·rin'go·my·e'li·a
syr'inx
  sy·rin'ges or syr'inx·es
syr'up
sys·sar·co'sis
  ·ses
sys·tal'tic
sys·tat'ic
sys'tem
sys·tem·at'ic or
  ·at'i·cal
sys'tem·a·ti·za'tion
sys'tem·a·tize'
  ·tized' ·tiz'ing
sys'tem'ic
sys'tem·i·za'tion
sys'tem·ize'
  ·ized' ·iz'ing
sys'to·le'
sys·tol'ic
sys'to·lom'e·ter
sys·trem'ma
sy·zy'gi·al
sy·zyg'i·ol'o·gy
syz'y·gy
  ·gies

T

tab'a·co'sis
tab'a·gism
tab'a·nid
ta·ba'ti·ère' a·na·to·mique'
ta·bel'la
  ·lae
ta'bes
ta·bes'cence
ta·bes'cent
tabes dor·sa'lis
ta·bet'ic
ta·bet'i·form'
tab'la·ture
ta'ble
ta'ble·spoon'
ta'ble·spoon'ful
  ·fuls
tab'let
ta·boo' or ·bu'
ta'bo·pa·re'sis
  ·ses
tab'u·lar
tac'a·ma·hac' or tac'a·ma·hac'a or tac'ma·hack'
tache
ta·chet'ic
ta·chis'to·scope'
ta·chis'to·scop'ic
ta·chom'e·ter
ta·chom'e·try
tach'y·car'di·a
tach'y·car'di·ac
ta·chym'e·ter
ta·chym'e·try
ta·chys'ter·ol'
tac'tile
tac·til'i·ty
tac'tion
tac'tu·al
tae'di·um vi'tae
tae'ni·a
  ·ni·ae'
tae'ni·a·ci'dal
tae'ni·a·cide'
tae·ni'a·sis
tae·ni'id
tag
  tagged tag'ging
tail'bone'
tail'gut'
taint
take
  took tak'en tak'ing
ta·lal'gi·a
ta'lar
talc
  talcked or talced talck'ing or talc'ing
tal'cum (powder)
ta'li
  (sing. ta'lus)
tal'i·ped'
tal'i·pes'
tal'i·pom'a·nus
ta·lo·cal·ca·ne·al or ·ca'ne·an
tal'o·nid
ta'lo·tib'i·al
ta'lus
  ·lus·es or ·li
tam'a·rind
tam'bour
tam'pon
tam'pon·ade' or ·age'
tam·pon'ment
tan
  tanned tan'ning
Tan·gier' disease
tan'nate
tan'nic
tan'nin
tan'ta·lum
T antigen
tan'trum
tap
  tapped tap'ping
tape
  taped tap'ing
ta·pe'tal
ta·pe'to·ret'i·nal
ta·pe'tum
  ·pe'ta
tape'worm'
tap'i·no·ceph'a·ly
  ·lies
ta'pir·oid
ta·pote·ment'
tar
tar'ant·ism
ta·ran'tu·la
  ·las or ·lae
tar'dive dyskinesia

tare
  tared tar′ing
  (weight deduction;
  SEE tear)
tar′get
tar′sal
tar·sal′gi·a
tar·sec′to·my
  ·mies
tar′si
  (sing. tar′sus)
tar′so·cla′si·a *or*
  tar′so·cla′sis
tar′so·met′a·tar′sal
tar·sot′o·my
tar′sus
  ·si
tar′tar
tar·tar′ic
tar′tar·ous
tar′trate
tar′trat·ed
taste
  tast′ed tast′ing
taste bud
tast′er
tat·too′
  ·tooed′ ·too′ing
tau
tau′rine
tau′ro·cho′lic acid
tau·to·mer′ic
tau·tom′er·ism
Ta·wa′ra's node
tax′is
  ·es

tax′on
  tax′a
tax′o·nom′ic
tax·on′o·mist
tax·on′o·my
Tay′–Sachs′
  disease
T cell
tear
  tore torn tear′ing
  (rip; SEE tare)
tear
  teared tear′ing
  (eye fluid)
tear′drop′
tear gas
tear′–gas′
  –gassed′ –gas′sing
tease
  teased teas′ing
tea′spoon′
tea′spoon′ful
  ·fuls
teat
tech·ne′ti·um
tech′nic
tech′ni·cal
tech·ni′cian
tech·nique′
tech·no·log′i·cal *or*
  ·log′ic
tech·nol′o·gist
tech·nol′o·gy
tec·ton′ic
tec·to′ri·al
tec′to·spi′nal

tec′tum
  tec′ta
teen′–age′ *or*
  teen′age′
teeth
  (sing. tooth)
teethe
  teethed teeth′ing
teeth′ridge′
Tef′lon
  (trademark)
teg′men
  ·mi·na
teg·men′tal
teg·men′tum
  ·ta
teg′mi·nal
teg′u·ment
teg′u·men′tal *or*
  ·men′ta·ry
tei·cho′ic acid
tei·chop′si·a
te′la
  ·lae
tel·an′gi·ec·ta′si·a
tel·an′gi·ec·ta′sis
  ·ses
tel·an′gi·ec·tat′ic
tel′e·car′di·o·gram′
  or
  ·e·lec′tro·car′di·o·
  gram′
tel′e·car′di·o·phone′
tel′e·cep′tive
tel′e·cep′tor
tel′e·cu′rie·ther′
  a·py

tel′e·ki·ne′sis
  ·ses
tel′e·ki·net′ic
te·lem′e·ter
tel′e·met′ric
te·lem′e·try
tel′en·ce·phal′ic
tel′en·ceph′a·lon′
  ·la
te′le·o·log′i·cal
te′le·ol′o·gy
tel′e·path′ic
te·lep′a·thist
te·lep′a·thy
tel′e·ra′di·og′ra·phy
tel′er·gy
tel′e·roent′gen·og′ra·phy
tel′es·the′si·a
tel′es·thet′ic
tel′e·sys·tol′ic
tel′e·tac′tor
tel′e·ther′a·py
tel·lu′ric
tel′lu·rite′
tel·lu′ri·um
tel′lu·rize′
  ·rized′ ·riz′ing
tel′lu·rous
tel′o·lem′ma
tel′o·phase′
tel′o·tism
tel′son
tem′peh
tem′per
tem′per·a·ment
tem′per·a·men′tal
tem′per·ance
tem′per·ate
tem′per·a·ture
tem′plate
tem′ple
tem′po·la′bile
tem′po·ral
tem′po·ra′lis
tem′po·ri·za′tion
tem′po·rize′
  ·rized′ ·riz′ing
tem′po·ro·max·il′lar′y
tem′po·ro–oc·cip′i·tal
te·na′cious
te·nac′i·ty
te·nac′u·lum
  ·la
tend′en·cy
  ·cies
ten′der
ten′di·ni′tis
ten′di·no·plas′ty
  ·ties
ten′di·nous
ten′do
  ten′di·nes′
ten·dol′y·sis
ten′don
ten′do·vag′i·ni′tis
te·nec′to·my
  ·mies
te·nes′mic
te·nes′mus

For words beginning teni– *see also* TAENI
te′ni·fuge′ *or* te′ni·a·fuge′
ten′nis elbow
te·nod′e·sis
  ·ses′
Te′non's space
ten·on′to·plas′ty
  ·ties
ten′o·plas′ty
  ·ties
te·nor′rha·phy
  ·phies
ten′o·syn′o·vi′tis
ten′o·tome′
te·not′o·mize′
  ·mized′ ·miz′ing
te·not′o·my
  ·mies
ten′o·vag′i·ni′tis
tense
  tens′er tens′est
  tensed tens′ing
ten′si·om′e·ter
ten′sion
ten′si·ty
  ·ties
ten′sive
ten′sor
  ten·so′res
tent
ten′ta·tive
ten·to′ri·al
ten·to′ri·um
  ·ri·a

te·nu'i·ty
· ties
ten'u·ous
tep'e·fac'tion
tep'e·fy'
· fied' ·fy'ing
teph'ro·ma·la'ci·a
teph'ro·my'e·li'tis
teph·ro'sis
tep'id
te·pid'i·ty
te'por
ter'as
· a·ta
ter·at'ic
ter'a·tism
ter'a·to·blas·to'ma
· mas or ·ma·ta
ter'a·to·gen
ter'a·to·gen'ic
ter'a·toid'
ter'a·to·log'i·cal
ter'a·tol'o·gy
ter'a·to'ma
· mas or ·ma·ta
ter'a·to'ma·tous
ter'a·to'sis
ter'bi·um
ter'e·bene'
ter'e·binth'
ter'e·bin'thi·nate'
ter'e·bin'thine
ter'e·bin'thi·nism
ter'e·brant or
· brat'ing
ter'e·bra'tion

te'res
 ter'e·tes
ter'gal
ter'gum
 ·ga
ter in di'e
term
ter'mi·nal
ter'mi·nate'
 ·nat'ed ·nat'ing
ter'mi·na'ti·o
 ·na'ti·o·nes'
ter'mi·na'tion
ter'mi·na'tion·al
ter'mi·no·log'i·cal
ter'mi·nol'o·gy
 ·gies
ter'mi·nus
 ·ni' or ·nus·es
ter'na·ry
ter'pene
ter·pin'e·ol'
ter'pin hydrate
ter'ra al'ba
ter'race
 ·raced ·rac·ing
Ter'ra·my'cin
 (trademark)
ter'ror
ter'tian
ter'ti·a·rism
ter'ti·ar'y
ter·tip'a·ra
tes'la
tes'sel·late adj.
tes'sel·late'
 ·lat'ed ·lat'ing

tes'sel·la'tion
test
tes·ta'ceous
tes·tal'gi·a
tes·tec'to·my
test'er
tes'tes
 (sing. tes'tis)
tes'ti·cle
tes'ti·cond
tes·tic'u·lar
tes'tis
 ·tes
tes'ti·tox'i·co'sis
tes'to·lac'tone
tes·tos'ter·one'
test'-tube' adj.
test tube
te·tan'ic
tet'a·ni·za'tion
tet'a·nize'
 ·nized' ·niz'ing
tet'a·node'
tet'a·noid'
tet'a·no·mo'tor
tet'a·nus
tet'a·ny
 ·nies
te·tar'ko·cone' or
 te'tar·cone'
te'tar·ta·no'pi·a or
 ·ta·nop'si·a
tet'ra·bas'ic
tet'ra·chlo'ride
tet'ra·cy'cline
tet'rad
tet'ra·eth'yl lead

tet′ra·gon′
te·trag′o·nal
tet′ra·hy′dro·can′na·bi·nol′
te·tral′o·gy
·gies
tet′ra·ploid′
tet′ra·ploi′dy
tet·ras′ter
tet′ra·tom′ic
tet′ra·va′lent
tet′ro·do·tox′in
te·trox′ide
tet′ter
tex′is
tex′ti·form′
tex′tur·al
tex′ture
·tured ·tur·ing
T′–group′
thal′a·men′ce·phal′ic
thal′a·men·ceph′a·lon′
·la
tha·lam′ic
thal′a·mot′o·my
·mies
thal′a·mus
·mi′
thal·as·se′mi·a
tha·las′so·ther′a·py
·pies
tha·lid′o·mide′
thal′lic
thal′lin·i·za′tion
thal′li·um

thal′loid
thal′lo·phyte′
thal′lo·phyt′ic
thal′lo·spore′
thal′lous
(of thallium)
thal′lus
thal′li or thal′lus·es
(growing plant)
tha·mu′ri·a
than′a·to·gno·mon′ic
than′a·tol′o·gist
than′a·tol′o·gy
than′a·tom′e·ter
than′a·to·pho′bi·a
than′a·tos
thau·ma·trope′
thau·mat′ro·py
the′a·ism
the′a·ter *or* ·tre
the·ba′ic
the′ba·ine′
the·be′si·an valve
the′ca
·cae
the′cal
the′cate
the·ci′tis
thee′lin
thee′lol
the′ine
the′in·ism *or* the′ism
the·lar′che
the·la·zi′a·sis

the′li·um
·li·a
the′lo·thism *or*
·tism
the′nad
the′nal
the′nar
the·o·bro′mine
the·o·ma′ni·a
the·o·phyl′line
the′o·rem
the′o·ret′i·cal *or*
·ret′ic
the′o·rize′
·rized′ ·riz′ing
the′o·ry
·ries
ther′a·peu′tic *or*
·peu′ti·cal
ther′a·peu′tics
ther′a·pist *or* ther′a·peu′tist
ther′a·py
·pies
ther′en·ceph′a·lous
therm
ther′mae
ther′mal
ther′mal·ge′si·a
therm′an·es·the′si·a
or ·an·aes·the′si·a
ther′ma·tol′o·gy
ther′mel
ther′me·lom′e·ter
therm′es·the′si·a *or*
·aes·the′si·a
ther′mic

therm·is′tor
ther′mo·chem′i·cal
ther′mo·chem′is·try
ther′mo·cou′ple
ther′mo·dy·nam′ics
ther′mo·e·lec′tric
ther′mo·gen′e·sis
ther′mo·ge·net′ic
ther′mo·gram′
ther′mo·graph′
ther′mo·graph′ic
ther·mog′ra·phy
ther′mo·la′bile
ther′mo·la·bil′i·ty
ther′mo·lu′mi·nes′cence
ther·mol′y·sis
 ·ses′
ther′mo·lyt′ic
ther·mom′e·ter
ther′mo·met′ric
ther·mom′e·try
ther′mo·neu·ro′sis
ther′mo·phile′
ther′mo·phil′ic
ther′mo·pile′
ther′mo·plas′tic
ther′mo·re·cep′tor
ther′mo·reg′u·la′tion
ther′mo·reg′u·la′tor
ther′mo·scope′
ther′mo·scop′ic
ther′mo·sta·bil′i·ty
ther′mo·sta′ble
ther′mo·stat′

ther′mo·stat′ic
ther′mo·tax′ic *or*
 ·tac′tic
ther′mo·tax′is
ther′mo·trop′ic
ther·mot′ro·pism
the′roid
the·sau′ris·mo′sis
the·sau·ro′sis
the′sis
 the′ses
thet′a
thews
 (sing. thew*)*
thi′a·mine *or* ·min
thi′a·zine′
thi′a·zole′
Thiersch′'s graft
thigh′bone′ *or*
 thigh bone
thig′mo·tac′tic
thig′mo·tax′is
 ·es
thig′mo·trop′ic
thig·mot′ro·pism
thi·mer′o·sal′
think
 thought think′ing
thin′–lay′er
 chromatography
thi′o·a·ce′tic acid
thi′o acid
thi′o·ar′se·nite′
thi′o·bac·te′ri·a
 (sing. ·bac·te′ri·um*)*
thi′o·car′ba·mide′
thi′o·cy′a·nate′

thi′o·cy·an′ic acid
thi′o·gen′ic
thi′ol
thi′o·nate′
thi′o·ne′ine
thi·on′ic
thi′o·nine′
thi′o·nyl
thi′o·pen′tal
 (sodium)
thi′o·phene′
thi′o·phil
thi′o·phil′ic
thi′o·sin·am′ine
thi′o·sul′fate
thi′o·sul·fu′ric acid
thi′o·u′ra·cil
thi′o·u·re′a
thi′o·xan′thene
thi′ram
third′–de·gree′
thirst
thirst′y
 ·i·er ·i·est
Thi′ry's fistula
thix′o·trop′ic
thix·ot′ro·py
thlip′sen·ceph′a·lus
Thomp′son's test
Thom′sen's disease
Thom′son's sign
tho·ra·cal′gi·a
tho·rac′ic
tho′ra·co·ce·los′
 chi·sis
tho′ra·co·dyn′i·a

tho´ra·co·lum´bar
tho´ra·cop´a·thy
tho´ra·co·scope´
tho´ra·cos´co·py
tho´ra·cot´o·my
　·mies
tho´rax
　·rax·es *or* ·ra·ces´
tho´ric
tho´ri·um
tho´ron
thread´worm´
thre´o·nine´
thresh´old
thrill
thrix an´nu·la´ta
throat
throb
　throbbed throb´bing
throe
throm´bin
throm´bo·an´gi·i´tis
throm·boc´la·sis
throm´bo·clas´tic
throm´bo·cyte´
throm´bo·cyt´ic
throm´bo·cy´to·crit
throm´bo·em·bol´ic
throm´bo·em´bo·lism
throm´bo·gen
throm´bo·gen´ic
throm´bo·ki´nase
throm´bo·ki·ne´sis
throm´bo·phle·bi´tis
throm´bo·plas´tic

throm´bo·plas´tin
throm´bose
　·bosed ·bos·ing
throm·bo´sis
　·ses
throm´bos´ta·sis
throm·bot´ic
throm·box´ane
throm´bus
　throm´bi
throt´tle
　·tled ·tling
throw´back´
thrush
thrust
　thrust thrust´ing
thryp´sis
thu´ja
thu´li·um
thumb´nail´
thumb´–print´ing
thus
thy·mec´to·mize´
　·mized´ ·miz´ing
thy·mec´to·my
　·mies
thy´mic
thy´mi·dine´
thy´mine
thy´mi·on
thy´mo·ke´sis
thy´mol
thy´mo·lize´
　·lized´ ·liz´ing
thy´mol´y·sis
thy´mo·lyt´ic

thy·mo´ma
　·mas *or* ·ma·ta
thy´mo·priv´al *or*
　thy´mo·priv´ic *or*
　thy´mop´ri·vous
thy´mo·sin
thy´mus
　·mus·es *or* ·mi
thy´mus–de·pend´ent
thy´ro·cal´ci·to´nin
thy´ro·gen´ic *or*
　thy·rog´e·nous
thy´ro·glos´sal
thy´ro·hy´oid
thy´roid
thy´roid·ec´to·mize´
　·mized´ ·miz´ing
thy´roid·ec´to·my
　·mies
thy´roid·ism
thy´roid·i´tis
thy´roid·ol´o·gy
thy´roid·ot´o·my
thy´roid·o·tox´in
thy´roid–stim´u·lat´ing hormone
thy´ro·nine´
thy´rop·to´sis
thy·ro´sis
　·ses
thy´ro·tox´i·co´sis
thy´rot´ro·phin *or*
　·pin
thy·rox´ine
tib´i·a
　·i·ae´ *or* ·i·as

tib′i·al
tic
  (spasm; SEE tick)
tic′ dou′lou·reux′
tick
  (insect; SEE tic)
tick′le
  ·led ·ling
tid′al
tide
tig′lic acid
tig′li·um
ti·gret′i·er′
ti′groid
ti·grol′y·sis
  ·ses′
til·tom′e·ter
tim′bre
time
  timed tim′ing
tim′er
tin
tin′cal
tinc′ta·ble
tinc·to′ri·al
tinc·tu·ra′tion
tinc′ture
  ·tured ·tur·ing
tin′e·a
tine test
tin′foil′
tinge
  tinged tinge′ing *or*
    ting′ing
tin′gi·bil′i·ty
tin′gi·ble

tin′gle
  ·gled ·gling
tin·ni′tus
tint
tin·tom′e·ter
tin′to·met′ric
tin·tom′e·try
tip
  tipped tip′ping
ti·queur′
tire
  tired tir′ing
tire-bal′
tire-fond′
ti·sane′
tis′sue
tis′sue–type′
  plasminogen
  activator
tis′su·lar
ti·ta′ni·um
ti′ter
tit′il·la′tion
ti′trate
  ·trat·ed ·trat·ing
ti·tra′tion
ti′tri·met′ric
ti·trim′e·try
tit′u·ba′tion
T lymphocyte
toad′skin′
toad′stool′
to·bac′co
  ·cos
to′co·dy′na·mom′e·
  ter
to′co·graph′

to·cog′ra·phy
to·col′o·gy
to·com′e·ter
to·coph′er·ol′
to·co·pho′bi·a
to′co·tri′e·nol′
to′cus
toe′nail′
to′ga·vi′rus
toi′let
tol·bu′ta·mide′
tol′er·ance
tol′er·ant
tol′er·ate′
  ·at·ed ·at·ing
tol′er·a′tion
tol′er·o·gen′
tol′er·o·gen′e·sis
tol′er·o·gen′ic
tol′u·ate′
tol′u·ene′
to·lu′ic acid
to·lu′i·dine′
tol′u·ol′
tol′yl
to′ma·tine′
to·men′tum
  ·ta
to′mo·gram′
to′mo·graph′
to·mog′ra·phy
  ·phies
to′mo·lev′el
to′mo·ma′ni·a
ton′al
tone
tone′–deaf′

ton'er
tongue
tongue'-tie'
  -tied' -ty'ing
ton'ic
to·nic'i·ty
ton'i·cize'
  ·cized' ·ciz'ing
ton'i·co·clon'ic *or*
  ton'o·clon'ic
to'nin
ton'ing
ton'o·fi'bril
to'no·gram'
to'no·graph'
to·nog'ra·phy
to·nom'e·ter
ton'o·met'ric
to·nom'e·try
ton'sil
ton'sil·lar
ton'sil·lec'to·my
  ·mies
ton'sil·lit'ic
ton'sil·li'tis
ton·sil'lo·lith'
ton'sil·lo·tome'
ton'sil·lot'o·my
  ·mies
ton'sure
to'nus
tooth
  teeth
tooth'ache'
tooth'-borne'
tooth'brush'
toothed

tooth'paste'
tooth powder
top'ag·no'sis
to·pec'to·my
top'es·the'si·a
to·pha'ceous
to'phus
  ·phi
top'i·cal
top'o·graph'ic *or*
  ·graph'i·cal
to·pog'ra·phy
  ·phies
top'o·log'i·cal
to·pol'o·gy
  ·gies
top'o·nar·co'sis
top'o·nym'
top'o·nym'ic *or*
  ·nym'i·cal
to·pon'y·my
top'o·ther'mes·the'-
  si·om'e·ter
tor'cu·lar
tore
  (*pt. of* tear)
to'ri
  (*sing.* to'rus)
to'ric
tor'mi·na
torn
  (*pp. of* tear)
to'rose *or* ·rous
tor'pent
tor'pid
tor·pid'i·ty
tor'por

tor'por·if'ic
torque
torr
tor're·fac'tion
tor're·fy'
  ·fied' ·fy'ing
tor'si
  (*sing.* tor'so)
tor'si·bil'i·ty
tor'sion
tor'sion·al
tor'sive
tor'si·ver'sion
tor'so
  ·sos *or* ·si
tor'so·clu'sion
tor'ti·col'lar
tor'ti·col'lis
tor'ti·pel'vis
tor'tu·os'i·ty
  ·ties
tor'tu·ous
tor'ture
  ·tured ·tur·ing
tor'u·lop'sis
tor'u·lo'sis
tor'u·lus
  ·li
to'rus
  ·ri
to'tal
to'ta·quine'
to·tip'o·ten·cy
  ·cies
to·tip'o·tent
touch
tour'ni·quet

tow'el·ette'
tox·al·bu'min
tox'a·phene'
tox·e'mi·a or
·ae'mi·a
tox·e'mic
tox'ic
tox'i·cant
tox'i·cide'
tox·ic'i·ty
·ties
tox'i·co·der'ma·ti'tis
tox'i·co·gen'ic
tox'i·coid
tox'i·co·log'ic or
·log'i·cal
tox'i·col'o·gist
tox'i·col'o·gy
tox'i·co'sis
·ses
tox'ic–shock'
syndrome
tox'i·gen'ic
tox'i·ge·nic'i·ty
·ties
tox·ig·nom'ic
tox'in
tox'in–an'ti·tox'in
tox'i·ne'mi·a
tox·in'i·cide'
tox'in·o·gen'ic
tox'i·no'sis
tox·ip'a·thy
tox'o·ca·ri'a·sis
tox'oid
tox'o·phil or ·phile'

tox'o·phore'
tox·oph'o·rous
tox'o·plas'min
tox'o·plas·mo'sis
·ses
tra·bec'u·la
·lae'
tra·bec'u·lar or
·late
tra·bec'u·la'tion
trabs
tra'bes
trace
trac'er
tra'che·a
·che·ae' or ·che·as
tra'che·a·ec'ta·sy
tra'che·al
tra'che·al'gi·a
tra'che·i'tis
tra'che·lag'ra
tra'che·lism or tra'che·lis'mus
tra'che·lo·cys·ti'tis
tra'che·lo·pex'y
tra'che·lot'o·my
tra'che·o·a'er·o·cele'
tra'che·o·path'i·a or
·che·op'a·thy
tra'che·o·pha·ryn'ge·al
tra'che·os'chi·sis
tra'che·os'co·py
tra'che·os'to·my
tra'che·o·tome'

tra'che·ot'o·my
·mies
tra·chi'tis
tra·cho'ma
·ma·ta
tra·cho'ma·tous
tra'chy·chro·mat'ic
tra'chy·pho'ni·a
tract
trac'tate
trac'tion
trac'tor
trac·tot'o·my
trac'tus
trag'a·canth'
tra'gal
tra'gi·on
trag'o·mas·chal'i·a
trag'o·pho'ni·a or
tra·goph'o·ny
tra'gus
·gi
train'a·ble
train'ing
trait
tra·jec'tor
trance
tran'quil·ize' or
·quil·lize'
·ized' or ·lized'
·iz'ing or ·liz'ing
tran'quil·iz'er or
·liz'er
trans·am'i·nase'
trans·am'i·na'tion
trans·an'i·ma'tion
trans·au'di·ent

trans·ca′lent
tran′scen·den′tal meditation
trans·cor′ti·cal
tran·scribe′
 ·scribed′ ·scrib′ing
tran·scrip′tion
trans′cu·ta′ne·ous
trans·duce′
 ·duced′ ·duc′ing
trans·duc′er
trans·duc′tant
trans·duc′tion
trans·fect′
trans·fec′tion
tran·sect′
tran·sec′tion
trans′fer n.
trans·fer′
 ·ferred′ ·fer′ring
trans′fer·ase′
trans·fer′ence
trans·fer′rin
trans·fix′
trans·fix′ion
trans′fo·rate′
 ·rat′ed ·rat′ing
trans′fo·ra′tion
trans·form′
trans·form′ant
trans′for·ma′tion
trans·form′er
trans·fuse′
 ·fused′ ·fus′ing
trans·fus′i·ble
trans·fu′sion
trans·fu′sive

tran′sient
trans·il′lu·mi·nate′
 ·nat′ed ·nat′ing
trans·il′lu·mi·na′tion
tran·sis′tor
tran·si′tion
tran·si′tion·al or
 ·tion·ar′y
tran′si·to·ry
trans·late′
 ·lat′ed ·lat′ing
trans·la′tion
trans·lo′cate
 ·cat·ed ·cat·ing
trans′lo·ca′tion
trans·lu′cence or
 ·lu′cen·cy
trans·lu′cent or
 ·lu′cid
trans·mi′grate
 ·grat·ed ·grat·ing
trans′mi·gra′tion
trans·mis′si·bil′i·ty
trans·mis′si·ble
trans·mis′sion
trans·mit′
 ·mit′ted ·mit′ting
trans·mit′tance
trans·mit′ter
trans·mit′ti·ble or
 ·ta·ble
trans·mut′a·ble
trans′mu·ta′tion
trans·mute′
 ·mut′ed ·mut′ing
trans′o·nance

trans·or′bit·al
trans·par′ent
tran·spir′a·ble
tran′spi·ra′tion
tran·spire′
 ·spired′ ·spir′ing
trans′plant′ n.
trans·plant′ v.
trans·plant′a·ble
trans′plan·ta′tion
trans′port n.
trans·port′ v.
trans′por·ta′tion
trans·pos′a·ble
trans·pose′
 ·posed′ ·pos′ing
trans′po·si′tion
trans·po′son
trans′seg·men′tal
trans·sex′u·al
trans·sex′u·al·ism
tran′sub·stan′ti·ate′
 ·at′ed ·at′ing
tran′sub·stan′ti·a′tion
tran′su·date′
tran′su·da′tion
tran·sude′
 ·sud′ed ·sud′ing
trans·verse′
trans·ver′so·u′re·thra′lis
trans·ver′sus
trans·ves′tism or
 ·ves′ti·tism
trans·ves′tite
tra·pe′zi·form′

tra·pe′zi·um
  ·zi·ums or ·zi·a
tra·pe′zi·us
trap′e·zoid′
trap′e·zoi′dal
trau′ma
  ·mas or ·ma·ta
trau·mat′ic
trau′ma·tism
trau′ma·tize′
  ·tized′ ·tiz′ing
trau′ma·tol′o·gy
trau′ma·top′ne·a
trau′ma·to·ther′
  a·py
trav′ail
trav′erse
tray
trea′cle
treat′a·bil′i·ty
treat′a·ble
treat′ment
tre·ha′la
tre′ha·lose′
trem′a·tode′
trem′a·toid′
trem′ble
  ·bled ·bling
trem′el·loid′ or
  ·lose′
trem′o·graph′
tre′mo·la′bile
trem′or
trem′or·ous
trem′u·lous or ·lant
trench fever
trench foot

trend
tre·pan′
  ·panned′ ·pan′ning
trep′a·na′tion
trep′a·nize′
  ·nized′ ·niz′ing
treph′i·na′tion
tre·phine′
  ·phined′ ·phin′ing
tre·phin′er
trep′i·dant
trep′i·da′tion
trep′o·ne′ma
  ·mas or ·ma·ta
trep′o·ne′mal or
  ·ne′ma·tous
trep′o·ne·mi′a·sis
trep′pe
tre′sis
tri·ac′e·tate′
tri·ac′id
tri′ad
tri·ad′ic
tri·age′
tri′al
tri·an′gle
tri·an′gu·lar
tri·an′gu·lar′i·ty
tri′a·tom′ic
trib′ad·ism
tri·bas′ic
tri′bo·lu′mi·nes′
  cence
tri′bo·lu′mi·nes′
  cent
tri·bro′mide

tri·bro′mo·eth′a·
  nol′
tri·car′box·yl′ic
tri′ceps
  ·cep·ses or ·ceps
trich·an′gi·ec·ta′si·a
  or ·ec′ta·sis
trich′a·tro′phi·a
trich·aux′is
tri·chi′a·sis
tri·chi′na
  ·nae
trich′i·ni·za′tion
trich′i·nize′
  ·nized′ ·niz′ing
trich′i·no′sis
  ·ses
trich′i·nous
trich′i·on
tri·chlo′ride
tri·chlo′ro·a·ce′tic
  acid
tri·chlo′ro·eth′yl·
  ene′
trich′o·cyst′
tri′choid
tri·chol′o·gist
tri·chol′o·gy
tri′chome
tri·chom′ic
trich′o·mon′ad
trich′o·mo·ni′a·sis
  ·ses′
trich′or·rhex′is
tri·cho′sis
  ·ses

tri·chot′o·mize′
  ·mized′ ·miz′ing
tri·chot′o·mous
tri·chot′o·my
tri·chro′ic
tri′chro·ism
tri′chro·mat′
tri′chro·mat′ic or
  tri·chro′mic
tri·chro′ma·tism
tri·cip′i·tal
tri′corn or ·corne
tri·crot′ic
tri′cro·tism
tri·cus′pid
tri·cus′pi·date′
tri·cy′clic
tri·dac′ty·lous
tri′dent
tri·den′tate
tri·der′mic
tri·dig′i·tate′
tri·el′con
tri·eth′y·lene·mel′a·mine
tri·fa′cial
tri′fid
tri·fo′cal *n.*
tri·fo′cal *adj.*
tri·fur′cate′
  ·cat′ed ·cat′ing
tri·fur′cat·ed
tri′fur·ca′tion
tri·gem′i·nal
tri·gem′i·ny
  ·nies
trig′ger

trigger finger
tri·glyc′er·ide′
trig′o·nal
tri′gone
trig′o·no·ceph′a·lus
tri·go′num
  ·na
tri·hy′drate
tri·hy′drat·ed
tri·i′o·do·thy′ro·nine′
tri′labe
tri·lat′er·al
tri·loc′u·lar
tril′o·gy
  ·gies
tri·men′su·al
tri′mer
tri·mer′ic
tri·mes′ter
tri·meth′a·di′one
tri·mor′phic or
  ·phous
tri·mor′phism
tri·ni′tro·cre′sol
tri·ni′tro·glyc′er·in
tri·ni′tro·tol′u·ene′
tri·no′mi·al
tri′ol
tri′ose
tri·ox′ide
trip
tri·par′tite
tri·phen′yl·meth′ane
tri′ple
  ·pled ·pling

tri′plet
tri′plex
trip′lo·blas′tic
trip′loid
trip′loi·dy
trip·lo′pi·a
tri′pod
trip′sis
tri·que′trous
tri·ra′di·ate
tri·sac′cha·ride′
tri·sect′
tri·sec′tion
tris′kai·dek·a·pho′bi·a
tris′mic
tris′moid
tris′mus
tri·so′di·um
tri·so′mic
tri·so′my
  ·mies
tri·splanck′nic
tri·sul′fide
tri·ta·no′pi·a
trit′i·um
trit′u·ra·ble
trit′u·rate′
  ·rat′ed ·rat′ing
trit′u·ra′tion
trit′u·ra′tor
tri·va′lence or
  ·va′len·cy
tri·va′lent
tri′valve
tro′car or ·char

tro·chan′ter
tro·chan′ter·ic
tro′che
troch′le·a
  ·le·ae′
troch′le·ar
troch′o·car′di·a
tro′choid or tro·choi′dal
troch′o·ri′zo·car′di·a
troi′lism
tro′land
trom·bic′u·li′a·sis
trom′bi·di′a·sis or ·di·o′sis
tro·meth′a·mine′
trom′o·ma′ni·a
troph′e·de′ma
troph′e·sy
  ·sies
troph′ic
troph′o·blast′
troph′o·blas′tic
troph′o·plasm
troph′o·zo′ite
tro′pi·a
trop′ic acid
trop′i·cal
tro′pine
tro′pism
tro·pis′tic
tro·po·col′la·gen
tro·pom′e·ter
tro′po·nin

trough
Trous·seau's′ symptom
troy
true
trun′cal
trun′cate
  ·cat·ed ·cat·ing
trun·ca′tion
trun′cus
trunk
tru′sion
truss
truth drug
try′–in′
try·pan′o·cide′ or ·pan′o·so′mi·cide′
tryp′a·no·some′
tryp′a·no·so·mi′a·sis
  ·ses′
try·pan′o·so·mid
tryp·ars′a·mide′
tryp′sin
tryp·sin′o·gen
tryp′tic
tryp′to·phan′ or ·phane′
tset′se fly
tsu′tsu·ga·mu′shi disease
tub
tu′ba
  ·bas or ·bae
tub′age
tub′al
tu′bate

tube
  tubed tub′ing
tu′ber
  ·bers or ·ber·a
tu′ber·cle
tu·ber′cu·lar
tu·ber′cu·late or ·lat′ed
tu·ber′cu·la′tion
tu·ber′cu·lid
tu·ber′cu·lig′e·nous
tu·ber′cu·lin
tu·ber′cu·lo·fi′broid
tu·ber′cu·loid′
tu·ber′cu·lo′ma
  ·mas or ·ma·ta
tu·ber′cu·lo′sis
tu·ber′cu·lot′ic
tu·ber′cu·lous
tu·ber′cu·lum
  ·la
tu′ber·os′i·tas
  ·os′i·ta′tes
tu′ber·os′i·ty
  ·ties
tu′ber·ous or ·ose′
tub′ing
tu′bo·ab·dom′i·nal
tu′bo·tor′sion
tu′bu·lar
tu′bu·lar′i·ty
  ·ties
tu′bu·late
tu′bule
tu′bu·lin
tu′bu·li·za′tion

tu′bu·lize′
  ·lized′ ·liz′ing
tu′bu·lo·cyst′
tu′bu·lo·der′moid
tu′bu·lous
tu′bu·lus
tu′bus
tuft
tug
  tugged tug′ging
tu·la·re′mi·a or
  ·rae′mi·a
tu·la·re′mic
tulle gras′
tu′me·fa′cient
tu′me·fac′tion
tu′me·fy′
  ·fied′ ·fy′ing
tu·men′ti·a
tu·mes′cence
tu·mes′cent
tu′mid
tu·mid′i·ty
tu′mor
tu′mor·af′fin
tu′mor·i·gen′e·sis
tu′mor·i·gen′ic
tu′mor·let
tu′mor·ous
tung′state
tung′sten
tu′nic
tu′ni·ca
  ·cae′
tu′ni·ca′ry
tu′ni·cin
tun′ing fork

tun′nel
tu′ra·nose
tur′bid
tur′bi·dim′e·ter
tur′bi·di·met′ric
tur′bi·dim′e·try
tur·bid′i·ty
tur′bi·nate or
  ·bi·nat′ed or
  ·bi·nal
tur′bi·nec′to·my
  ·mies
tur′bi·not′o·my
  ·mies
Türck's bundle
tur·ges′cence or
  ·ges′cen·cy
tur·ges′cent
tur′gid
tur·gid′i·ty
tur′gid·i·za′tion
tur′gom′e·ter
tur′gor
Türk's cell
tur′mer·ic
turn′o′ver
tur′pen·tine′
tu·run′da
  ·dae
tus′sal
tus·sic′u·lar
tus·sic′u·la′tion
tus′sis
tus′sive
T wave
tweez′ers

twen′ty–twen′ty
  (or 20/20) vision
twig
twi′light′ sleep
twin
  twinned twin′ning
twinge
  twinged twing′ing
twitch
ty′le
ty·lec′to·my
  ·mies
tyl′i·on
  ·i·a
ty·lo′ma
ty·lo′sis
  ·ses
ty·lot′ic
tym′pa·nal
tym·pan′ic
tym′pa·nic′i·ty
tym′pa·nism
tym′pa·ni′tes
  (abdominal
    distention; SEE
    tympanitis)
tym′pa·nit′ic
tym′pa·ni′tis
  (ear inflammation;
    SEE tympanites)
tym′pa·no·mas′toid·
  i′tis
tym′pa·no·plas′ty
  ·ties
tym′pa·no′sis
tym′pa·not′o·my
tym′pa·nous

tym'pa·num
  ·nums *or* ·na
tym'pa·ny
  ·nies
type
  typed typ'ing
ty·phe'mi·a
typh·lit'ic
typh·li'tis
typh'lo·dic'li·di'tis
typh'lo·li·thi'a·sis
  ·ses'
typh·lol'o·gy
ty'pho·bac'ter·in
ty'phoid
ty·phoi'dal
ty'pho·ma·lar'i·al
ty'phous
  (of typhus)
ty'phus
  (infectious disease)
typ'i·cal *or* typ'ic
typ'i·fy'
  ·fied' ·fy'ing
typ'o·scope'
ty'ra·mine'
ty'ro·ci'dine
ty·rog'e·nous
ty'roid
ty'ro·si·nase
ty'ro·sine'
ty'ro·si·no'sis
ty'ro·syl·u'ri·a
ty'ro·tox'ism
tzet'ze fly

U

u'ber·ous
u'ber·ty
u·biq'ui·none'
u·lag'a·nac·te'sis
u·lat'ro·phy
ul'cer
ul'cer·ate'
  ·at'ed ·at'ing
ul'cer·a'tion
ul'cer·a'tive
ul'cer·ous
ul'cus
  ·cer·a
u'le·gyr'i·a
u'ler·y·the'ma
u·li'tis
ul'na
  ·nae *or* ·nas
ul'nad
ul'nar
u'lo·car'ci·no'ma
  ·mas *or* ·ma·ta
u'lo·der·ma·ti'tis
u'loid
u'lor·rha'gi·a
u·lot'o·my
ul'tra·cen·trif'u·gal
ul'tra·cen'tri·fuge'
  ·fuged' ·fug'ing
ul'tra·fil·tra'tion
ul'tra·mi'cro·scope'
ul'tra·mi'cro·
  scop'ic

ul'tra·mi·cros'co·py
  ·pies
ul'tra·mi'cro·tome'
ul'tra·red'
ul'tra·son'ic
ul'tra·son'ics
ul'tra·son'o·gram'
ul'tra·son'o·
  graph'ic
ul'tra·so·nog'ra·phy
  ·phies
ul'tra·sound'
ul'tra·struc'tur·al
ul'tra·struc'ture
ul'tra·vi'o·let
ul'tra·vi'rus
um'ber
um·bi'lec·to·my
um·bil'i·cal
um·bil'i·cate *or*
  ·cat'ed
um·bil'i·ca'tion
um·bil'i·cus
  ·ci'
um'bo
  um·bo'nes *or* um'
  bos
um·bo'nal *or* um'
  bo·nate *or* um·
  bon'ic
un·bal'ance
  ·anced ·anc·ing
un'ci·form'
un'ci·na'ri·a'sis
  ·ses'
un'ci·nate
un'ci·na'tum

un'ci·pres'sure
un·cir'cum·cised'
un'con·di'tioned
un·con'scious
un'co·or'di·nat·ed
un'co-os'si·fied
un'co·ver'te·bral
unc'tion
unc'tu·ous
un·cur'a·ble
un'cus
un'dec·y·le'nic acid
un'der·a·chieve'
　·chieved'　·chiev'ing
un'der·a·chieve'ment
un'der·a·chiev'er
un'der·bite'
un'der·cut'
un'der·horn'
un'der·lip'
un'der·nour'ish
un'der·nu·tri'tion
un'der·sexed'
un'der·shot'
un'der·slung'
un'der·stain'
un'der·toe'
un'der·weight'
un·de·scend'ed
un'dif·fer·en'ti·at·ed
un'dif·fer·en'ti·a'tion
un'di·gest'ed
un'di·lut'ed
un·dine'

un'din·ism
un'dis·solved'
un·do'ing
un'du·lant
un'du·late'
　·lat'ed　·lat'ing
un'du·late or
　·lat'ed
un'du·la'tion
un'gual
un'guent
un'guen·tar'y
un'guen·tum
un'guis
　un'gues
un'gu·la
un·health'y
　·i·er　·i·est
un·hy'gi·en'ic
u'ni·ar·tic'u·lar
u'ni·ax'i·al
u'ni·cam'er·al
u'ni·cel'lu·lar
u'ni·ceps
u'ni·corn'
u'ni·cor'nous
u'ni·di·rec'tion·al
u'ni·fi'lar
u'ni·form'
u'ni·form'i·ty
　·ties
u'ni·lat'er·al
u'ni·lo'bar
u'ni·loc'u·lar
u'nin·hib'it·ed
u'ni·nu'cle·ar
u'ni·oc'u·lar

un'ion
u'ni·o'val or u'ni·ov'u·lar
u'ni·po'lar
u'ni·po·lar'i·ty
u'ni·port
u'ni·por'ter
u·nip'o·tent
u'ni·sen'so·ry
u'ni·sex'u·al
u'ni·sex'u·al'i·ty
u'nit
u'ni·tar'i·an
u'ni·tar'y
u·nite'
　·nit'ed　·nit'ing
u'ni·va'lence or
　·len·cy
u'ni·va'lent
u'ni·ver'sal
un·joint'
un·med'i·cat'ed
un·med'ul·lat'ed
un·mod'i·fied'
un·my'e·li·nat'ed
un·nat'u·ral
un'of·fi'cial
un·or'gan·ized'
un·o'ri·en·ta'tion
un·paired'
un·phys'i·o·log'i·cal
　or ·log'ic
un·pure'
un·pu'ri·fied'
un're·al'i·ty
un·rest'

253

un·san'i·tar'y
un·sat'u·rat'ed
un·sex'
un·sound'
un·spec'i·fied'
un·sta'ble
un·stri·at·ed
un·u·nit·ed
un·well'
up'per
up·set'
  ·set' ·set'ting
up'set' n.
up'si·loid'
up'si·lon'
up'take'
u'ra·chal
u'ra·chus
u'ra·cil
u'ra·cra·si·a
u'ra·cra·ti·a
u'ra·gogue'
ur'a·nal
u'ra·nis'co·plas'ty
u'ra·nis'cor·rha·phy
u'ra·nis'cus
u·ra'ni·um
u'ra·no'ma
u'ra·no·plas'ty
u'ra·no·ple'gi·a
u'ra·nor'rha·phy
u'rase
u'rate
u'ra·te'mi·a
u'ra·tu'ri·a
ur'ce·o·late'

ur'–de·fense'
u·re'a
u·re'al or ·re'ic
u're·ase'
u're·do
u're·ide'
u·re'mi·a
u·re'mic
u·re'o·tel'ic
u·rc'si·cs·the'si·a or ·sis
u·re'ter
u·re'ter·al or u·re·ter'ic
u·re'ter·i'tis
u·re'ter·o·ne'o·py'e·los'to·my
  ·mies
u·re'ter·os'to·my
  ·mies
u're·thane' or ·than'
u·re'thra
  ·thrae or ·thras
u·re'thral
u·re'threm·phrax'is
u·re·thri'tis
u·re'thro·cele'
u·re'thro·graph'
u·re·throg'ra·phy
u·re·throm'e·ter
u·re'thro·m·e·try
u·re'thro·scope'
u·re'thro·scop'ic
u·re'thro·tome'
u're·throt'o·my
  ·mies

u·ret'ic
ur'gen·cy
  ·cies
ur·hi·dro'sis
u'ric
u'ri·case'
u'ri·ce'mi·a
u'ri·cos·u'ric
u'ri·dine'
u'ri·dro'sis
u'ri·nal
u'ri·nal'y·sis
  ·ses
u'ri·nar'y
u'ri·nate'
  ·nat·ed ·nat·ing
u'ri·na'tion
u'ri·na'tive
u·rine
u'ri·nif'er·ous
u'ri·nip'a·rous
u'ri·no·gen'i·tal
u'ri·nog'i·nous
u'ri·no'ma
  ·mas or ·ma·ta
u'ri·nom'e·ter
u'ri·no·phil'
u'ri·nous or ·nose'
u'ri·sol'vent
u'ro·chrome'
u'ro·fus'co·hem'a·tin
u'ro·gen'i·tal
u·rog'e·nous
u'ro·ki'nase
u'ro·lith'

u′ro·lith′ic
u′ro·log′ic or
　·log′i·cal
u·rol′o·gist
u·rol′o·gy
u′ro·lu′te·in
u·rom′e·ter
u′ro·scop′ic
u·ros′co·py
　·pies
u′ro·tox·ic′i·ty
u′ro·tox′in
u′ro·xan′thin
ur′ti·ca
　·cae
ur′ti·cant
ur′ti·car′i·a
ur′ti·car′i·cal
ur′ti·cate′
　·cat′ed ·cat′ing
ur′ti·ca′tion
u·ru′shi·ol′
us′ti·lag′in·ism
us′tion
us′tu·late
us′tu·la′tion
us′tus
u·surp′
u·sur·pa′tion
u′ta
u′ter·ine
u′ter·o·ges·ta′tion
u′ter·o·ton′ic
u′ter·o·vag′i·nal
u′ter·o·ves′i·cal
u′ter·us
　u′ter·i′

u′tri·cle
u·tric′u·lar
u·tric′u·lo·sac′cu·lar
u·tric′u·lus
　·li′
u′tri·form′
u′ve·a
u′ve·al
u′ve·it′ic
u′ve·i′tis
u′ve·o·neu′rax·i′tis
u′ve·o·par′o·ti′tis
u′vi·form′
u′vi·o·fast′
u′vi·o·lize′
　·lized′ ·liz′ing
u′vi·om′e·ter
u′vu·la
　·las or ·lae′
u′vu·lar
u′vu·la′ris
u′vu·lec′to·my
　·mies
u′vu·lop·to′sis or
　·lap·to′sis
　·ses
u′vu·lo·tome′ or u′
　vu·la·tome′
u′vu·lot′o·my
　·mies
U wave

V

vac′ci·nal

vac′ci·nate′
　·nat′ed ·nat′ing
vac′ci·na′tion
vac′ci·na′tor
vac·cinc′
vac·cin′i·a
vac·cin′i·al
vac·cin′i·form′
vac·cin·i·o′la
vac′ci·noid′
vac′cin·o·ther′a·
　peu′tics
vac′u·a
　(sing. vac′u·um)
vac′u·o·lar
vac′u·o·late′
　·lat′ed ·lat′ing
vac′u·o·lat′ed
vac′u·o·la′tion
vac′u·ole′
vac′u·ome′
vac′u·um
　·u·ums or ·u·a
va′gal
va′gi
　(sing. va′gus)
va·gi′na
　·nas or ·nae
vag′i·nal
vag′i·na·li′tis
vag′i·nate′
　·nat′ed ·nat′ing
vag′i·nate adj.
vag′i·nec′to·my
　·mies
vag′i·nis′mus or
　vag′in·ism

vag′i·ni′tis
vag′i·no·ab·dom′i·nal
vag′i·no·cele′
va·gi′tis
(nerve inflammation)
va·gi′tus
(newborn cry)
va·go·de·pres′sor
va′go·mi·met′ic
va·got′o·my
·mies
va′go·to′ni·a
va′go·ton′ic
va′go·trop′ic
va′go·va′gal
va′grant
va′gus
·gi
va′gus·stoff′
va′lence
va′len·cy
·cies
val′er·ate′
va·le′ri·an
va·ler′ic acid
val′e·tu′di·nar′i·an
val′e·tu′di·nar′i·an·ism
val′e·tu′di·nar′y
·nar′ies
val′gus
val′id
val′i·date′
·dat′ed ·dat′ing
val′i·da′tion

va·lid′i·ty
·ties
val′ine
val′in·o·my′cin
val′late
val·lec′u·la
·lae′
val·lec′u·lar
Val·sal′va maneuver
val′ue
val′va
·vae
val′vate
valve
val·vot′o·my
·mies
val′vu·la
·lae
val′vu·lar *or* val′var
val′vule *or* valve′let
val′vu·li′tis
van′a·date′
va·nad′ic acid
va·na′di·um
van′a·dous
van′co·my′cin
van den Bergh's test
va·nil′la
va·nil′lic acid
va·nil′lin
van′il·lyl·man·del′ic acid
va′por

va′por·if′ic
va·po′ri·um
va′por·i·za′tion
va′por·ize′
·ized′ ·iz′ing
va′por·iz′er
va·por·os′i·ty
va′por·ous *or* ·por·y
va′po·ther′a·py
var′i·a·bil′i·ty
var′i·a·ble
var′i·ance
var′i·ant *or* ·ate
var′i·a′tion
var′i·ca′tion
var′i·ce′al
var′i·cel′la
var′i·cel′loid
var′i·ces′
(sing. var′ix)
var·ic′i·form′
var′i·co·cele′
var′i·cog′ra·phy
·phies
var′i·cose′
var′i·co′sis
·ses
var′i·cos′i·ty
·ties
var′i·cot′o·my
·mies
va·ric′u·la
var′i·e·gat′ed
var′i·e·ga′tion
va·ri′e·ty
·ties

var′i·form′
va·ri′o·la
va·ri′o·lar
var′i·o·late′
  ·lat′ed ·lat′ing
var′i·o·loid′
va·ri′o·lous
var′ix
  ·i·ces′
var′nish
va·ro′li·an
var′us
vas
  va′sa
va′sal
vas′cu·lar
vas·cu·lar′i·ty
  ·ties
vas·cu·lar·i·za′tion
vas′cu·lar·ize′
  ·ized′ ·iz′ing
vas′cu·la·ture
vas·cu·li′tis
vas de′fe·rens′
  va′sa de′fe·ren′ti·a
va·sect′
vas·ec′to·mize′
  ·mized′ ·miz′ing
vas·ec′to·my
  ·mies
Vas′e·line′
  (trademark)
vas′o·con·stric′tion
vas′o·con·stric′tor
vas′o·dil′a·ta′tion
  or ·di·la′tion
vas′o·di′la·tor

vas′o·in·hib′i·tor
vas′o·in·hib′i·to′ry
vas′o·mo′tor
vas′o·pres′sin
vas′o·pres′sor
vas′o·spasm′
vas·ot′o·my
  ·mies
vas′o·va′gal
vas′tus
vault
V′–bends′
vec′tion
vec′tis
vec′tor
vec′tor·car′di·o·gram′
vec′tor·car′di·og′ra·phy
vec·to′ri·al
veg′an
veg′an·ism
veg′e·ta·ble
veg′e·tal
veg′e·tar′i·an
veg′e·tar′i·an·ism
veg′e·tate′
  ·tat′ed ·tat′ing
veg′e·ta′tion
veg′e·ta′tive *or* ·e·tive
ve′hi·cle
veil
veil′lo·nel′la
vein
veined

vein′let
vein′ule
ve′la
  (sing. ve′lum*)*
ve·la′men
  ·lam′i·na
vel′a·men′tous
ve′lar
ve′late
veld *or* veldt
vel′li·cate′
  ·cat′ed ·cat′ing
vel′li·ca′tion
vel′lus
ve·loc′i·ty
  ·ties
vel′o·syn′the·sis
ve′lum
  ·la
ve′na
  ve′nae
vena ca′va
  venae ca′vae
ve·na′tion
ven′e·na′tion
ven′e·nous
ven′e·punc′ture
ve·ne′re·al
ve·ne′re·ol′o·gist
ve·ne′re·ol′o·gy
ven′er·y
ven′e·sec′tion
ven′in
ven′i·punc′ture
ve′no·a′tri·al
ven′om
ven′om·ous

ve′nose
ve·nos′i·ty
  ·ties
ve′nous
vent
ven′ter
ven′ti·late′
  ·lat′ed ·lat′ing
ven′ti·la′tion
ven′ti·la′tor
ven′ti·la·to′ry
ven′trad
ven′tral
ven′tri·cle
ven·tric′u·lar
ven·tric′u·li′tis
ven·tric′u·lo·nec′tor
ven·tric′u·lus
  ·u·li′
ven′tro·dor′sal
ven′tro·lat′er·al
ven·tros′co·py
ven·tu′ri (tube)
ven′u·lar
ven′ule
ven′u·lose′
ve·rat′ri·dine′
ver′a·trine′ or ve·ra′tri·a
ve·ra′trum
ver′bal
ver′bal·ize′
  ·ized′ ·iz′ing
ver·big′er·a′tion
ver′do·he′mo·glo′bin

verge
ver′gence
ver′mi·cide′
ver·mic′u·lar or ·u·late or ·lat′ed
ver·mic′u·la′tion
ver′mi·form′
ver′mi·fuge′
ver·mil′ion
ver′min
ver′min·ous
ver′mis
  ·mes
ver′nal
ver′ni·er acuity
ver′nix
ver·ru′ca
  ·cae
ver′ru·cose′ or ·ru·cous
ver′si·col′or
ver′sion
ver′te·bra
  ·brae′ or ·bras
ver′te·bral
ver′te·brate
ver′te·bro·cos′tal
ver′te·bro·mam′ma·ry
ver′te·bro·ster′nal
ver′tex
  ·tex·es or ·ti·ces′
ver′ti·cal
ver′ti·ces′
  (sing. ver′tex)
ver′ti·cil

ver·tic′il·late or ·lat′ed
ver·tig′i·nous
ver′ti·go
  ver′ti·goes′ or ver·tig′i·nes′
very low–density lipoproteins
ve·si′ca
  ·cae
ves′i·cal
  (of the bladder; SEE vesicle)
ves′i·cant
ves′i·cate′
  ·cat′ed ·cat′ing
ves′i·ca′tion
ves′i·ca·to′ry
  ·to′ries
ves′i·cle
  (sac; SEE vesical)
ves′i·co·cer′vi·cal
ves′i·co·u′ter·o·vag′i·nal
ve·sic′u·lar or ·u·late
ve·sic′u·late′
  ·lat′ed ·lat′ing
ve·sic′u·la′tion
ve·sic′u·lo·cav′er·nous
ve·sic′u·lot′o·my
  ·mies
ves′sel
ves·tib′u·lar
ves′ti·bule′

ves·tib′u·lo·coch′le·ar
ves·tib′u·lot′o·my
  ·mies
ves·tib′u·lum
  ·la
ves′tige
ves·tig′i·al
ves·tig′i·um
  ·i·a
vi′a
  vi′ae
vi′a·bil′i·ty
vi′a·ble
vi′al
vi′bex
  vi·bi′ces
vi′bra·punc′ture
vi′brate
  ·brat·ed ·brat·ing
vi′bra·tile
vi′bra·til′i·ty
vi·bra′tion
vi′bra·tor
vi′bra·to′ry or ·tive
vib′ri·o′
  ·ri·os′
vi·bris′sa
  ·sae
vi′bro·car′di·o·gram′
vi′bro·mas·sage′
vi·bur′num
vi·car′i·ous
vice
vi′cious
vi·dar′a·bine′

view
vig′il
vig′il·am′bu·lism
vig′i·lance
vig′or
vig′or·ous
vil′li
  (sing. vil′lus)
vil′li·form′
vil·los′i·ty
  ·ties
vil′lous or ·lose
  (of villi)
vil′lus
  vil′li
  (hairlike projection)
vin·blas′tine sulfate
Vin′cent's angina
vin·cris′tine
vin′cu·lum
  ·la
vin′e·gar
vi′nic
vi′nous
vi′nyl
vi·nyl′i·dene′
vi′o·la′ceous
vi′o·late′
  ·lat·ed ·lat·ing
vi′o·la′tion
vi′o·lence
vi′o·les′cent
vi′o·let
vi·os′ter·ol′
vi′per
vi′per·ine
vi′ra·gin′i·ty

vi′ral
vi·re′mi·a
vi′res
  (sing. vis)
vir′gin
vir′gin·al
vir·gin′i·ty
vi′ri·ci′dal
vi′ri·cide′
vir′ile
vir′i·les′cence
vir′i·lism
vi·ril′i·ty
vir′i·li·za′tion
vir′i·lize′
  ·lized′ ·liz′ing
vi′ri·on′
vi′roid
vi′ro·log′ic or ·log′i·cal
vi·rol′o·gist
vi·rol′o·gy
vi′rose or ·rous
vi·ro′sis
  ·ses
vir′tu·al
vi·ru′ci·dal
vi·ru′cide′
vir′u·lence or ·len·cy
vir′u·lent
vi′rus
vis
  vi′res
vis′age
vis′cance

vis'cer·a
(sing. vis'cus)
vis'cer·ad
vis'cer·al
vis'cer·o·meg'a·ly
vis'cer·o·mo'tor
vis'cer·op·to'sis
vis'cid
vis·cid'i·ty
  ·ties
vis·co·e·las'tic
vis'coid or vis·coi'dal
vis·com'e·ter
vis'co·sim'e·ter
vis·co·sim'e·try
vis·cos'i·ty
  ·ties
vis'cous
(viscid)
vis'cus
  vis'cer·a
(visceral organ)
vis'i·bil'i·ty
  ·ties
vis'i·ble
vis'ile
vi'sion
vis'u·al
vis'u·al·i·za'tion
vis'u·al·ize'
  ·ized' ·iz'ing
vis'u·o·au'di·to'ry
vis'u·o·sen'so·ry
vi'tal
vi'tal·ism
vi'tal·ist

vi'tal·is'tic
vi·tal'i·ty
  ·ties
vi'tal·i·za'tion
vi'tal·ize'
  ·ized' ·iz'ing
vi'ta·mer
vi'ta·min
vi'ta·min'ic
vi·tel'lin
(a protein in egg yolk)
vi·tel'line
(of egg yolk)
vi·tel'lus
  ·li
vi'ti·a'tion
vit'i·lig'i·nous
vit'i·li'go
  ·i·lig'i·nes
vit'i·li·goi'de·a
vit'i·um
  vit'i·a
vit're·ous
vi·tres'cence
vit'ric
vit'ri·ol
vit'ri·o·lat'ed
vit'ri·ol'ic
vi·var'i·um
  ·i·ums or ·i·a
viv'i·di·al'y·sis
  ·ses
viv'i·fi·ca'tion
viv'i·par'i·ty
vi·vip'a·rous
viv'i·sect'

viv'i·sec'tion
viv'i·sec'tion·ist
viv'i·sec'tor
vo'cal
voice
  voiced voic'ing
voice'print'
void
vo'la
vo'lar
vo·la'ris
vol'a·tile
vol'a·til'i·ty
vol'a·til·i·za'tion
vol'a·til·ize'
  ·ized' ·iz'ing
vo·le'mic
vo·li'tion
vo·li'tion·al
Volk'mann's canal
vol'ley
  ·leys
vol·sel'la
volt'age
vol·ta'ic
vol'ta·ism
vol·tam'e·ter
vol·ta·met'ric
volt'am'me·ter
volt'–am'pere
volt'me·ter
vol'ume
vo·lu'me·ter
vol'u·met'ric or
  ·met'ri·cal
vol'u·mom'e·ter or
  ·u·me·nom'e·ter

vol'un·tar'y
vo·lup'tu·ous
vo·lute'
vol'u·tin
vol'vox
vol'vu·late'
  ·lat'ed ·lat'ing
vol'vu·lo'sis
vol'vu·lus
vo'mer
vo'mer·ine'
vo'mer·o·na'sal
vom'i·ca
  i·cae
vom'it
vom'it·ing
vom'i·tive
vom'i·to'ry
  ·ries
vom'i·tu·ri'tion
vom'i·tus
Voor'hees' bag
vo·ra'cious
vor'tex
  ·tex·es or ·ti·ces'
vor'ti·cal
vor'ti·cel'la
  ·lae
vor'ti·cose'
vous·sure'
vox
  vo'ces
vox'el
vo·yeur'
vo·yeur'ism
vo·yeur·is'tic
vu'er·om'e·ter

vul'can·ite'
vul'can·i·za'tion
vul'can·ize'
  ·ized' ·iz'ing
vul·ga'ris
vul'ner·a·bil'i·ty
vul'ner·a·ble
vul'ner·ant
vul'ner·ar'y
  ·ar'ies
vul'ner·ate'
  ·at'ed ·at'ing
vul'nus
  ·ner·a
vul·sel'la or vul·sel'lum
vul'va
  ·vae or ·vas
vul'val or ·var
vul·vec'to·my
vul·vis'mus
vul·vi'tis
vul·vop'a·thy
vul'vo·u'ter·ine
vul'vo·vag'i·nal
vul'vo·vag'i·ni'tis

W

Wach'en·dorf's' membrane
wad
  wad'ded wad'ding
wad'dle
  ·dled ·dling

wa'fer
waist
wake'ful
walk'er
wal·le'ri·an degeneration
wall'eye'
wall'eyed'
wam'ble
  ·bled ·bling
wan
  wan'ner wan'nest
wan'der·ing
ward
war'far·in
warm'blood'ed
wart
wash
wash'ing soda
wasp
Was'ser·mann test
waste
  wast'ed wast'ing
wa'ter
wa'ter·borne'
water cure
wa'ter·fall'
wa'ters
wa'ter·shed'
watt'age
watt'me·ter
wave'length'
wax'y
  ·i·er ·i·est
weak
weal
wean'ling

wea′sand
web
  webbed web′bing
we′ber
wedge
  wedged wedg′ing
weep
  wept weep′ing
weigh
weight
weight′less
weis′mann·ism
well
welt
wen
Wen′ke·bach′ phenomenon
wet
  wet′ter wet′test
wet brain
wet′–nurse′
  –nursed′ –nurs′ing
wet nurse
wet pack
wet′ta·ble
Wet′zel grid
Whar′ton's duct
wheal
wheat
wheel′chair′
wheeze
  wheezed wheez′ing
whelk
whelp
whey
whiff

whine
  whined whin′ing
whip′lash′
whip′worm′
whirl′bone′
whirl′pool′ bath
whis′ker
whis′key
  ·keys *or* ·kies
whis′per
whis′tle
white
  whit′er whit′est
white′head′
white room
whit′low
who′lism
who·lis′tic
whoop′ing cough
whorl
wid′ow's peak
width
wild′fire′
will′pow′er
Wil′son's disease
wince
  winced winc′ing
wind′age
wind′–borne′
wind′burn′
wind′chill′ factor
wind′ed
wind′lass, Span′ish
win′dow
win′dow·ing
wind′pipe′
wine

wing
wink
wink′ing
win′ter·green′
wire
  wired wir′ing
wir′y
wis′dom tooth
witch ha′zel
with·draw′
  ·drew′ ·drawn′
  ·draw′ing
with·draw′al
wit′kop
wit′zel·sucht′
wohl·fahr′ti·a
Wolff′i·an body
Wolff′–Park′in·son–White′ syndrome
wolf′ram
wolfs′bane′
womb
won′der drug
wood alcohol
Wood's rays
wool′sort′er's disease
word′–as·so′ci·a′tion
word sal′ad
work′out′
work′up′
wor′mi·an bones
worm′seed′
wort
wound′wort′

W–plas′ty
wrap
  wrapped *or* wrapt
  wrap′ping
wreath
  wreaths
wrench
wrin′kle
  ·kled ·kling
Wris′berg's nerve
wrist′drop′
writ′er's cramp
writhe
  writhed
  writh′ing
writ′ing
wry′neck′
wu′cher·e·ri′a·sis
  ·ses′

X

xan′than
xan′thate
xan′thel·as′ma
  ·mas *or* ·ma·ta
xan·the·las·moi′de·a
xan·the′mi·a
xan′thene
  (molecular structure; SEE xanthine)
xan′thic
xan′thine
  (nitrogenous compound; SEE xanthene)
xan′thin·u′ri·a
xan′thism
xan′tho·chroid′
xan′tho·chro·mat′ic *or* ·chro′mic
xan′tho·chro′mi·a *or* ·chroi′a
xan′thoch′ro·ous
xan′tho·gran′u·lo′ma
  ·mas *or* ·ma·ta
xan′tho·gran′u·lom′a·tous
xan·tho′ma
  ·mas *or* ·ma·ta
xan′tho·ma·to′sis
  ·ses
xan′tho·ma·tous
xan′thone
xan′tho·phyll
xan′tho·phyl′lous
xan′tho·pro·te′ic
xan′tho·pro′te·in
xan·thop′si·a
xan·tho·sine′
xan·tho′sis
  ·ses
xan′thous
xan′thu·re′nic acid
xan·thu′ri·a
xan′thyl
xan·thy′lic
X chromosome
xen′o·bi·ot′ic
xen′o·gen′e·sis
xen′o·ge·net′ic *or* ·gen′ic
xen′o·graft′
xe′non
xen′o·phobe′
xen′o·pho′bi·a
xen′o·pho′bic
xen′o·pho′ni·a
xen′yl
xe·ran′sis
xe·ran′tic
xe·ra′si·a
xe′ro·chei′li·a
xe′ro·der′ma
xe′ro·der·mat′ic
xe′ro·gram′
xe′ro·graph′ic
xe·rog′ra·phy
  ·phies
xe·ro′ma
  ·mas *or* ·ma·ta
xe′ro·pha′gi·a *or* xe·roph′a·gy
xe′roph·thal′mi·a
xe′roph·thal′mic
xe·ro·ra′di·og′ra·phy
xe·ro′sis
  ·ses
xe′ro·sto′mi·a
xer′o·tes′
xe·rot′ic
xi
xiph′i·ster′nal

263

xiph′i·ster′num
  ·na
xiph′o·cos′tal
xiph′o·dyn′i·a
xiph′oid
xiph′oid·i′tis
xi·phop′a·gus
X′–ray′ or X ray
  or x′–ray′ or x
  ray
xy′lene
xy′li·dine′
xy′li·tol′
xy′lol
xy′lo·met′a·zo′line
xy′lose
xy′lu·lose′
xy′lyl
xy′ro·spasm
xys′ma
xys′ter

Y

yawn
yaws
Y chromosome
yeast
yel′low enzyme
yer′ba bue′na
yer·sin′i·a
yer′sin·i·o′sis
Yer′sin's serum
yin′–yang′
yo′ga
yo′gic
yo′gurt or ·ghurt
  or ·ghourt
yo·him′bé
yo·him′bine
yoke
  *(connecting
  structure)*
yolk
  (egg nutrient)
youth
yp·sil′i·form
yt·ter′bic
yt·ter′bi·um
yt′tric
yt′tri·um

Z

ze′a
ze′a·xan·thin
zed′o·a′ry
Zee′man effect
ze′in
ze′ism
zen′ker·ism
ze′o·lite′
ze′o·lit′ic
ze′ro
  ·ros or ·roes
zero gravity
zes′to·cau′sis
ze′ta
zig′zag·plas′ty
zinc
zinc′a·lism
zinc·if′er·ous
zinc′oid
Zinn's ligament
Zins′ser
  inconsistency
zir·con′ic
zir·co′ni·um
zit
zo′ac·an·tho′sis
  ·ses
zo·an′thro·py
zo·et′ic
zo′ic
Zol′lin·ger–El′li·
  son syndrome
zo′na
  ·nae
zon′al
zona pel·lu′ci·da
zon′a·ry
zon′ate or ·at·ed
zone
zon′es·the′si·a
zo·nif′u·gal
zon′u·la
  ·lae
zon′u·lar
zon′ule
zo′o·der′mic
zo′o·gen′ic or zo·
  og′e·nous
zo′o·gloe′a
zo′o·gloe′al or
  ·gloe′ic
zo′o·graft′
zo′oid

zo·oi′dal
zo′o·log′i·cal or ·log′ic
zo·ol′o·gist
zo·ol′o·gy
zoom
zo·on′o·sis
·ses′
zo′o·not′ic
zo′o·par′a·site
zo′o·par′a·sit′ic
zo·oph′a·gous
zo′o·phil′ic or zo·oph′i·lous
zo′o·oph′i·lism or zo·oph′i·ly or zo′o·phil′i·a
zo′o·pho′bi·a
zo′o·plas′ty
·ties
zo·op′si·a
zo′os·mo′sis
zo′o·spore′
zo′o·spor′ic or zo·os′po·rous
zo·os′ter·ol′
zo′o·tom′ic or ·tom′i·cal
zo·ot′o·mist
zo·ot′o·my
zo′o·tox′in

zos′ter
zos·ter′i·form′
zos′ter·oid′
zox′a·zol′a·mine
Z–plas′ty
·ties
zwit′ter·i′on
zwit′ter·i·on′ic
zy′gal
zyg′ap·o·phys′e·al
zyg·a·poph′y·sis
·ses′
zyg′i·on′
·i·a
zy·go·dac′tyl or ·tyl·ous
zy′go·dac′ty·ly
zy′goid
zy·go′ma
·ma·ta or ·mas
zy′go·mat′ic
zy′go·mat′i·co·fa′cial
zy′go·mat′i·co–or′bit·al
zy′go·mat′i·cum
zy′gon
zy·gos′i·ty
zy′go·spore′ or ·sperm

zy′go·spor′ic
zy′go·style′
zy′gote
zy′go·tene′
zy·got′ic
zy′mase
zyme
zy′mic
zy′mo·ex′ci·ta′tor or ·ex·cit′er
zy′mo·gen
zy′mo·gene′
zy′mo·gen·e′sis
zy′mo·gen′ic or zy·mog′e·nous
zy′mo·log′ic or ·log′i·cal
zy·mol′o·gist
zy·mol′o·gy
zy·mol′y·sis
·ses′
zy·mo·lyt′ic
zy·mom′e·ter
zy′mose
zy·mo′sis
·ses
zy′mos·then′ic
zy·mot′ic